Electronic Circuit Design Ideas

Asheville-Buncombe
Technical Community College
Learning Resources Center
340 Victoria Rd.
Asheville, NC 28801

DISCARDED

JUN 25 2025

# Electronic Circuit Design Ideas

V. Lakshminarayanan

## NEWNES

Newnes
An imprint of Butterworth-Heinemann Ltd
Linacre House, Jordan Hill, Oxford OX2 8DP

A member of the Reed Elsevier plc group

OXFORD  LONDON  BOSTON
MUNICH  NEW DELHI  SINGAPORE  SYDNEY
TOKYO  TORONTO  WELLINGTON

First published 1995

© Butterworth-Heinemann Ltd 1995

All rights reserved. No part of this publication
may be reproduced in any material form (including
photocopying or storing in any medium by electronic
means and whether or not transiently or incidentally
to some other use of this publication) without the
written permission of the copyright holder except in
accordance with the provisions of the Copyright,
Designs and Patent Act 1988 or under the terms of a
licence issued by the Copyright Licensing Agency Ltd,
90 Tottenham Court Road, London, England W1P 9HE.
Applications for the copyright holder's written permission
to reproduce any part of this publication should be addressed
to the publishers

**British Library Cataloguing in Publication Data**
Lakshminarayanan, Venkataraman
  Electronic Circuit Design Ideas
  I. Title
  621.3815

ISBN 0 7506 2047 1

**Library of Congress Cataloguing in Publication Data**
Lakshminarayanan, V.
  Electronic circuit design ideas/V. Lakshminarayanan.
  p.  cm.
  Includes bibliographical references and index.
  ISBN 0-7506-2047-1
  1. Electronic circuit design.  I. Title.
  TK7867.L344
  621.3815–dc20                                94–19585
                                                    CIP

NOTE
The author and publishers, while exercising the greatest care in compiling this publication, do not
hold themselves responsible for the consequences arising from any inaccuracies therein.

Composition by Genesis Typesetting, Laser Quay, Rochester, Kent
Printed and bound in Great Britain

# Contents

*Preface* xi

*Acknowledgements* xiii

1 Digital circuits 1
   1.1 Monostable multivibrators 1
   1.2 Digitally programmable monostable uses a PLD 6
   1.3 Use a counter as a pulse-stretcher 8
   1.4 Digital noise canceller 9
   1.5 Build a low-cost delay line 10
   1.6 Digital phase-detector 12
   1.7 Digital frequency mixer 12
   1.8 Digital attenuator 13
   1.9 Binary frequency divider 15
   1.10 Digital edge-detectors 15
   1.11 One IC doubles frequency 16
   1.12 Inexpensive frequency counter/tachometer 17
   1.13 Missing pulse detector 18
   1.14 Digital quadrature phase-shifter 19
   1.15 Pulse-width discriminator 20
   1.16 Power-on reset circuit using monostable 74121 21
   1.17 Quadrature waveform decoder 22
   1.18 2 to 1 digital multiplexer using spare tristate buffers 23
   1.19 Switch debouncers 23
   1.20 Spare buffers debounce push-button switch 25
   1.21 Programmable digital frequency comparator 25
   1.22 Programmable frequency comparator for analog signals 27
   1.23 AC line synchronizer 28
   1.24 AC line synchronizer 28
   1.25 Time-base generator 29
   1.26 Bit rate generators 30
   1.27 Single pulse extractor 32
   1.28 Use a spare EX–OR gate as an inverter 33
   1.29 Monolithic decade counter decoder display driver 34
   1.30 Digital glitch detector 37
   1.31 Phase lead–lag indicator 38
   1.32 Digital synchronizer 39
   1.33 Electronic alternate-action switch 41
   1.34 Programmable undervoltage/overvoltage detector 42

2 Interface circuits  44
    2.1 Logic interfacing techniques  44
    2.2 RS 232 line driver/receiver  47
    2.3 Optically isolated RS 232 interface  49
    2.4 Low-power 5 V RS 232 driver/receiver  50
    2.5 Programmable micropower level translator/receiver/driver  52
    2.6 RS 232 and RS 423 line driver  55
    2.7 Monolithic relay driver  57

3 Timer circuits  59
    3.1 Astable multivibrator using 555 timer  59
    3.2 Low-cost appliance timer  60
    3.3 Sequential timer  61
    3.4 Long duration timer using quad timer 558  62
    3.5 CMOS programmable 0–99 seconds/minutes timer  63
    3.6 Retriggerable monostable using 555 timer  64
    3.7 Low-power monostable using 555 timer  65

4 Op-amp circuits  66
    4.1 Integrator  66
    4.2 Differentiator  66
    4.3 Voltage follower  67
    4.4 Simulated inductor  68
    4.5 Op-amp power booster  68
    4.6 High current booster  69
    4.7 Window comparator  70
    4.8 Voltage comparator with hysteresis  71
    4.9 Voltage-to-current converter  72
    4.10 Supply-frequency reject filter  73
    4.11 Visible voltage indicator  73
    4.12 Precision half-wave rectifier  74
    4.13 Precision full-wave rectifier  75
    4.14 Op-amp supply splitter  75
    4.15 Four-quadrant analog multiplier  76
    4.16 Programmable positive and negative voltage references  81
    4.17 Negative voltage reference using a positive voltage reference  82

5 Amplifier circuits  83
    5.1 Inverting amplifier  83
    5.2 Non-inverting amplifier  84
    5.3 Summing amplifier  84
    5.4 Absolute value amplifier  85
    5.5 Differential amplifier  85
    5.6 Bridge transducer amplifier  86
    5.7 Monolithic logarithmic amplifier  87
    5.8 Antilog amplifier  90

    5.9   Low-distortion audio amplifier   91
    5.10  Low-noise, high-speed precision op-amp   92
    5.11  Low-cost 50 W per channel audio amplifier   94
    5.12  Power amplifier   95
    5.13  Single-supply instrumentation amplifier   98

6  Waveform generators   101
    6.1   Monolithic precision waveform generator   101
    6.2   Single-supply function generator   104
    6.3   Triangle and square wave generator   106
    6.4   Ramp generators   107
    6.5   Sawtooth and pulse generator   108
    6.6   Square wave tone-burst generator   108
    6.7   Single-tone-burst generator   109
    6.8   Linear ramp voltage generator   110
    6.9   Staircase waveform generator   111
    6.10  Two-phase sine wave generator   112
    6.11  Get ±15 V square waves from +5 V   113
    6.12  Programmable pulse generator   114
    6.13  Nanoseconds pulse generator   115
    6.14  3-phase clock generator   116
    6.15  4-phase clock generator   117
    6.16  Digital phase-shifted clock generator   118
    6.17  Wien-bridge oscillator   119
    6.18  Wien-bridge oscillator using spare logic inverters   121
    6.19  *RC* phase-shift oscillator   121
    6.20  Crystal oscillator   122
    6.21  Gated oscillator   123

7  Phase-locked loop circuits   124
    7.1   Monolithic phase-locked loop   124
    7.2   Monolithic tone decoder   125
    7.3   Dual-tone decoder   130
    7.4   Go/no-go frequency meter   132
    7.5   High-speed, narrow-band tone decoder   133
    7.6   One-shots tame tone decoder   133
    7.7   PLL lock indicator   134
    7.8   Frequency multiplication using PLL 565   136
    7.9   Reduce distortion in mod–demod circuit   137
    7.10  Phase modulator   138
    7.11  Phase detector   139
    7.12  Frequency doubler   140

8  Power supply circuits   141
    8.1   Transformer-isolated +5 V power supply   141
    8.2   Uninterruptible +5 V power supply   142

- 8.3 +5 V to ±10 V voltage converter  143
- 8.4 ±5 V power supply from a 9 V battery  145
- 8.5 +3 V battery to +5 V DC–DC converter  146
- 8.6 Low-cost DC voltage booster  147
- 8.7 DC voltage splitter  148
- 8.8 Supply voltage splitter  148
- 8.9 Step-up negative converter  149
- 8.10 Positive voltage doubler  149
- 8.11 Circuit gives constant DC output with selectable AC input  150
- 8.12 Power-fail alarm  151
- 8.13 AC power fail and brownout detector  152
- 8.14 Power-fail warning and power-up/power-down reset  153
- 8.15 Blown-fuse indicator  154

9 Voltage regulator circuits  155
- 9.1 Positive adjustable voltage regulator  155
- 9.2 5 A positive adjustable voltage regulator  158
- 9.3 Negative adjustable voltage regulator  161
- 9.4 Automobile voltage regulator  164
- 9.5 Voltage/current regulator  165
- 9.6 Voltage inverting switching regulator  165
- 9.7 Voltage converter  170

10 Battery circuits  171
- 10.1 Simple battery charger  171
- 10.2 Wind-powered battery charger  171
- 10.3 Battery status indicator  172
- 10.4 9 V battery life extender  173
- 10.5 Battery switchover circuit  174
- 10.6 Automatic battery back-up switch  175
- 10.7 Polarity insensitive battery-powered supply  176
- 10.8 Combination low-battery warning and low-battery disconnect  177

11 Motor control circuits  178
- 11.1 DC servomotor phase-locked loop  178
- 11.2 Simple bidirectional DC motor speed controller  179
- 11.3 Constant current motor drive  179
- 11.4 Control a bidirectional 4-phase stepper motor  180
- 11.5 PLD adds flexibility to motor controller  182

12 Encoders/decoders  187
- 12.1 Revolution sensor  187
- 12.2 Photo-diode detector  187
- 12.3 Detector for magnetic transducer  188
- 12.4 FSK demodulator using PLL 565  188
- 12.5 SCA decoder  189

13   Tester circuits   191
    13.1   On-board transistor tester   191
    13.2   Zener diode tester and unmarked Zener identifier   192
    13.3   FET tester   193
    13.4   Crystal tester   194
    13.5   Coaxial cable tester   195

14   Miscellaneous circuits   197
    14.1   High-side switch   197
    14.2   Power MOSFET driver   198
    14.3   DTMF filter detects tones in exchanges   202
    14.4   Four channels on a single channel scope   203
    14.5   Light-activated alarm   204
    14.6   Remote light monitor   205
    14.7   Low-cost electronic lamp dimmer   206
    14.8   Incandescent lamp dimmer and protector   206
    14.9   Analog switch needs no supply   208
    14.10  Simple high/low temeprature alarm   208

*Appendix 1* Reference index of integrated circuits and their sources   210

*Appendix 2* Addresses of manufacturers   213

*Appendix 3* Datasheets of commonly used integrated circuits   214

*Bibliography*   247

*Circuits index*   248

# Preface

This book grew out of a collection of circuit ideas designed by me over a number of years, many of which were published in electronics design magazines; to this collection I have added a number of useful circuits taken from the applications literature of a few device manufacturers. The compendium consists of a wide variety of electronic circuits, each one of which can be used as a building block for a larger system design or, in some cases, the short design idea is an independent application by itself.

Each design idea consists of a circuit diagram (and waveforms, where applicable) and an explanation of how the circuit works. For better understanding of the design, in most cases the relevant design equations/formulae used in the design for computing the component values are given. There are 14 chapters in the book and each chapter consists of a collection of circuits which belong to a particular category.

A book of this type, I must admit, obviously cannot cover all areas of electronic circuit design. The book covers certain areas of circuit design often encountered in design work and should prove useful to electronics professionals, hobbyists and students. Students taking courses in electronic circuits should find this compendium useful as supplemental reading for locating certain useful application circuits, and practising engineers should find this compilation useful as a ready reference book, saving considerable time and effort which would normally be required in locating certain often-used designs.

Chapter 1 of the book focuses on digital circuits. Chapter 2 gives examples of interface circuits frequently used in circuit design. These include methods of interfacing different types of logic circuits with one another, RS 232 drivers and receivers of different types, level translators/drivers and receivers and relay drivers. Chapter 3 is devoted to the description of commonly used timer circuits. Chapter 4 describes frequently used op-amp circuits. Chapter 5 gives examples of amplifier circuits. Chapter 6 gives examples of waveform generators and oscillators commonly used in circuit design. Phase-locked loop circuits are described in Chapter 7. The popularly used devices PLL 565 and tone decoder 567 are described in detail and examples of their application in circuit design are presented. Power-supply circuits form the subject of Chapter 8. Chapter 9 describes a few voltage regulator circuits. Battery-related circuits are covered in Chapter 10. Methods of controlling DC servomotors and stepper motors are covered in Chapter 11. The examples given include a simple 4-phase bidirectional stepper motor controller, a programmable logic device (PLD) based stepper motor controller, a constant current motor driver, a PLL-based servomotor controller and a simple bidirectional DC motor controller. Chapter 12 describes encoding and decoding circuits. The design ideas described include a slotted optical sensor based revolution detector, a photodiode detector, a magnetic transducer detector and a FSK circuit. Tester circuits are described in Chapter 13. The circuits described, include an on-board transistor tester, a coaxial-cable tester, a FET tester, a crystal tester and a circuit which can identify unmarked Zener diodes and also test them for functionality. Chapter 14 describes miscellaneous circuits of different types.

An alphabetically arranged circuits index lists the circuits covered in the book and should prove useful when searching for a circuit for a specific application.

Datasheets of a few commonly used ICs have been appended. A reference index of most of the ICs used in the circuits described in the book, and their sources, is given in the Appendices.

V. Lakshminarayanan

# Acknowledgements

I was helped in my effort by a number of magazines and companies who kindly gave me permission to reproduce copyrighted material and generously helped me with copies of technical literature; I am extremely grateful to the following magazines and companies for their help:

*Electronics World* and *Wireless World*, Quadrant House, The Quadrant, Sutton, Surrey SM2 5AS, UK

*Electronic Engineering*, 30 Calderwood Street, London SE18 6QH, UK

*EDN Magazine*, Cahners Publishing Company, Cahners Building, 275 Washington Setreet, Newton, MA-02158, USA

Electronic Design, 611 Route 46 West Hasbrouck Heights, New Jersey 07604, USA

Harris Semiconductor, Sector P.O. Box 883, Melbourne, FL 32902–0883, USA

Maxim Integrated Products, 120 San Gabriel Drive, Sunnyvale, California 94086, USA

Linear Technology Corporation, 1630 McCarthy Blvd, Milpitas, California 95035, USA

Philips, Building BAF – 1, P.O. Box 218, 5600 MD Eindhoven, The Netherlands

I am indebted to Duncan Enright, Commissioning Editor, Butterworth-Heinemann, for his constant encouragement and guidance throughout the preparation of the final manuscript and the publication of the book.

I would like to thank Diane Chandler, Editorial Projects Manager and Alison Boyd, Desk Editor, Butterworth-Heinemann, for their excellent cooperation and help during the publication process. Finally, I would like to thank the Executive Director and the Directors of the Centre for Development of Telematics for their encouragement which made this work possible.

V. Lakshminarayanan

# 1 Digital circuits

## 1.1 Monostable multivibrators

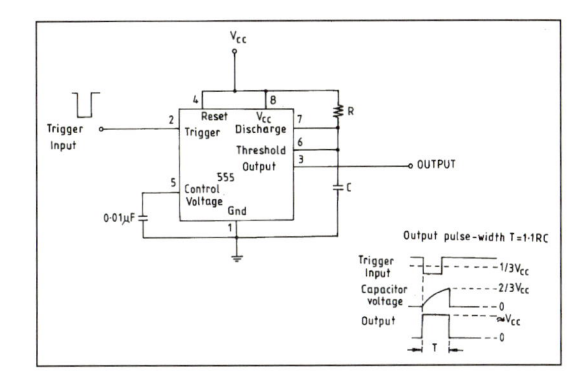

**Figure 1.1.1** Monostable using 555 timer

One of the frequently used building blocks in electronic circuit design is a monostable multivibrator. A monostable or one-shot is a circuit which remains in its stable state until it is triggered; on being triggered it goes into a temporary state for a duration depending on the timing components used and then returns to its stable state. The output of the one-shot is a pulse of specific width which depends on the timing components used.

Monostable multivibrators are of two basic types – non-retriggerable monostable and retriggerable monostable. A non-retriggerable monostable can't be retriggered by applying a trigger input when it is in the process of timing out, after being triggered once. A retriggerable monostable can be retriggered by applying a trigger input even before it has timed out; by repeatedly retriggering a retriggerable monostable long pulses can be obtained. A retriggerable one-shot remains in its triggered state until one pulse-width after the last trigger input was applied.

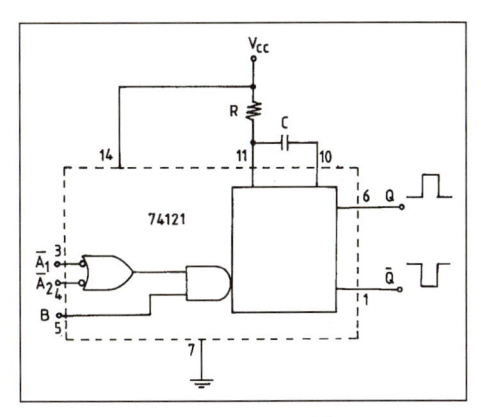

**Figure 1.1.2** 74121 monostable

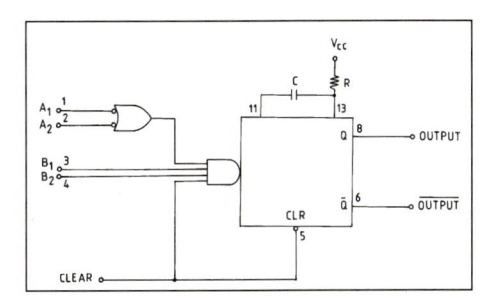

**Figure 1.1.3** Monostable using 74122

## 2 Digital circuits

Monostable multivibrators can be of edge-triggered type or voltage-level triggered type. An edge-triggered monostable could be positive-edge triggered or negative-edge triggered. A positive-edge triggered one-shot is triggered by a low-to-high going edge of the trigger input and a negative-edge triggered one-shot is triggered by a high-to-low going edge of the trigger input. Voltage-level triggered one-shots are triggered when their trigger input encounters a specific voltage level. Each of these types of one-shots has its own applications.

The operation of some of the commonly used monostable multivibrators is explained below:

### Monostable using a 555 timer

One of the simplest and most widely used operating modes of the 555 timer is the monostable (one-shot). This configuration requires only two external components for operation (see Figure 1.1.1). The sequence of events starts when a voltage below $(1/3)V_{CC}$ is sensed by the trigger comparator. The trigger is normally applied in the form of a short negative-going pulse. On the negative-going edge of the pulse, the device triggers, the output goes high and the discharge transistor turns OFF. Note that prior to the input pulse, the discharge transistor is ON, shorting the timing capacitor to ground. At this point the timing capacitor $C$ starts charging through the timing resistor $R$. The voltage on the capacitor increases exponentially with a time constant $T = RC$. Ignoring capacitor leakage, the capacitor will reach the $(2/3) V_{CC}$ level in 1.1 time constants, or

$$T = 1.1RC$$

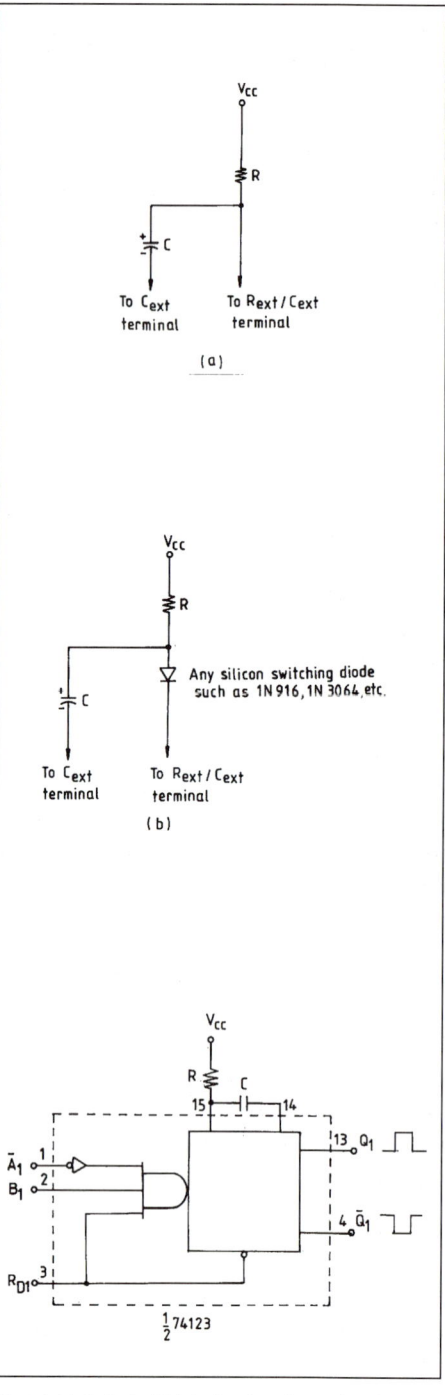

**Figure 1.1.4** (a,b) Methods of connecting $RC$ timing components in 74123. (c) Typical monostable using 74123

where $T$ is in seconds, $R$ is in ohms and $C$ is in farads. This voltage level trips the threshold comparator, which in turn, drives the output low and turns ON the discharge transistor. The transistor discharges the capacitor $C$ rapidly; the timer has now completed its cycle and will await another trigger pulse.

## 74121

This is a TTL monostable multivibrator. It has two low-going edge inputs ($\overline{A_1}$ and $\overline{A_2}$) and a high-going edge input (B) which can be used as an enable input. Complementary outputs Q and $\overline{Q}$ are available. The B input has a Schmitt-trigger circuit which allows the triggering of the one-shot even by slowly varying waveforms. This monostable is non-retriggerable, i.e. once it is triggered, further application of trigger input has no effect on the output until the one-shot has timed out. The output pulse duration can be varied from 20 ns to 28 s by choosing suitable timing components. The output pulse width is given by

$$T = RC \ln 2 = 0.7 RC$$

In applications where pulse cutoff is not critical capacitance up to 1000 μF and resistance as low as 1.4k can be used for timing. Figure 1.1.2 shows the typical connection diagram.

## 74122

This is a TTL retriggerable monostable multivibrator. It has a pair of low-going edge inputs, a pair of high-going edge inputs and a clear input. The B inputs are provided with Schmitt-trigger circuitry to ensure jitter-free

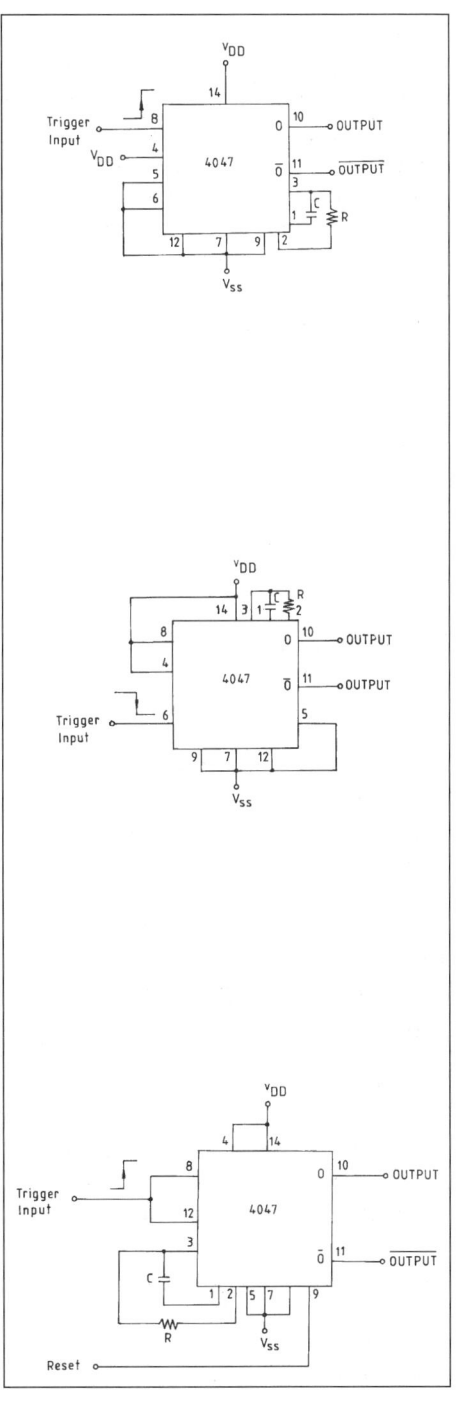

**Figure 1.1.5** Monostables using 4047: (a) positive-edge triggered; (b) negative-edge triggered; (c) retriggerable

triggering with transition rates as slow as 0.1 mV/ns. For $C \leq 1000\,\text{pF}$ the output pulse width can be obtained from the graph given in the datasheet of the 74122 in any TTL catalogue; for $C > 1000\,\text{pF}$ the output pulse width is given by $T = 0.45\,RC$. Figure 1.1.3 shows the typical connection diagram.

### 74221

This is a TTL dual monostable multivibrator identical to the 74121 in performance. The timing pulse width is given by the same equation as in the case of the 74121. The 74221 is non-retriggerable. Once triggered, the output can be reset by applying an active low input to the Reset terminal.

### 74123

This is a TTL dual retriggerable monostable multivibrator. Once the one-shot is triggered, the pulse length can be extended by applying the trigger input before the one-shot has timed out. The

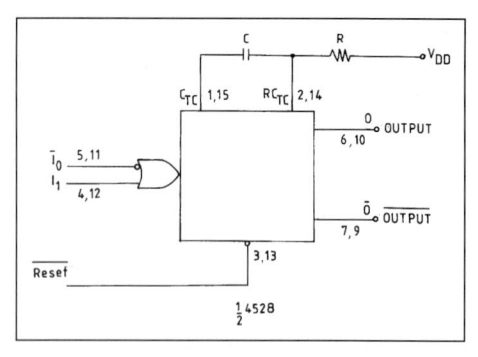

**Figure 1.1.6** Monostable using 4528

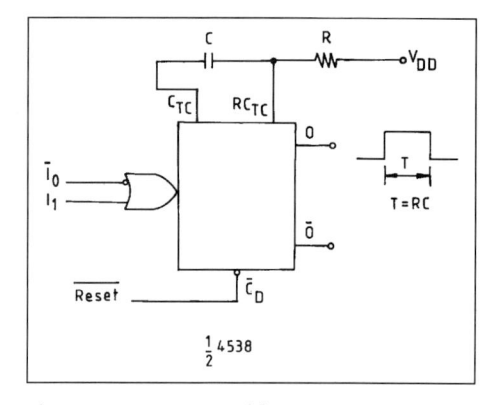

**Figure 1.1.7** Monostable using 4538

monostable can be triggered by a low-going edge input at $\overline{A}$ or a high-going edge input at B. After being triggered, the pulse can be terminated by applying a logic low input to the Reset terminal. For $C \leq 1000\,\text{pF}$, the monostable pulse width can be obtained from the nomograph given in the datasheet of 74123 in any TTL catalogue. When $C > 1000\,\text{pF}$, the output pulse width is given by

$$T = 0.28\,R/\{C[1 + (0.7/R)]\}$$

The timing components are connected as shown in Figure 1.1.4(a); if, however, an electrolytic capacitor with an inverse voltage rating of less than 1 V is to be used, then the connection diagram shown in Figure 1.1.4(b) should be used, in which case the output pulse width is given by

$$T = 0.25\,R/\{C[1 + (0.7/R)]\}$$

### 4047

This is a CMOS astable/monostable multivibrator. In the monostable mode, the device can be used as a positive-edge triggered (low-to-high transition at the positive trigger input) and as a negative-edge triggered (high-to-low transition at the negative trigger input) monostable. The device can be retriggered by applying a simultaneous low-to-high transition to both the positive trigger input and the retrigger input. The one-shot is reset by applying a high-level input to the Reset terminal, which makes Q low and $\overline{Q}$ high. In the monostable mode, the output pulse width is given by

$$T = 2.48\,RC$$

at the Q and $\overline{Q}$ outputs, where $R$ and $C$ are the timing components connected as shown in Figure 1.1.5 ($R$ between pins 2 and 3 and $C$ between pins 1 and 3).

## 4528

This is a CMOS dual retriggerable monostable multivibrator. Each one-shot has an active low input ($\bar{I}_O$), an active high input ($I_O$), an active low clear (or reset) input ($\overline{C}_D$) and complementary outputs O and $\overline{O}$. The monostable is triggered by a high-to-low transition at the $\bar{I}_O$ input when $I_O$ is low, or a low-to-high transition at the $I_O$ input when $\bar{I}_O$ is high. The monostable can be reset by applying a logic low signal to $\overline{C}_D$, which makes O low and $\overline{O}$ high and inhibits the monostable until $\overline{C}_D$ is made high.

The pulse width of the monostable depends on the $RC$ timing components used and is given by

$$T = KRC$$

where $K = 0.42$ for $V_{DD} = 5\,\text{V}$, $K = 0.32$ for $V_{DD} = 10\,\text{V}$ and $K = 0.30$ for $V_{DD} = 15\,\text{V}$, provided $C > 0.01\,\mu\text{F}$. For $C < 0.01\,\mu\text{F}$ the pulse width can be obtained from the nomogram given in the datasheet of 4528 in any CMOS 4000 series databook. Figure 1.1.6 shows the connections for a monostable using the 4528.

## 4538

This is a CMOS dual retriggerable monostable multivibrator. Each one-shot has an active low input ($\bar{I}_O$), an active high input ($I_O$), an active low reset input ($\overline{C}_D$) and complementary outputs (O and $\overline{O}$). Each monostable can be triggered by either a positive-going edge input or a negative-going edge input. The pulse width depends on the $RC$ timing components used and is given by $T = RC$. The range of pulse widths obtainable varies from 10 µs to infinity. A low level on the reset input $\overline{C}_D$ makes the O output low and $\overline{O}$ output high. The trigger inputs have Schmitt-trigger circuitry and hence slowly changing waveforms do not affect the triggering. Figure 1.1.7 shows the connections for a monostable using the 4538.

*Courtesy of Philips, The Netherlands*

6  Digital circuits

## 1.2 Digitally programmable monostable uses a PLD

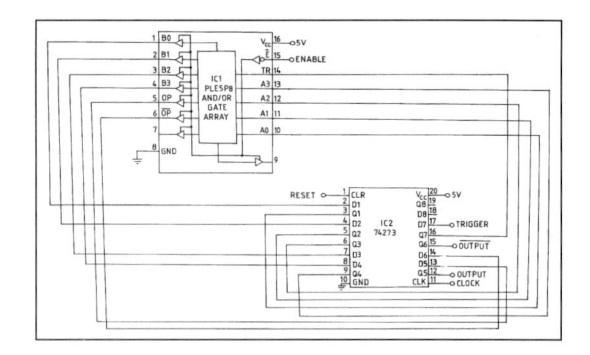

**Figure 1.2.1** Digitally programmable monostable using a PLD

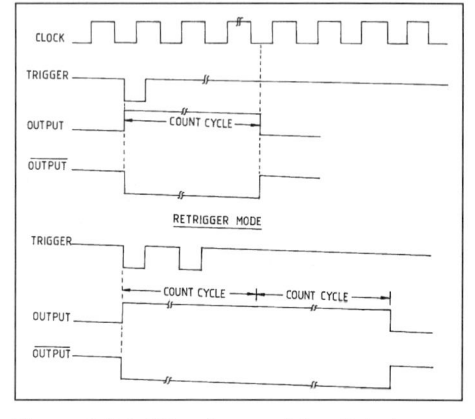

**Figure 1.2.2** Waveforms of the digital monostable

Commonly used ICs such as 74121/123 or 555 timer for monostable applications are limited in their flexibility as far as the timing that can be obtained is concerned.

In addition, inaccuracy of timing due to drift and tolerance in the values of RC components used, temperature drift inherent in the IC itself, time delays due to the switching speeds of the comparator in the IC (such as in the 555) and similar factors contribute to timing errors which could be a critical factor in some applications.

Longer time delays require large values of resistor and capacitor and in such cases the input bias current of the comparator and the leakage currents associated with the timing capacitor or the internal discharge transistor in the IC may limit the timing accuracy of the circuit.

The circuit shown in Figure 1.2.1 overcomes these disadvantages of conventional monostables and can give a wide range of pulse widths according to the clock frequency used. The stability and accuracy of the timing is decided by the stability and accuracy of the clock frequency input. The timing circuit does not use any RC components.

The monostable consists of a PLE 5P8 Programmable Logic Element (Monolithic Memories) and an octal latch, 74273.

The reference clock input to the 74273 may be part of the system in which the one-shot is to work. Four of the five inputs of the PLE are used for the state incrementing control function and the fifth input serves as the trigger input for the one-shot. The clear input (CLR) of the octal latch functions as the reset input of the one-shot.

Both true as well as complement (active high and active low) outputs are available from this monostable which makes it easier to interface to certain types of logic without an additional inverter.

Digital circuits 7

When the trigger input goes LOW, the true output of the one-shot goes HIGH and the monostable starts timing out the $n$ clock cycles it is programmed for. After the completion of one timing cycle, the true output of the monostable becomes logic LOW. In the retrigger mode, the output timing continues for another $n$ cycles and if the retrigger continues further, the output timing also continues. In this monostable, the number of clock cycles that can be programmed as the monostable timing period can vary from 1 to 16. By suitably selecting the clock frequency, a wide selection of timing pulses can be obtained. By selecting a PLD with a larger number of inputs, a larger number of timing cycle combinations can be obtained; for example, a PLD with 9 inputs can give 1 to 256 clock cycles as the timing period (one input being used as the trigger input and eight inputs for timing sequence generation). Figure 1.2.2 shows the waveforms of the circuit Figure 1.2.1. If the trigger pulse width happens to be greater than the period of the monostable, the one-shot will time out as usual; however, at the end of the $n$ clock cycles another timing cycle will begin without changeover.

**Table 1.2.1** Truth table: timing generator for digital monostable

| State | Next state | Timing waveforms | | | Comments |
|---|---|---|---|---|---|
| AAAA | BBBB | Trigger input | Monostable outputs | | |
| 3210 | 3210 | T | OP | COP | |
| LLLL | LLLH | H | L | H | |
| LLLH | LLHL | H | L | H | |
| LLHL | LLHH | L | H | L | ; Timing starts |
| LLHH | LHLL | H | H | L | ; for |
| LHLL | LHLH | H | H | L | ; $n$ ( = 4) clock |
| LHLH | LHHL | H | H | L | ; cycles |
| LHHL | LHHH | L | H | L | ; Retriggered, |
| LHHH | HLLL | H | H | L | ; another timing |
| HLLL | HLLH | H | H | L | ; cycle starts |
| HLLH | HLHL | H | H | L | ; Timing over |
| HLHL | HLHH | H | L | H | |
| HLHH | HHLL | H | L | H | |
| HHLL | HHLH | H | L | H | |
| HHLH | HHHL | H | L | H | |
| HHHL | HHHH | H | L | H | ; Loop here until reset |

The PLD programming table is given in Table 1.2.1 and the Boolean equations used to program the PLD are given in Table 1.2.2.

Since this circuit does not use any *RC* components, and the monostable is digitally programmable over a wide range, it can be very useful for precision timing applications.

8  Digital circuits

**Table 1.2.2** PLD equations

| ADD | A0 | A1 | A2 | A3 | T | |
|---|---|---|---|---|---|---|
| DAT | B0 | B1 | B2 | B3 | OP | COP |

;NEXT STATE GENERATOR

B0 = /A0   ;INCREMENT LSB

B1 = /A1*A0 + A1*A0   ;INCREMENT BIT1

B2 = A2*/A1 + A2*/A0 + /A2*A1*A0   ;INCREMENT BIT2

B3 = A3*/A2 + A3*/A0 + A3*/A1 + /A3*A2*A1*A0   ;INCREMENT BIT3

;TIMING WAVEFORMS

OP = /A3*A1*/A0*/T + /A3*A1*A0*T + /A1*T   ;OUTPUT

COP = /OP = A3*T + /A3*/A2*/A1*T + A3*A2*/A0*T   ;COMPLEMENTARY OUTPUT

*Electronic Engineering, February 1989. Reprinted with permission*

## 1.3  Use a counter as a pulse-stretcher

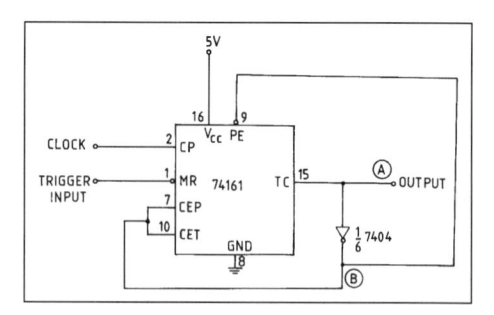

**Figure 1.3.1** Monostable using a counter

The circuit in Figure 1.3.1 shows a technique to use a 4-bit binary counter, 74161, as a programmable pulse stretcher, i.e. a digital monostable. It does not use any *RC* components and the timing can be as stable as the clock used.

In the circuit shown in Figure 1.3.1, the terminal count output TC of the counter 74161 is connected to the count enable inputs CET and CEP through a

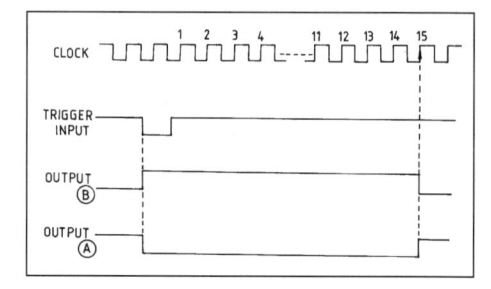

**Figure 1.3.2** Waveforms of circuit Figure 1.3.1

spare inverter. The trigger input of the monostable is connected to the master reset input MR. The counter is reset by the fall of an input pulse, making terminal count output TC LOW. This makes CET and CEP HIGH, enabling the counter. When the outputs of the counter become HHHH, i.e. at the 15th count, the terminal count output TC goes HIGH which makes the inverter output LOW, thereby pulling CET and CEP to LOW level. This inhibits further counting.

Figure 1.3.2 shows the waveforms of the monostable. The two output waveforms from this circuit at points A and B give complementary timing outputs. By suitably changing the clock frequency as required, the monostable timing can be made programmable. Conventional monostables use the time required to charge a capacitor through a resistor to derive the timing whereas this monostable uses an external clock (which could be the clock available on board) as the timing reference and the counter to accumulate the delay count. The timing accuracy of this circuit is, therefore, independent of leakage current, drift, tolerance and similar problems associated with conventional monostables. A wide range of timing pulses can be obtained using this technique.

## 1.4 Digital noise canceller

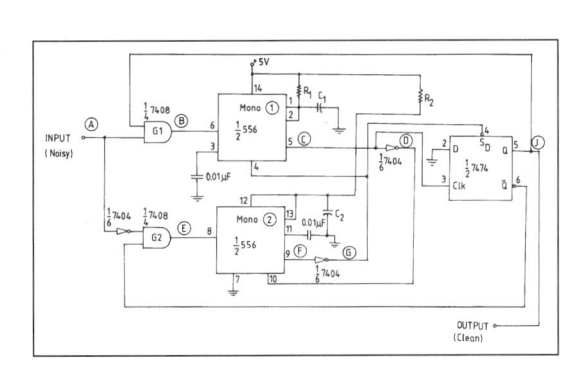

**Figure 1.4.1** Digital noise canceller

The circuit shows in Figure 1.4.1 can stop noise pulses giving a false digital output. It can clean up noise from any digital signal provided the timing details of the signal are known.

The corrupted digital signal is input to an AND gate and the output of the gate is used to trigger a negative-edge triggered monostable built using a 556 timer. When the logic signal goes LOW, the monostable is triggered and the output stays HIGH for a time $T_A$. The rising edge of one-shot 1 resets the output of the arbitration D flip-flop to zero.

When the output of the flip-flop goes LOW, the AND gate $G_1$ is disabled. Until one-shot 1 times out, further transitions at the input due to noise cannot cause any change in the output. After a time $T_A$, the positive edge of the input triggers one-shot 2 at the negative edge and the inverted output of one-shot 2 pre-sets the flip-flop to a HIGH state.

This disables $G_1$ and the inverted output from one-shot 2 inhibits AND gate $G_2$. Since both gates and the flip-flop are disabled, spurious noise cannot give a false

## 10 Digital circuits

output at J, which is the flip-flop output.

The times $T_A$ and $T_B$ of the one-shots should be chosen such that both are less than the time when the digital signal is low but their sum is higher.

*First published in Electronics and Wireless World, May 1990. Reprinted with permission*

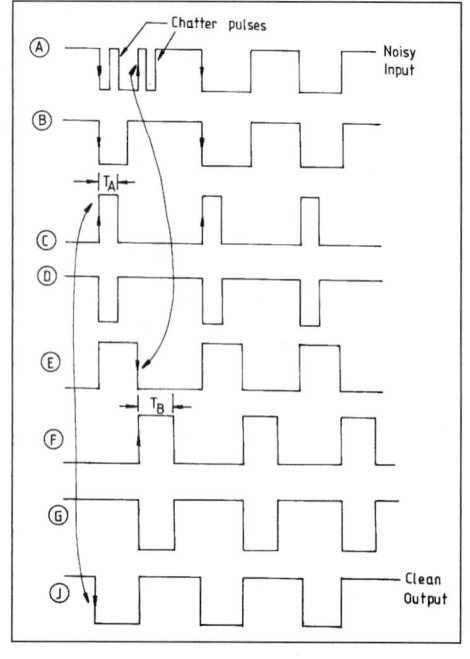

**Figure 1.4.2** Waveforms of circuit in Figure 1.4.1

## 1.5 Build a low-cost delay line

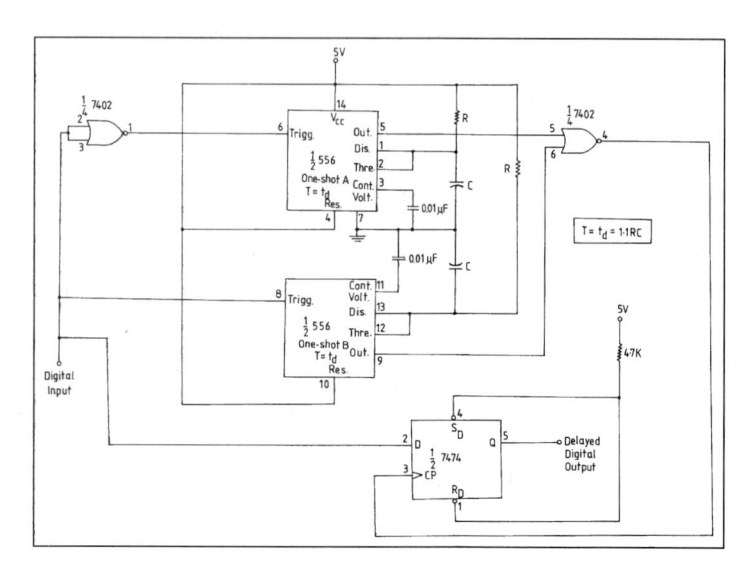

**Figure 1.5.1** Digital delay line

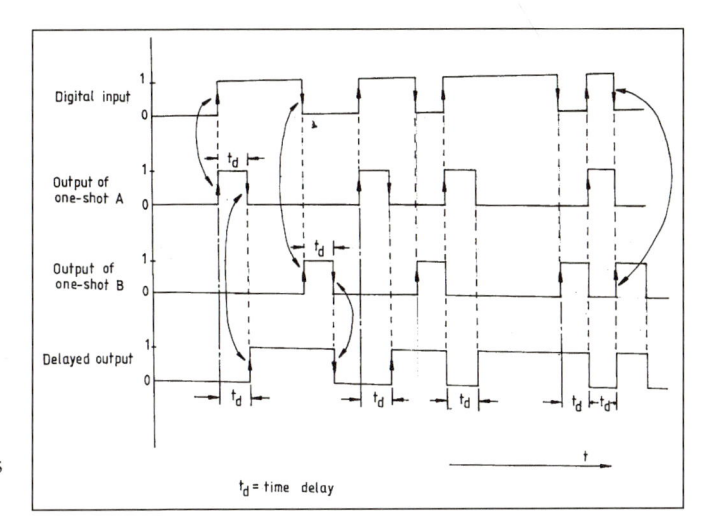

**Figure 1.5.2** Waveforms of digital delay line

$t_d$ = time delay

An inexpensive programmable delay-line circuit for TTL logic signals can supply long delays for complex digital waveforms and still preserve the input pulse spacing at its output (see Figure 1.5.1). Just two *RC* networks determine the time delay.

The circuit uses two negative-edge triggered monostable multivibrators, built around a dual 556 timer. The delay of both the multivibrator circuits is the same and equal to $t_D$, as follows:

$$t_D = 1.1\ RC$$

A NOR gate (1/4 of a 7402) inverts the input to one of the multivibrators with respect to the other input. Therefore, one of the multivibrators triggers at the leading edge and the other at the trailing edge of the input waveform. The outputs of the two multivibrators feed another NOR gate (also 1/4 of a 7402), whose output, in turn, clocks an arbitration flip-flop (1/2 of a 7474). Also, the input waveform steers the D-input of this flip-flop. Finally, a time-delayed version of the input waveform appears at the output of the flip-flop (see Figure 1.5.2).

This circuit will work for any waveform where the input pulse widths are less than the delay time and where the period between input pulses is more than the delay time.

*Reprinted with permission from Electronic Design, 37(15) 13 July, 1989. Copyright 1989, Penton Publishing Inc.*

◆ ◆

## 1.6 Digital phase detector

**Figure 1.6.1** Digital phase detector

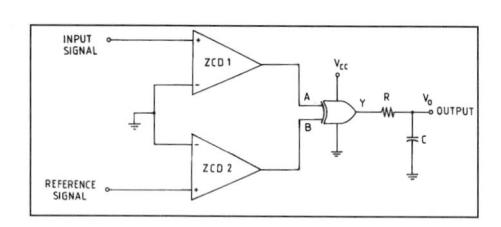

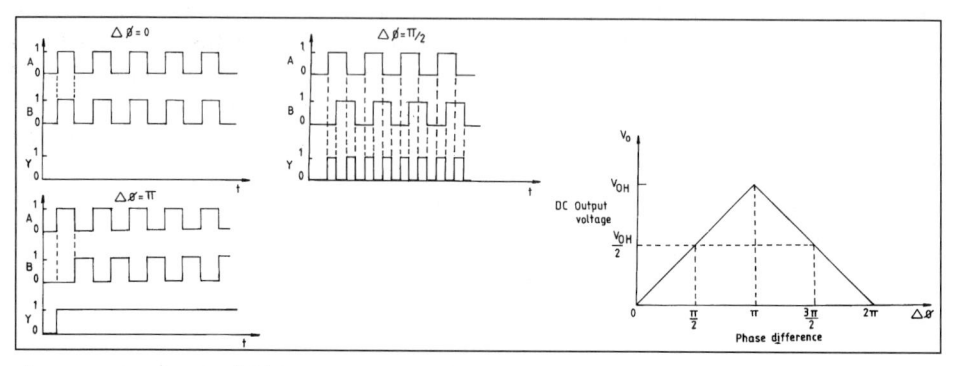

**Figure 1.6.2** (a) Digital phase comparator waveforms. (b) Plots of phase difference versus DC output for the digital phase detector

A simple digital phase detector is shown in Figure 1.6.1. It consists of a pair of zero-crossing detectors to which the input signal and the reference signal are connected. The outputs of the zero-crossing detectors are connected to an EX-OR gate whose output is averaged by a simple $RC$ low-pass filter. As seen from the waveforms of Figure 1.6.2(a), the EX-OR gate acts as an edge detector, detecting the rising and falling edges of the reference and the input waveforms. The output of the $RC$ filter consists of a DC voltage varying with phase difference between the input and reference signals as shown in the graph of Figure 1.6.2(b). The DC voltage is maximum at 180° phase difference and 0 at 0° phase difference; at 90° phase difference, the DC voltage is mid-way, i.e. $V_{OH}/2$, where $V_{OH}$ is the logic HIGH level output voltage of the EX-OR gate used.

Such a phase detector can be used for lock detection in PLL circuits.

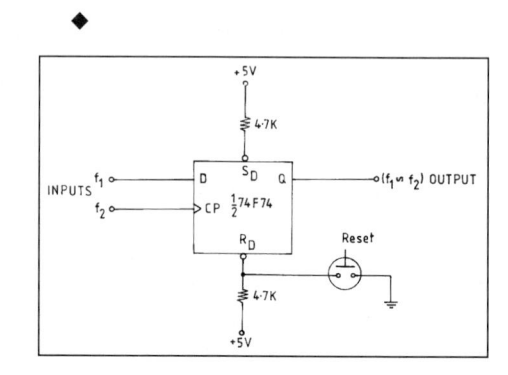

## 1.7 Digital frequency mixer

**Figure 1.7.1** Digital frequency-difference extractor

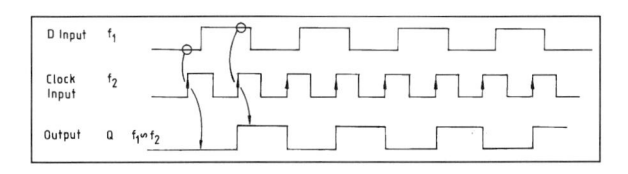

**Figure 1.7.2** Digital frequency mixer waveforms

A D flip-flop can be used as a digital frequency-mixer, i.e. to extract the difference between the frequencies of two digital pulse trains. One of the pulse trains is connected to the D input of the flip-flop and the other to the clock input. The Q and $\overline{Q}$ outputs of the flip-flop will have a frequency which is equal to the difference between the frequencies of the inputs, i.e. $f_Q = f_{\overline{Q}} = f_1 \sim f_2$ (see Figures 1.7.1 and 1.7.2).

Such a digital frequency mixer can find application where it is required to compare two frequencies, e.g. in PLL lock indication.

◆ ◆

## 1.8 Digital attenuator

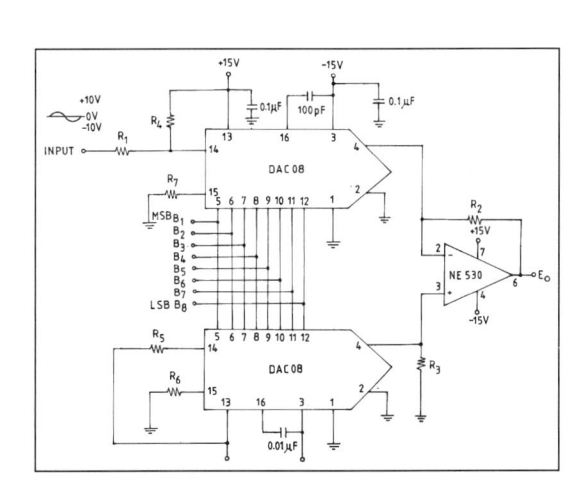

**Figure 1.8.1** Digital attenuator

Figure 1.8.1 shows the circuit for a DC-coupled digital attenuator or programmable gain amplifier.

In the circuit pin 14 of the DAC is a virtual ground. Current must always flow into pin 14, so the current through $R_4$ must be greater than that through $R_1$ when the input signal is at its most negative usable value. If the input signal value goes low enough to cause the current through $R_1$ to be greater than that through $R_4$, output clipping will occur.

To extend the operating frequency range, the compensation capacitor $C_C$ needs to be minimized, which implies that the resistance at pin 14 ($R_1$ and $R_4$) must be minimized. If the voltage to which $R_4$ and $R_5$ are returned has any noise on it at all, $R_4$ and $R_5$ should be formed of two resistors with their junction bypassed with 0.1 µF to ground. Pin 15 could be grounded with a small sacrifice in accuracy and temperature drift; $R_6$ and $R_7$ compensate for reference amplifier input offset.

## 14 Digital circuits

$R_1$ and $R_4$ should be chosen such that, when the input is at peak usable signal, the total current into pin 14 does not exceed 4 mA. When the input is most negative, $R_1$ current must be less than $R_4$ current (remember, pin 14 is always at 0 V). Also when the input is at the absolute positive peak value, current into pin 14 should not exceed 5 mA. Minimum compensation capacitor ($C_C$), in pF is 15 times the parallel combination of $R_1$ and $R_4$ in kohms. With a single DAC there is a DC offset at the circuit output that varies with the digital word input. To eliminate this, we use a second DAC to subtract this offset at the sum node of the op amp.

## Example

Input signal is to be 20 V p-p, centred at 0 V. Maximum input frequency is to be 15 kHz. Power supplies available are ± 15 V, both regulated. Determine values of all resistors for maximum gain of unit.

## Solution

At minimum input (–10 V), reference current, $I_{REF}$ is given by

$$I_{REF} = (15/R_4) + [(-10)/R_1]$$

If minimum $I_{REF} = 0$, then,

$$15/R_4 = 10/R_1$$

and $R_4 = (1.5)(R_1)$. Therefore, 60% of $I_{REF}$ comes through $R_4$. If we let $I_{REF}$ go to about 3.9 mA (4 mA is the maximum recommended), $R_4$ current is found to be

$$I_{R_4} = (0.6)(3.9) = 2.34 \text{ mA}$$

and

$$R_4 = 6.4k$$

The balance of the reference current $I_{R_1}$ is found to be

$$I_{R_1} = 3.9 - I_{R_4} \text{ mA}$$

i.e. $I_{R_1} = 3.9 - 2.34 = 1.56$ mA and $R_1 = 6.4$k

Using commonly available values, and remembering that $R_4$ current must exceed $R_1$ current, we set

$$R_1 = 6.8k$$

and

$$R_4 = 6.2k$$

Maximum reference current is now

$$I_{REF(max)} = (15/6.2k) + (10/6.8k) = 3.9 \text{ mA}$$

The parallel combination of $R_1$ and $R_4$ is found to be 3.24k, so minimum compensation capacitor value is

$$C_{C(min)} = (3.24)(15) = 48.6\,pF$$

If we use 50 pF, from the graph (ref. 26, pp. 6–79) we find $f_{MAX}$ to be 370 kHz. For unity gain,

$R_2 = R_1 = 6.8k$

$R_3 = R_2 = 6.8k$

$R_5 = R_1 = 6.8k$

$R_6 = R_7 = (R_1)(R_4)/(R_1 + R_4) = 3.24k$ (use 3.3k)

*Courtesy of Philips, The Netherlands*

## 1.9 Binary frequency divider

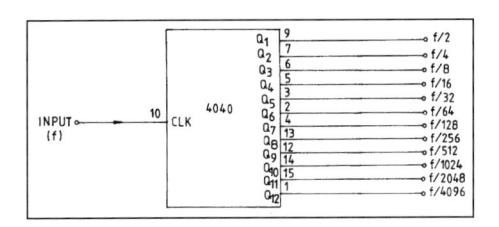

**Figure 1.9.1** Binary frequency divider

There is, sometimes, a need to divide an available clock frequency in powers of two. One technique to accomplish this using the CD 4040, which is a 12-stage binary ripple counter, is shown in Figure 1.9.1. The outputs $Q_1$–$Q_{12}$ of the counter are at frequencies of $f/2$, $f/4$, $f/8$, $f/16$ and so on up to $f/4096$ (where $f$ is the frequency of the input clock) as shown in Figure 1.9.1.

## 1.10 Digital edge-detectors

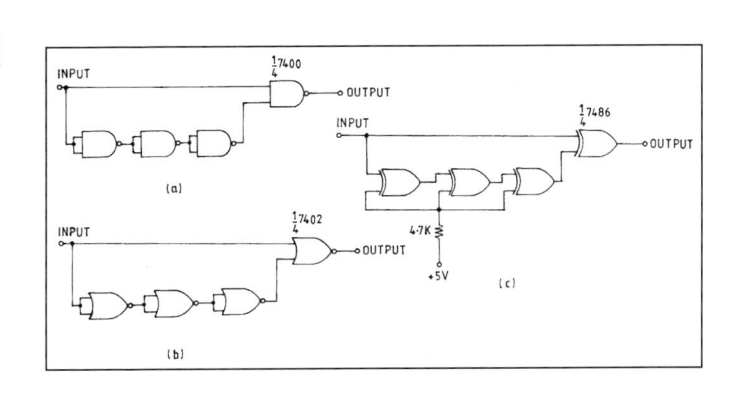

**Figure 1.10.1** Digital edge detectors

## 16  Digital circuits

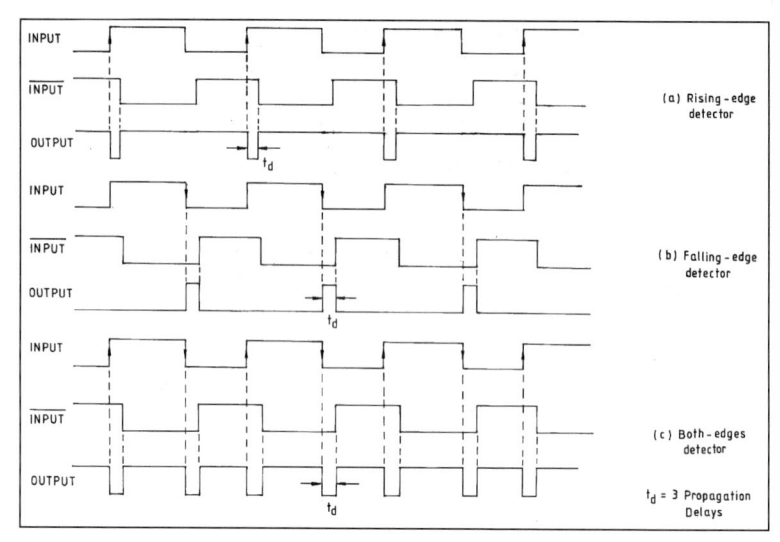

**Figure 1.10.2** Waveforms of digital edge detectors

A technique to detect the edges of a digital waveform is useful in applications such as frequency doubling, multiplication, etc. Three techniques to detect rising edges, falling edges and both rising and falling edges are shown in Figure 1.10.1. Figure 1.10.1(a) detects rising edges, Figure 1.10.1(b) detects falling edges and Figure 1.10.1(c) detects both rising and falling edges. Each edge detector consists of a logic gate which is fed by the waveform whose edge is to be detected, in true and complement forms. The complement signal is derived from the true signal by configuring three gates from the quad package as logic inverters. The three inverters are connected in tandem in each case in order to delay the input waveform by three propagation delays. As the waveforms in Figure 1.10.2 show, the propagation delay of the logic used is usefully employed in the circuits to detect the edges of the waveforms.

◆ ◆

## 1.11  One IC doubles frequency

**Figure 1.11.1** Single-chip frequency doubler

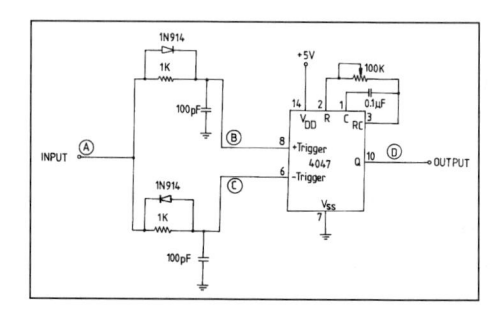

An inexpensive frequency doubler and duty-cycle-variation circuit can be designed around one IC – a monostable 4047 that's triggered directly by a LOW-to-HIGH or HIGH-to-LOW transition. The circuit uses two *RC* differentiators to detect the leading and trailing edges of a digital input signal (see Figure 1.11.1). The differentiator's transition spikes trigger the 4047 at both edges, effectively doubling the input signal frequency (see Figure 1.11.2). The external potentiometer–capacitor combination at pins 1, 2 and 3 of the monostable IC can be varied to adjust the output pulse train's duty cycle up to 100%.

*Reprinted with permission from Electronic Design, Vol.38, No.8, 26 April, 1990. Copyright 1990, Penton Publishing Inc.*

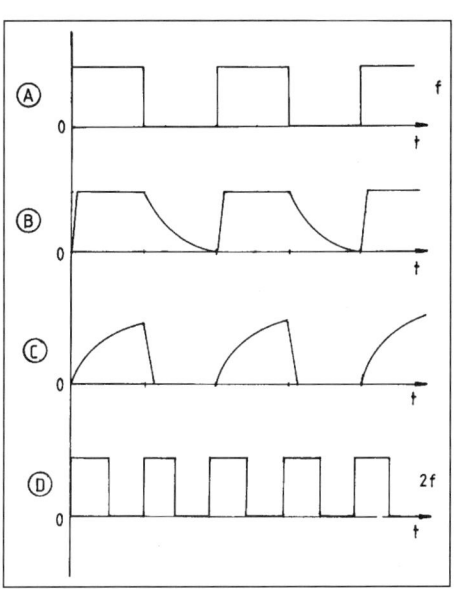

**Figure 1.11.2** Waveforms of the circuit Figure 1.11.1

## 1.12 Inexpensive frequency counter/tachometer

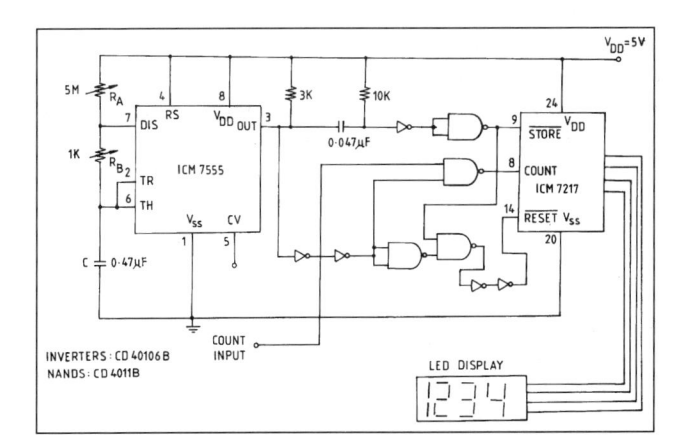

**Figure 1.12.1** Inexpensive frequency counter

The ICM 7217 is a four-digit, presettable up/down decade counter with an on-board presettable register continuously compared to the counter. It provides multiplexed 7-segment LED display outputs with common-anode configuration. Digit and segment drivers are provided to directly drive displays of up to 0.8"

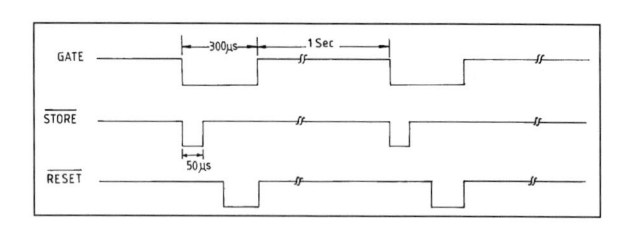

**Figure 1.12.2** Waveforms of circuit Figure 1.12.1

character height (common-anode) at a 25% duty cycle. The frequency of the on-board multiplex oscillator may be controlled with a single capacitor or the oscillator may be allowed to free run. Leading zeros can be blanked. The data appearing at the 7-segment and BCD outputs is latched; the content of the counter is transferred into the latches under external control by means of the STORE pin. The ICM 7217A is a common-cathode version. The maximum count with 7217 and 7217A is 9999. To permit operation in noisy environments and to prevent multiple triggering with slowly changing inputs, the count input is provided with a Schmitt trigger. Input frequency is guaranteed to 2 MHz, although the device will run typically with $f_{in}$ as high as 5 MHz. The device operates on a single 5 V supply.

The Digit and SEGment drivers provide a decoded 7-segment display system, capable of directly driving common-anode LED displays at typical peak currents of 40 mA/segment. This corresponds to average currents of 10 mA/segment at a 25% multiplex duty cycle.

The circuit shown in Figure 1.12.1 can be used as an inexpensive frequency counter and tachometer. It uses the low-power timer ICM 7555 (CMOS 555) to generate the gating, store and reset signals for the ICM 7217. To provide the gating signal, the timer is connected as an astable multivibrator, using $R_A, R_B$ and $C$ to provide an output that is positive for approximately one second and negative for approximately 300–500 μs. The positive waveform time is given by $t_{wp} = 0.693 (R_A + R_B)C$ while the negative waveform time is given by $t_{wn} = 0.693 R_B C$. The system is calibrated by using a 5 Mohm potentiometer for $R_A$ as a 'coarse' control and a 1k potentiometer for $R_B$ as a 'fine' control. CD 40106B inverters are used as a monostable multivibrator and for reset time delay.

The circuit can be used as a simple frequency counter and tachometer.

*Courtesy of Harris Semiconductor, Melbourne, FL*

◆ ◆

## 1.13 Missing pulse detector

You can use the retriggerable monostable 74123 as a missing pulse detector. The circuit shown in Figure 1.13.1 gives a steady logic HIGH output as long as the input pulses arrive at a predetermined rate. If a single pulse fails to arrive, as the waveforms in Figure 1.13.2 show, the output of the monostable goes LOW; the output then goes HIGH as the pulses continue to arrive at the input. The principle of operation is quite simple: 74123 is retriggerable, i.e. if the monostable is triggered before it times out, the Q output remains HIGH. In the case of the missing pulse detector, the pulse train to be monitored is used as the trigger input for the monostable. The time period of the monostable is kept about 25% longer than the time interval between pulses. The monostable pulse width is given by $T = 0.28\, R/C\,(1 + 0.7/R)$ for $C > 1000\,\text{pF}$. For values of $C \leq 1000\,\text{pF}$ use the monostable design chart given in the data sheet of 74123. The circuit can be used to monitor a data line and flag an error condition when one or more pulses are missed. Optionally you can connect an LED through a suitable current-limiting resistor for visual indication of an error condition.

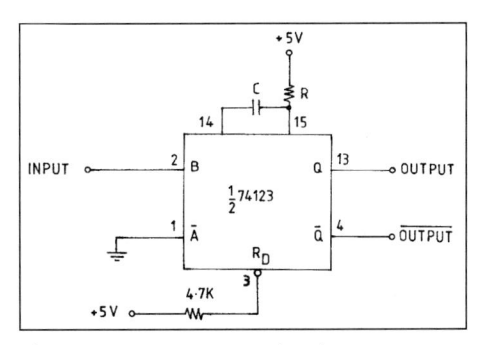

**Figure 1.13.1** Missing pulse detector

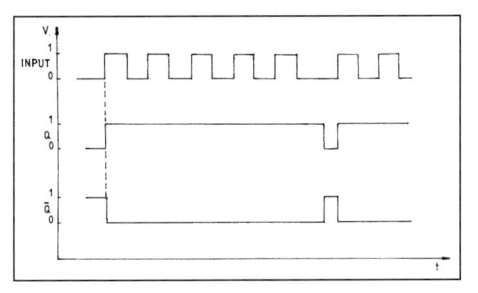

**Figure 1.13.2** Waveforms of circuit Figure 1.13.1

## 1.14 Digital quadrature phase-shifter

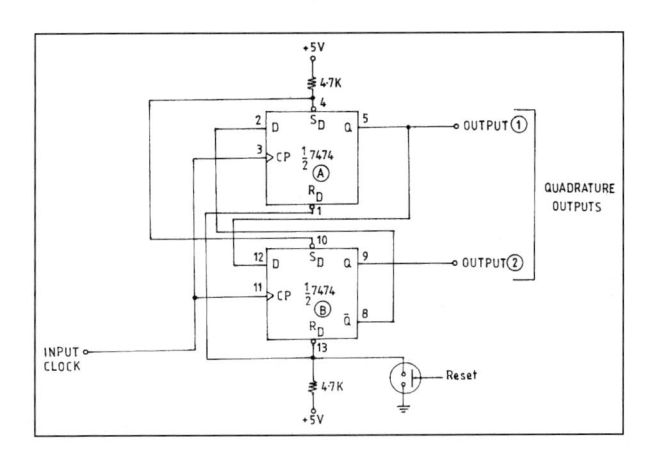

**Figure 1.14.1** Digital 90° phase-shifter

The circuit shown in Figure 1.14.1 can be used to generate quadrature clocks from an available system clock. The circuit essentially consists of a pair of synchronously clocked D flip-flops. The Q output of flip-flop A is tied to the D input of flip-flop B and the $\overline{Q}$ output of flip-flop B is tied to the D input of flip-flop A. The Set inputs of both the flip-flops are tied to a logic HIGH level and the Reset inputs to a push-button switch as shown. The Q outputs of flip-flops A and B are in quadrature, i.e. out of phase by 90° with respect to each other as shown in Figure 1.14.2. The time period of the quadrature outputs is $4T$ if $T$ is the period of the input clock. In other words, the frequency of the quadrature outputs is $f/4$ if $f$ is the frequency of the input clock. This single chip circuit should prove useful where a low-cost quadrature pulse-train generation technique is required.

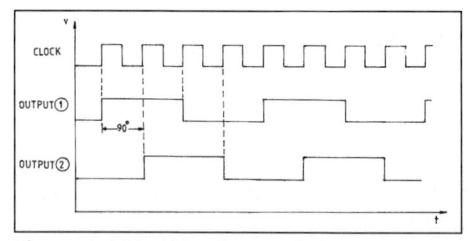

**Figure 1.14.2** Waveforms of circuit Figure 1.14.1

## 1.15 Pulse-width discriminator

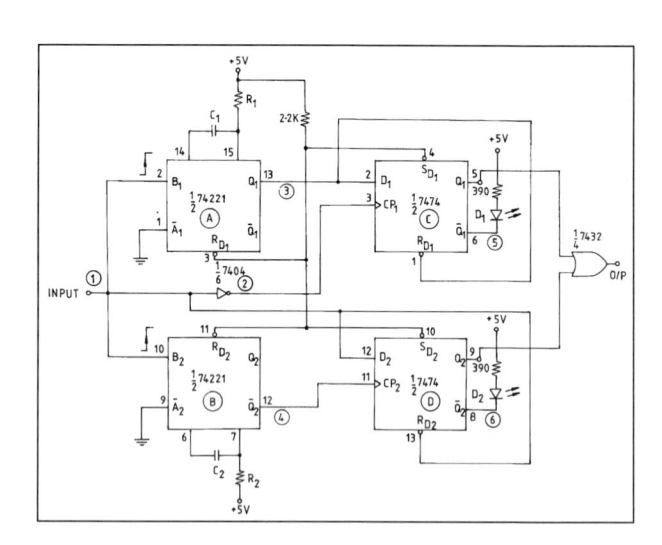

**Figure 1.15.1** Pulse-width discriminator

The circuit shown in Figure 1.15.1 can be used to monitor the width of pulses in a data stream. It produces an output pulse whenever the width of the input pulse is either less than a set minimum or greater than a set maximum limit. The circuit consists of a dual monostable multivibrator, 74221, whose halves A and B are configured as one-shots with time periods equal to the minimum and maximum pulse widths, i.e. $T_{min}$ and $T_{max}$, respectively. The minimum and maximum values of the time periods of the one-shots can range from 35 ns to 28 s ($T = RC \ln 2$). The input pulse train triggers both the one-shots at the rising edges of the waveform. As seen from the waveforms in Figure 1.15.2, the Q outputs of flip-flops C and D go HIGH (and the $\overline{Q}$ outputs go LOW) only if a pulse outside the pulse-width appears at their trigger inputs. The LED $D_1$ glows when a pulse of width less than the minimum set limit appears at the input and the LED $D_2$ glows when a pulse of width greater than the maximum set limit appears at

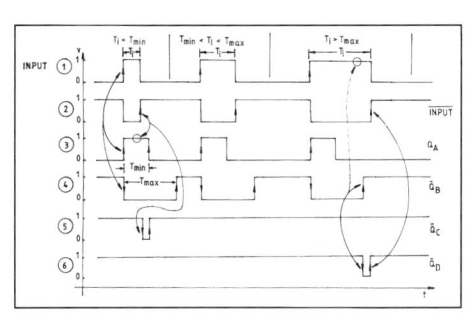

**Figure 1.15.2** Waveforms of circuit Figure 1.15.1

the input. Instead of LED indication, if required, an error flag can be derived by combining the Q outputs of the flip-flops through an OR gate.

The circuit, therefore, effectively monitors input data for conformance to predetermined pulse-width limits.

*Printed with the permission of Electronic Engineering*

◆ ◆

## 1.16 Power-on reset circuit using monostable 74121

**Figure 1.16.1** Power-on reset circuit

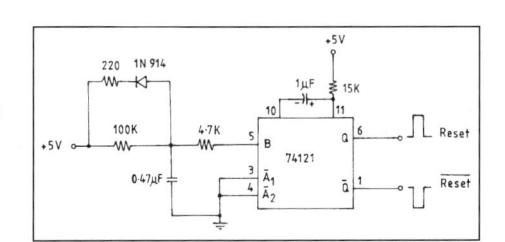

Circuits using counters, flip-flops, etc. need a reset signal on power-up so that the system does not enter a wrong logic state when power is first applied. The circuit shown in Figure 1.16.1 generates glitch-free complementary reset signals using a 74121 monostable multivibrator. When power is first applied, the 0.47 μF capacitor charges through the 100K resistor towards the supply voltage of 5 V, and the instantaneous voltage across the capacitor at any instant $t$ is given by $V(1 - e^{-t/RC})$ where $V$ is the supply voltage. When the capacitor voltage exceeds the positive-going threshold for 74121, which is 2 V, the monostable gets triggered and produces complementary pulses of duration $T = RC \ln 2$, which with $RC$ values of 1 μF and 15 k works out to about 10 mS, at the Q and $\overline{Q}$ outputs. This pulse width is sufficient to reset the logic circuits in the system.

The 1N914 diode provides a discharge path for the 0.47 μF capacitor when power is turned off; otherwise the capacitor will try to discharge through the B trigger input of the monostable. Since 74121 has Schmitt trigger inputs, even the slowly charging capacitor voltage is converted to a sharp positive edge for jitter-free triggering of the monostable thereby providing high noise immunity which is typically of the order of 1.2 V. Once the monostable is triggered the outputs are independent of further transitions of the inputs and are dependent only on the *RC* timing components used. One half of a 74221 dual monostable can also be used instead of a 74121.

◆ ◆

## 1.17 Quadrature waveform decoder

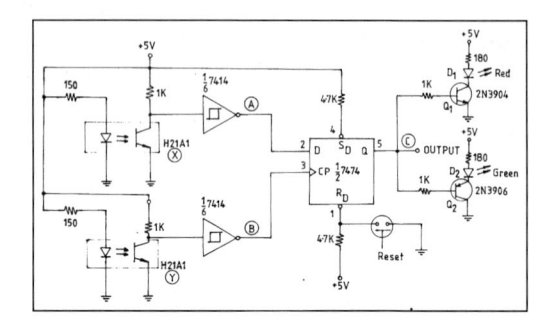

**Figure 1.17.1** Direction of rotation indicator

It is sometimes required to decode the information obtained from a pair of optically coupled interrupter modules mounted in quadrature (see Figure 1.17.2) in order to determine the direction of rotation of an object such as a motor shaft. A rotating transparent disc with an opaque pattern in it or conversely, an opaque disc with a transparent pattern is generally used with a photo-coupled interrupter to sense the rotation of the motor shaft. As the disc

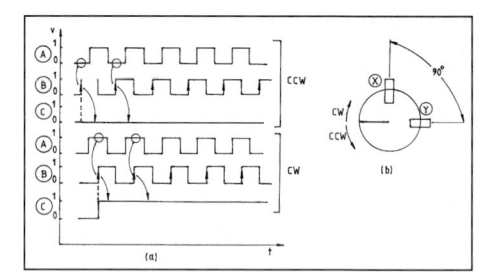

**Figure 1.17.2** (a) Waveforms of circuit Figure 1.17.1. (b) Mounting arrangement for direction sensing

rotates, the light beam from the LED in the photo-coupled interrupter module is interrupted before it falls on the photo-detector depending on the pattern in the disc. The interruptions of the light beam are converted to a digital output by the detector in the interrupter module. If a pair of photo-coupled interrupter modules is used, which is the case if the direction of rotation is to be sensed, the digital output from each of the modules is wave-shaped using a pair of Schmitt triggers (7414) as shown in Figure 1.17.1. One of the wave-shaped outputs is connected as the D input of an arbitration flip-flop (7474) and the other wave-shaped output is used to clock the D input waveform. The Q output of the arbitration flip-flop drives a pair of complementary transistors with LEDs connected as shown in Figure 1.17.1.

Now assume that the disc rotates in a clockwise (CW) direction. Under this condition, the wave-shaped output from the interrupter module X leads the wave-shaped output from the interrupter module Y by 90° (since the modules are mounted in quadrature along the disc as shown in Figure 1.17.2). Therefore, a logic 1 is clocked for every rising edge of the waveform at point A in Figure 1.17.1, and the Q output of the arbitration flip-flop is HIGH as long as the disc rotates clockwise, which is clear from the waveforms in Figure 1.17.2. This makes the transistor $Q_1$ ON and $Q_2$ OFF, lighting up the red LED $D_1$ indicating a clockwise rotation. The opposite effect occurs if the disc rotates counter-clockwise, i.e. the output of Y leads that of X and a logic O is clocked into the Q output of the arbitration flip-flop, turning ON transistor $Q_2$ and lighting the green LED $D_2$. If required, the Q output of the arbitration flip-flop can be used as a logic sense output, instead of LED indication. Since the outputs of the interrupter modules are in quadrature and this circuit decodes the outputs, this circuit can also be used to decode any pair of quadrature waveforms, i.e. the circuit functions as a quadrature waveform decoder.

## 1.18  2-to-1 digital multiplexer using spare tristate buffers

**Figure 1.18.1** 2 line to 1 line multiplexer using spare tristate buffers

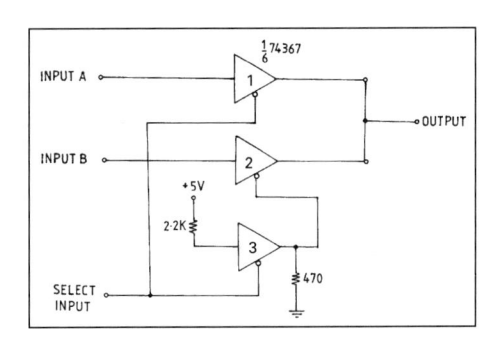

The arrangement shown in Figure 1.18.1 can be used to configure a simple 2-to-1 digital multiplexer using spare tristate buffers on board such as the 74367. The multiplexer enables one of two inputs A and B to the output depending on which tristate buffer is enabled. One of the tristate buffers (labelled 3 in Figure 1.18.1) is used as a logic inverter to derive complementary select signals for the other two tristate buffers. If the applied select signal is logic 0, the input A is directed to the output since buffer 1 is enabled and buffer 2 is disabled. The reverse happens if a logic 1 select signal is applied. This technique avoids the need for a multiplexer chip in a deisgn where spare tristate buffers are available.

## 1.19  Switch debouncers

Mechanical switch contacts generally bounce for a short duration after they are closed. This usually lasts from a few milliseconds to tens of milliseconds. Digital logic circuits can interpret such jitter or bounce pulses as valid logic transitions

## 24 Digital circuits

and give erratic performance. To avoid such invalid interpretation of contact bounce, debouncing circuits are used when interfacing mechanical switches to digital logic. The circuits in Figures 1.19.1(a) to (d) show the commonly used techniques for debouncing a mechanical switch.

The debouncer in Figure 1.19.1(a) uses a pair of cross-coupled NOR gates to form a Set–Reset flip-flop. Bounce of jitter inputs from the switch will not produce multiple output transitions of the S–R flip-flop since the first application of a logic high to the S (or R) input will set (or reset) the flip-flop and multiple application of the same logic level will not have any effect on the flip-flop.

The debouncer in Figure 1.19.1(b) uses a pair of cross-coupled NAND gates to form a S–R flip-flop instead of the NOR gates used in Figure 1.19.1(a). A similar explanation holds in this case as in Figure 1.19.1(a).

Figure 1.19.1(c) uses a D flip-flop to achieve debouncing action by configuring the D flip-flop as a S–R flip-flop.

In Figure 1.19.1(d) a Schmitt trigger inverter and a $RC$ network at its input achieve debouncing action of the switch S. The $RC$ network at the input to the Schmitt trigger provides an integrating or low-pass (smoothing) effect on the multiple pulses at the switch contact and when the capacitor voltage exceeds the threshold voltage level of the Schmitt trigger, it changes state. The $RC$ time constant chosen (62k $\times$ 0.47 µF) $\simeq$ 30 mS should be adequate for most mechanical switches. This method is suitable for CMOS type logic elements since they have a high input impedance, but is not recommended for TTL.

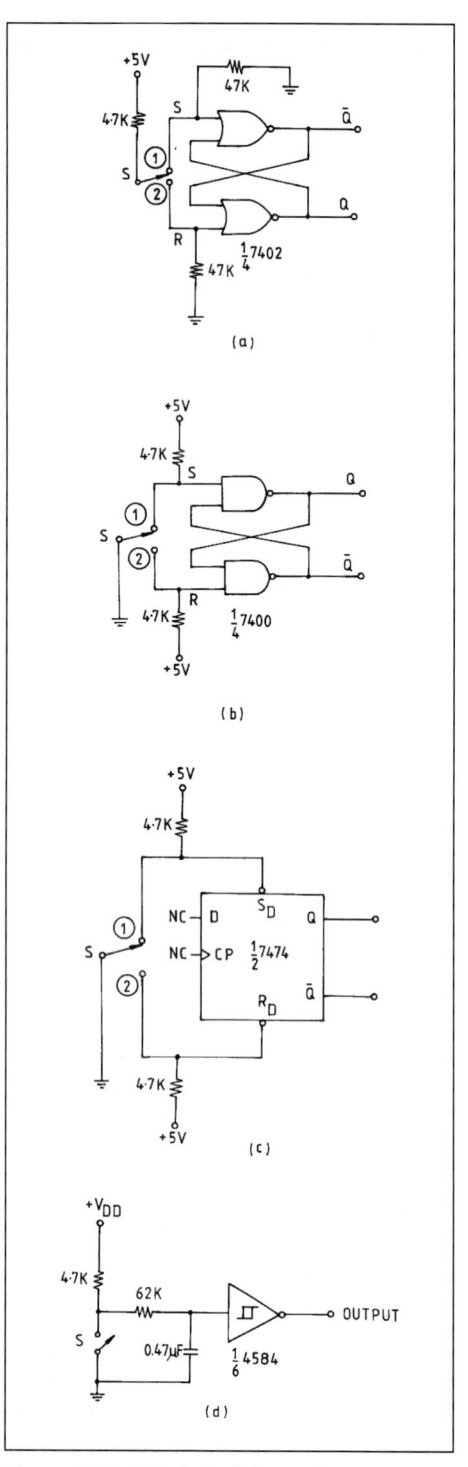

**Figure 1.19.1** Switch debouncing circuits

Variations of these circuits can also be configured for debouncing mechanical switches depending on the availability of components and the characteristics of the switch used.

◆ ◆

## 1.20 Spare buffers debounce pushbutton switch

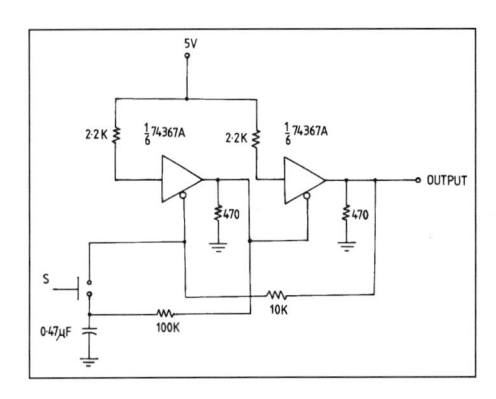

**Figure 1.20.1** Spare buffers debounce push-button switch

If you have a pair of spare tristate buffers in your design and need to debounce a pushbutton switch, you can use the circuit configuration in Figure 1.20.1. The key to the circuit is the ingenious use of a pair of tristate buffers (74367) as a pair of inverters. When the enable line of the tristate buffer is HIGH, the buffer is in its HIGH IMPEDANCE state and the output is pulled LOW by the 470 ohms resistor; when the enable line is pulled LOW, the buffer's output is HIGH because of the 2.2k pull-up resistor. In effect, therefore, the circuit uses the buffer's enable line as an input line for inverter operation. The two inverters are cross-coupled to form a latch (standard configuration) which, in turn, is used to debounce the pushbutton switch.

The circuit has very low cost since it uses spare buffers on board and a few passive components.

◆ ◆

## 1.21 Programmable digital frequency comparator

The simple circuit shown in Figure 1.21.1 can be used for the comparison of frequencies of a pair of digital pulse trains. The frequency comparator gives a logic HIGH output whenever the pulse trains are within $(1/T)$ Hz of each other, where $T$ is the output pulse width of the monostable, 74123. The principle of working of the circuit is quite simple: a D flip-flop can be used as a frequency mixer for digital signals (refer to design idea entitled 'digital frequency mixer' for more information). In other words, the frequency of the output of a D flip-flop whose D input and the clock input are fed by two pulse trains is equal to the difference between the frequencies of the input waveforms.

As seen from the waveforms in Figures 1.21.2 and 1.21.3, the waveform at B clocks the logic state of the waveform at A at its rising edges into the Q output of the D flip-flop. The monostable 74123 gets triggered at the rising edges of the digital frequency mixer waveform and the output of the one-shot has a pulse width given by

$$T = 0.28R/[C(1 + 0.7/R)]$$ for values of $C > 1000\,\text{pF}$

For $C \leq 1000\,\text{pF}$ the pulse width can be obtained from the nomograph given in the datasheet of 74123 in any TTL databook.

The output of the mixer flip-flop FF1 is used to clock another D flip-flop, FF2, whose D input is connected to the output of the monostable. The output of the frequency mixer, FF1, therefore, clocks the output state of the one-shot at its rising edges into the Q output of FF2. As seen from the waveforms in Figure 1.21.2 and Figure 1.21.3, which depict the situations when $f_1$ and $f_2$ are within $(1/T)$ Hz and not within $(1/T)$ Hz of each other, $T$ being the one-shot output pulse width, the steady state $\overline{Q}$ output of FF2 is in a logic HIGH state if $f_1$ and $f_2$ are within $(1/T)$ Hz and low if $f_1$ and $f_2$ are not within $(1/T)$ Hz. This is because in the case of $f_1$ being close to $f_2$, i.e. within $(1/T)$ Hz, the one-shot does not get re-triggered and it times out, enabling a logic 0 to be clocked into the Q output of FF2 making its $\overline{Q}$ output HIGH; if $f_1$ and $f_2$ are not close,

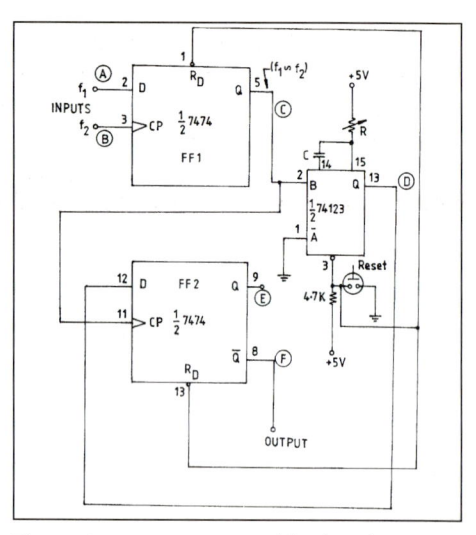

**Figure 1.21.1** Programmable digital frequency comparator

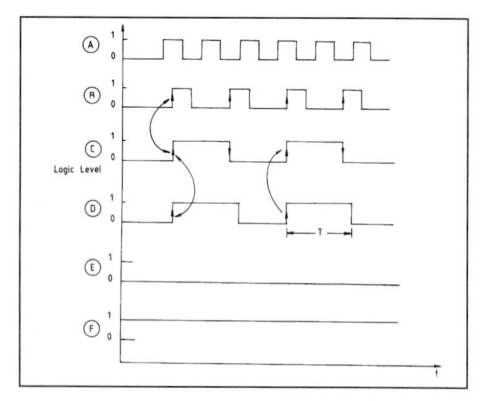

**Figure 1.21.2** Frequency comparator waveforms: $f_1$ and $f_2$ close

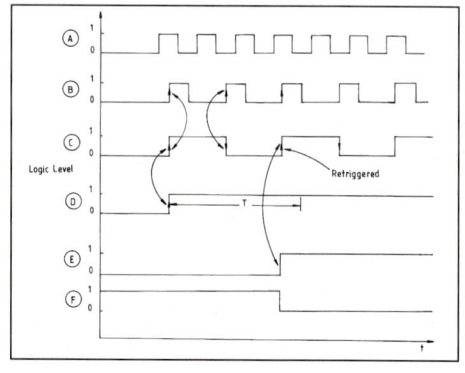

**Figure 1.21.3** Frequency comparator waveforms: $f_1$ and $f_2$ not close

i.e. not within $(1/T)$ Hz of each other, the monostable gets re-triggered and does not time out, which sets FF2 making its Q output HIGH and $\overline{Q}$ output LOW.

The circuit is programmable by choosing the RC components of the monostable. For convenience, $R$ can be made variable and $C$ can be kept fixed. This circuit can be useful for applications requiring a comparison of digital frequencies without reference to their absolute values.

## 1.22 Programmable frequency comparator for analog signals

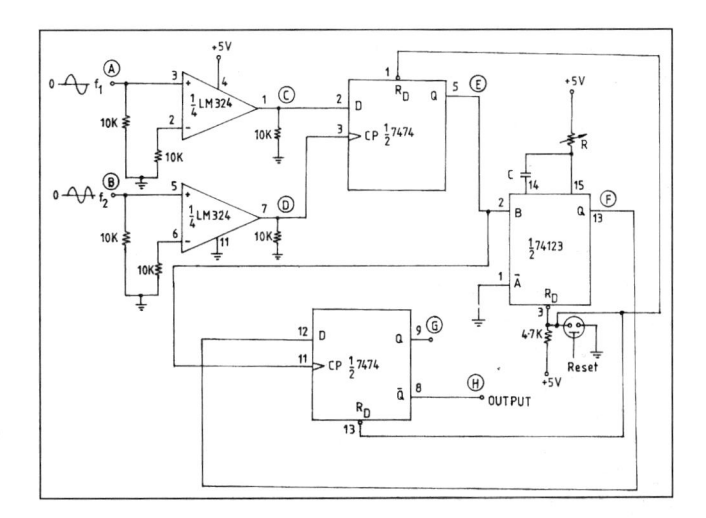

**Figure 1.22.1** Programmable frequency comparator for analog signals

The circuit shown in Figure 1.22.1 is similar to the programmable digital frequency comparator but uses a pair of zero-crossing detectors to 'square-up' the input analog signals into logic-compatible pulse trains. The zero-crossing detectors consist of a pair of LM 3424 op amps connected as shown. The outputs of the zero-crossing detectors are logic-compatible pulse trains and the same waveforms and explanation as in the case of the programmable digital frequency comparator apply in this case too. This circuit can be used to compare the frequencies of a pair of periodic analog signals such as sine waves. By making the timing resistor $R$ of the 74123 monostable variable, the timing of the one-shot can be made variable and hence the frequency window for comparison can be made programmable. In other words, the circuit functions as a programmable frequency comparator for analog periodic signals.

28  Digital circuits

## 1.23 AC line synchronizer

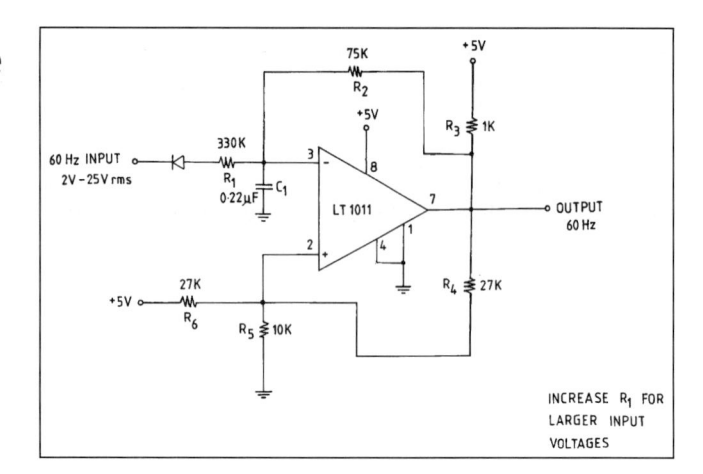

**Figure 1.23.1** Noise immune 60 Hz line sync

The circuit shown in Figure 1.23.1 produces a square wave TTL compatible output at 60 Hz line frequency. It uses the LT 1011 voltage comparator. The comparator oscillates at approximately 60 Hz, causing it to 'lock' on to the incoming line signal. The input AC voltage can be in the range 2 V to 25 V rms. The output can be used for synchronizing digital logic to the AC mains waveform.

*Courtesy of Linear Technology Corporation, Milpitas, CA*

◆ ◆

## 1.24 AC line synchronizer

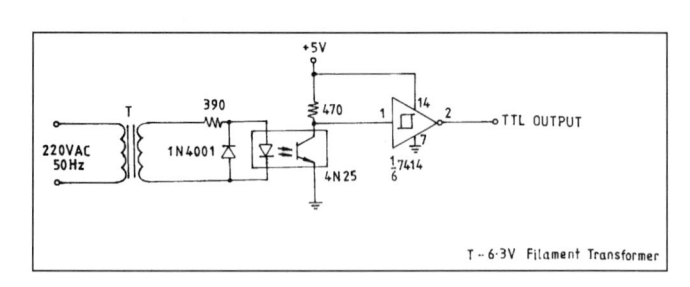

**Figure 1.24.1** AC line synchronizer

The circuit shown in Figure 1.24.1 can be used to derive TTL compatible output from AC mains for synchronization purposes. The circuit gives a square wave output corresponding to the sinusoidal AC input as shown in Figure 1.24.2, changing state at points near the zero crossings of the AC waveforms (near zero crossings and not exactly at the zero crossings due to the threshold voltage of the digital logic used). The circuit consists

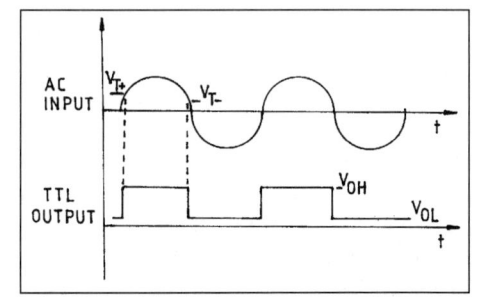

**Figure 1.24.2** Waveforms of circuit in Figure 1.24.1

of a 6.3 V filament transformer whose secondary is connected to the input diode of the opto-coupler 4N 25 through a current limiting resistor of 390 ohms. The output transistor of the opto-coupler has a collector load of 470 ohms and is connected to the input of a Schmitt trigger inverter (1/6 7414) which shapes the waveform at its input into square waves as shown in Figure 1.24.2. The 1N 4001 diode connected across the input diode of the opto-coupler with a reverse polarity protects the opto-coupler diode from reverse breakdown during the negative half-cycle transitions of the AC waveform; this is required since the reverse breakdown voltage rating of opto-coupler diodes is generally low (in this case 3 V).

This method is useful for synchronization of digital logic to the AC mains supply, e.g. for power control and similar applications.

## 1.25 Timebase generator

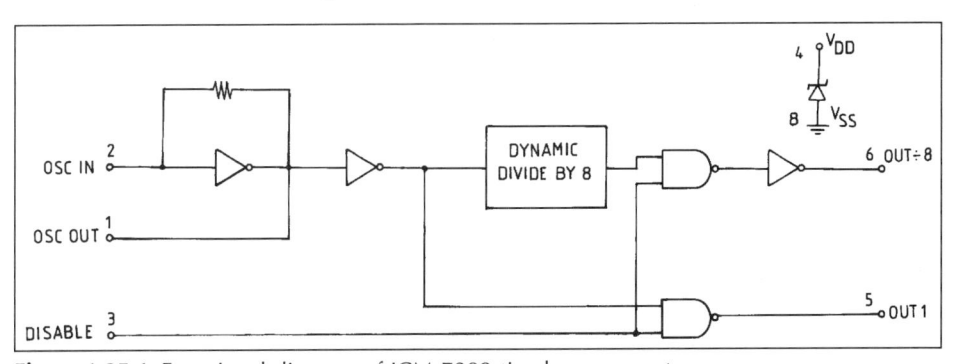

**Figure 1.25.1** Functional diagram of ICM 7209 timebase generator

The ICM 7209 is a versatile CMOS clock generator capable of driving a number of 5 V systems with a variety of input requirements. When used to drive up to 5 V TTL gates, the typical rise and fall times are 10 nS.

The ICM 7209 consists of an oscillator, a buffered output equal to the oscillator frequency and a second buffered output having an output frequency one-eighth that of the oscillator. The guaranteed maximum oscillator frequency is 10 MHz. Connecting the disable terminal to the negative supply forces the 8 output into the '0' state and the output 1 into the '1' state.

The oscillator consists of a CMOS inverter with a non-linear resistor connected between the oscillator input and output to provide DC biasing. Using commercially available quartz crystals the oscillator will operate from low frequencies (10 kHz) to 10 MHz.

The oscillator circuit consumes about 500 µA of current using a 10 MHz crystal with a 5 V supply, and is designed to operate with a high impedance tank circuit.

It is therefore necessary that the quartz crystal be specified with a load capacitance ($C_L$) of 10 pF instead of the standard 30 pF. To maximize the stability of the oscillator as a function of supply voltage and temperature, the motional capacitance of the crystal should be low (5 mpF or less). Using a fixed input capacitor of 18 pF on the output will result in oscillator stabilities of typically 1 ppm per volt change in supply voltage.

### The ÷ 8 output

A dynamic divider is used to divide the oscillator frequency by 8. Dynamic dividers use small nodal capacitances to store voltage levels instead of latches (which are used in static dividers). The dynamic divider has advantages in high speed operation and low power but suffers from limited low frequency operation. This results in a window of operation for any oscillator frequency.

### Output drivers

The output drivers consists of CMOS inverters having active pull ups and pull downs. Thus the outputs can be used to directly drive TTL gates, other CMOS gates operating with a 5 V supply, or TTL compatible MOS gates. The guaranteed fanout is 5 TTL loads although typical fanout capability is at least 10 TTL loads with slightly increased output rise and fall times.

### Device power consumption

At low frequencies the principal component of the power consumption is the oscillator. At high oscillator frequencies the major portion of the power is consumed by the output drivers; thus by disabling the outputs (activating the DISABLE INPUT) the device power consumption can be drastically reduced.

Figure 1.25.1 shows the functional diagram of ICM 7209.

*Courtesy of Harris Semiconductor, Melbourne, FL*

◆   ◆

## 1.26   Bit rate generators

Digital data transmission systems employ a wide range of standardized bit rates, ranging from 50 baud (for electromechanical devices) to 9600 baud (for high speed modems). Modern electronic systems commonly use universal asynchronous receiver and transmitter circuits (UARTs) to convert parallel data inputs into a serial bit stream (transmitter) and to reconvert the serial bit stream into

Digital circuits 31

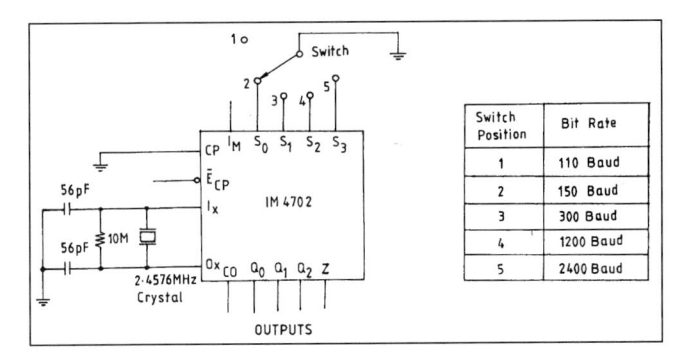

**Figure 1.26.1** Single-channel bit rate generator

| Switch Position | Bit Rate |
|---|---|
| 1 | 110 Baud |
| 2 | 150 Baud |
| 3 | 300 Baud |
| 4 | 1200 Baud |
| 5 | 2400 Baud |

parallel outputs (receiver). In order to resynchronize the incoming serial data, the receiver requires a clock rate which is a multiple of the incoming bit rate. Popular MOS LSI UART circuits use a clock that is 16 times the transmitted bit rate.

The IM 4702/12 can generate 14 standard clock rates from one common high frequency input, using a 2.4576 MHz crystal oscillator. They control up to eight output channels and can be cascaded for output expansion. The output rate is controlled by four digital input lines, and with the specified crystal, is selectable from 'zero' through 9600 baud. In addition, 19200 baud is possible via hardwiring. Multichannel operation is facilitated by making the clock frequency and the divide-by-8 prescaler outputs available externally. This allows up to eight simultaneous baud rates to be generated. The IM 4712 is identical to the IM 4702 with the exception that the IM 4712 integrates the oscillator feedback resistor and two load capacitors on-chip.

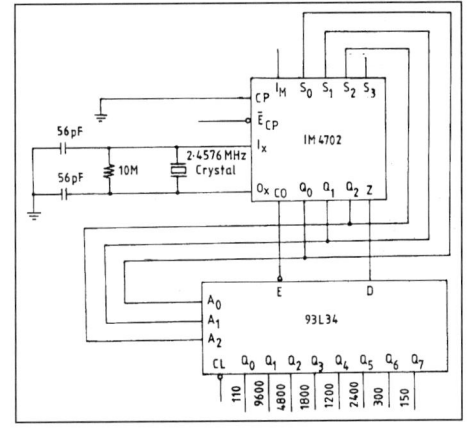

**Figure 1.26.2** Simultaneous generation of several bit rates

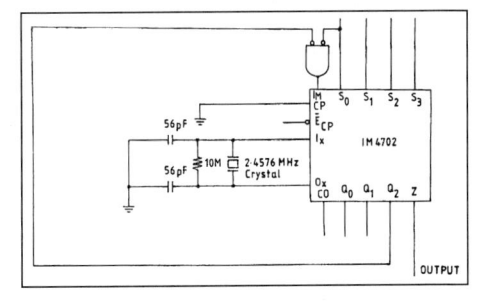

**Figure 1.26.3** 19200 baud rate generator

## Single-channel bit rate generator

Figure 1.26.1 shows the simplest application of the IM 4702/12. This circuit generates one of five possible bit rates as determined by the setting of a single pole, five-position switch. The bit rate output (Z) drives one standard TTL load or four low-power Schottky loads over the full temperature range. The possible output frequencies correspond to 100, 150, 300, 1200 and 2400 or 3600 baud. For many low cost terminals, these five bit rates are adequate.

This mode of operation is commonly chosen for applications using industry standard 1402/6402 UARTs.

## Multiple bit rate generators

Figure 1.26.2 shows a simple scheme that generates eight bit rates on eight output lines, using one IM 4702/12 and one 93L34 8-bit addressable latch. This design takes advantage of the built-in scan counter (prescaler) outputs. As shown in the block diagram, these ouputs ($Q_0$ to $Q_2$) go through a complete sequence of eight states for every half-period of the highest output frequency (9600 baud). Feeding these scan counter outputs back to the select inputs of the multiplexer causes the IM 4702/12 to sequentially interrogate the state of eight different frequency signals. The 93L34 bit-addressable latch, addressed by the same scan counter outputs, reconverts the multiplexed single output (Z) into eight parallel output frequency signals. In the simple scheme of Figure 1.26.2, input $S_3$ is left open (HIGH) and the following bit rates are generated:

$Q_0$: 110 baud     $Q_3$: 1800 baud    $Q_6$: 300 baud
$Q_1$: 9600 baud    $Q_4$: 1200 baud    $Q_7$: 150 baud
$Q_2$: 4800 baud    $Q_5$: 2400 baud

Other bit rate combinations can be generated by changing the scan counter to selector interconnection or by inserting logic gates into this path.

## 19200 baud bit rate generator

A 19200 baud signal is available on the $Q_2$ output, but is not internally connected to the multiplexer. This signal can be generated on the Z output by connecting the $Q_2$ output to the IM input and applying select code. An additional 2-input NOR gate can be used to retain the 'zero baud' feature on select code 1 for the IM 4702/12 (see Figure 1.26.3).

*Courtesy of Harris Semiconductor, Melbourne, FL*

◆  ◆

# 1.27 Single pulse extractor

The circuit shown in Figure 1.27.1 can be used to extract a single pulse from a pulse train. The circuit consists of a dual retriggerable monostable multivibrator, 74123, one half of which is configured as a positive-edge triggered monostable and the other half as a negative-edge triggered monostable. The pulse train from which a single pulse is to be sampled is used to trigger both the one-shots. The time-period of each monostable is given by

$T = 0.28\, R/C(1 + 0.7/R)$ for $C > 1000\,\text{pF}$

For $C \leq 1000\,\text{pF}$ the pulse-width can be obtained from the graph given in the data-sheet of 74123 in any TTL catalogue.

The time-periods of the monostables are made equal and the values of the RC components chosen such that for the repetition rate of the pulse train to be sampled, the monostables are made retriggerable. This requires that the time-period of each monostable is greater than the pulse repetition rate so that by the time each one-shot times out, it is re-triggered by the input pulse train to be sampled. The outputs of the one-shots are input to an EX–OR gate which, as the waveforms in Figure 1.27.2 show, gives a single extracted pulse at its output.

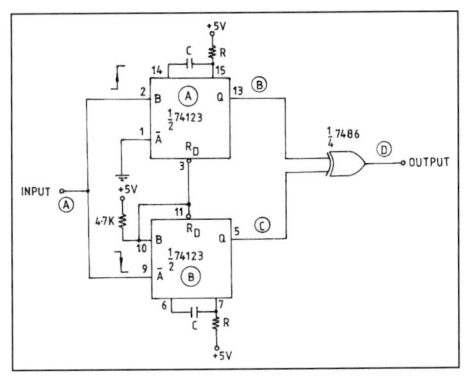

**Figure 1.27.1** Circuit extracts a single pulse from a pulse train

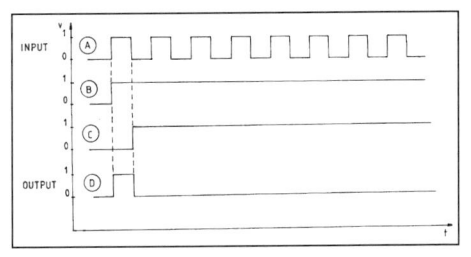

**Figure 1.27.2** Waveforms in the circuit of Figure 1.27.1

## 1.28 Use a spare EX–OR gate as an inverter

**Figure 1.28.1** Spare EX-OR gate used as inverter

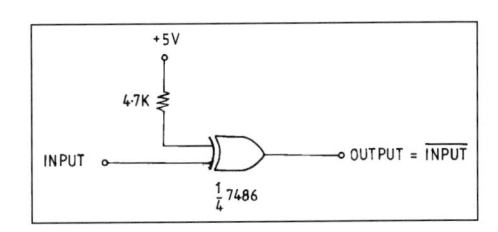

If you need a logic inverter in your design and have a spare EX–OR gate on board, you can save on chip count and save board space by configuring the spare EX–OR gate as an inverter, as shown in Figure 1.28.1. This can be accomplished by tying one input of the EX–OR to a logic HIGH level. Since one input of the EX–OR gate is already tied HIGH, if the other input is in a logic HIGH state, the output of the EX–OR goes to a logic LOW state and if the second input of the EX–OR is logic LOW, the output becomes HIGH, in accordance with the truth table of EX–OR gate – thus the EX–OR gate configured as in Figure 1.28.1 acts as an inverter.

## 1.29 Monolithic decade counter decoder display driver

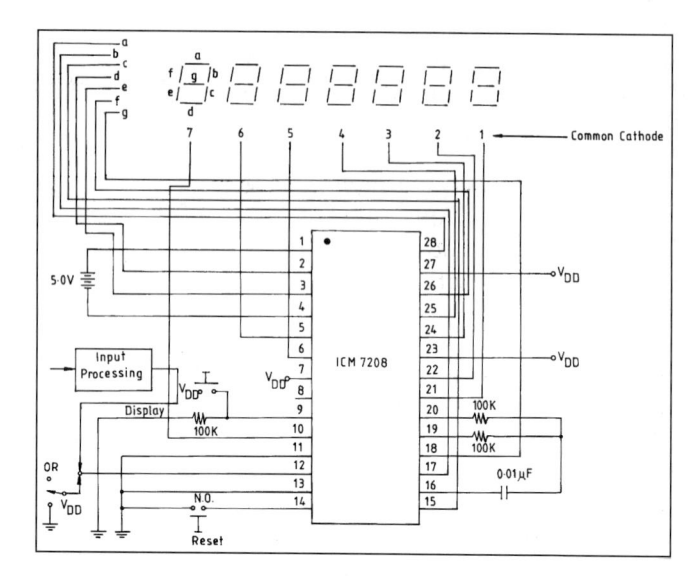

**Figure 1.29.1** Schematic of unit counter

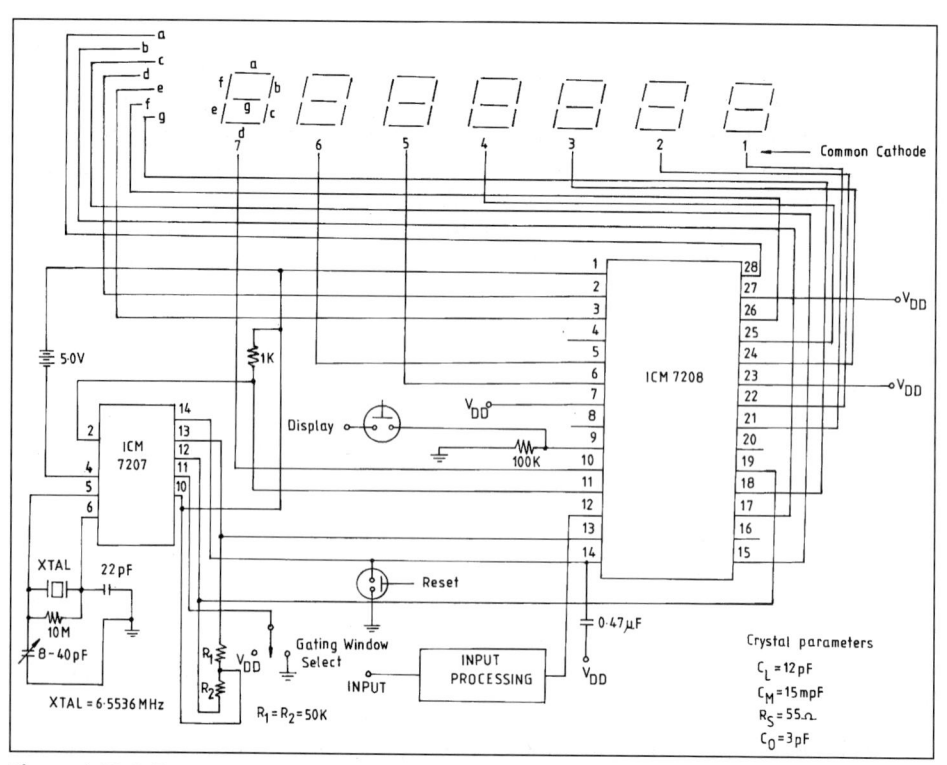

**Figure 1.29.2** Frequency counter

Digital circuits 35

The ICM 7208 is a fully integrated seven decade counter–decoder–driver providing the following on-chip functions: a 7 decade counter, multiplexer, 7 segment decoder, digit and segment driver, plus additional logic for display blanking, reset, input inhibit and display on/off.

For unit counter applications the only additional components are a 7 digit common cathode display, three resistors and a capacitor to generate the multiplex frequency reference, and the control switches.

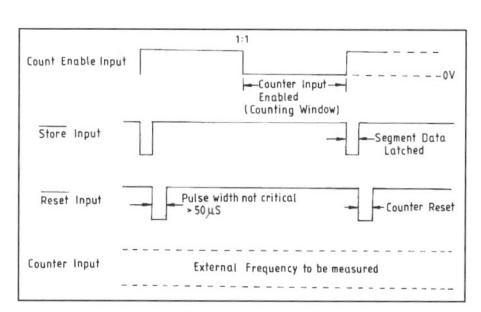

**Figure 1.29.3** Frequency counter input waveforms

The ICM 7208 is intended to operate over a supply voltage of 2 to 6 V as a medium speed counter, or over a more restricted voltage range for high frequency applications.

The internal counters of the ICM 7208 index on the negative edge of the input signal at terminal 12.

## Format of signal to be counted

The noise immunity of the COUNTER INPUT terminal is approximately 1/3 the supply voltage. Consequently, the input signal should be at least 50% of the supply in peak-to-peak amplitude and preferably equal to the supply.

The optimum input signal is a 50% duty cycle square wave equal in amplitude to the supply. However, as long as the rate of change of voltage is not less than approximately $10^{-4}$ V/µs, at 50% of the power supply voltage, the input waveshape can be sinusoidal, triangular, etc.

When driving the input of the ICM 7208 from TTL a 1k–5k ohm pull-up resistor to the positive supply must be used to increase peak-to-peak input signal amplitude.

## Display considerations

Any common-cathode multiplexable LED display may be used. However, if the peak digit current could exceed 150 mA for any prolonged time, it is recommended that resistors be included in series with the segment outputs to limit digit current to 150 mA.

The ICM 7208 is specified with 500 µA of possible digit leakage current. With certain new LED displays that are extremely efficient at low currents, it may be necessary to include resistors between the cathode outputs and the positive supply to bleed off this leakage current.

## Display multiplex rate

The ICM 7208 has approximately 0.5 µs overlap between output drive signals. Therefore, if the multiplex rate is very fast, digit ghosting will occur. The ghosting determines the upper limit for the multiplex frequency. At very low multiplex rates flicker becomes visible.

It is recommended that the display multiplex rate be within the range 50 Hz to 200 Hz, which corresponds to 400 Hz to 1600 Hz for the multiplex frequency input. For stand-alone systems, two inverters are provided so that a simple but stable RC oscillator may be built using only two resistors and a capacitor.

The multiplex oscillator is eight times the multiplex rate. The frequency is given by:

$$f = 1/2.2\, R_x C_x$$

$R_s$ should always be < 1 Mohm and $R_s = k R_x$ where $k$ is in the range 2–10.

An external generator may be used to provide the multiplex frequency input. This signal, applied to terminal 19 (terminals 16 and 20 open circuit) should be approximately equal to the supply voltage, and should be a square wave for minimum of power dissipation.

## Application examples

*Unit counter* Figure 1.29.1 shows the schematic of an extremely simple unit counter that can be used for remote traffic counting, to name one application. The power cell stack should consist of 3 or 4 nickel cadmium rechargeable cells (nominal 3.6 V or 4.8 V). If 4 × 1.5 V cells are used it is recommended that a diode be placed in series with the stack to guarantee that the supply voltage does not exceed 6 V.

The input switch is shown to be a single-pole double-throw (SPDT). A single-pole single-throw switch (SPST) could also be used (with a pull-up resistor), however, anti-bounce circuitry must be included in series with the counter input. In order to avoid contact bounce problems due to the SPDT switch the ICM 7208 contains an input latch on chip.

The unit counter updates the display for each negative transition of the input signal. The information on the display will count, after reset, from 00 to 9999999 and then reset to 0000000 and begin to count up again. To blank leading zeros, actuate reset at the beginning of a count. Leading zero blanking affects two digits at a time.

For battery operated systems the display may be switched off to conserve power.

*Frequency counter* The ICM 7208 may be used as a frequency counter when used with an external frequency reference and gating logic. This can be achieved using the ICM 7207 oscillator controller (Figure 1.29.2). The ICM 7207 uses a crystal controlled oscillator to provide the store and reset pulses together with the counting window. Figure 1.29.3 shows the recommended input gating waveforms to the ICM 7208. At the end of a counting period (50% duty cycle) the counter input is inhibited. The counter information is then transferred and stored in latches, and can be displayed. Immediately after this information is stored, the counters are cleared and are ready to start a new count when the counter input is enabled.

Using a 6.5536 MHz quartz crystal and the ICM 7207 driving the ICM 7208, two ranges of counting may be obtained using either 0.01 s or 0.1 s counter enable windows.

The ICM 7207 provides the multiplex frequency reference of 1.6 kHz. Previous comments on leading zero blanking, etc., apply as per the unit counter.

*Courtesy of Harris Semiconductor, Melbourne, FL*

## 1.30 Digital glitch detector

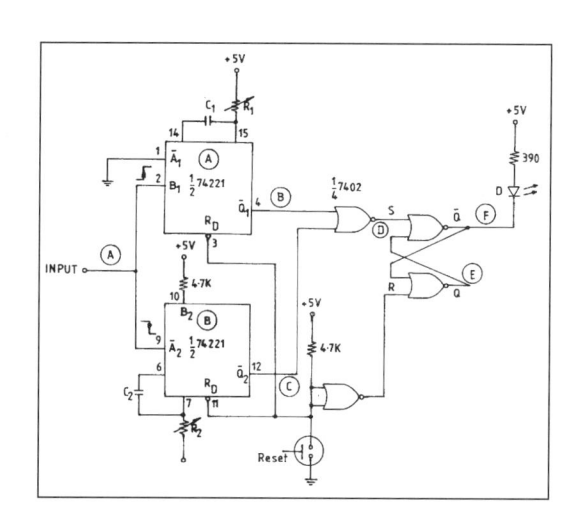

**Figure 1.30.1** Circuit flags glitches in digital logic circuits

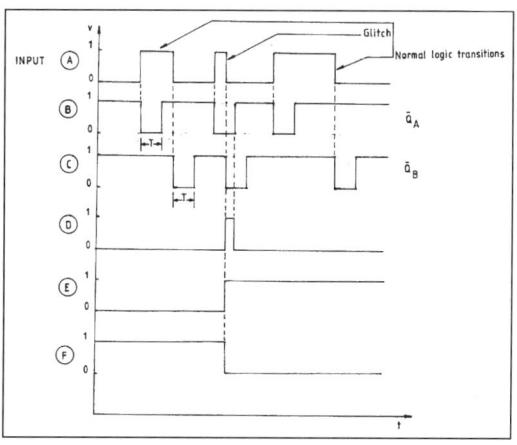

**Figure 1.30.2** Waveforms of circuit in Figure 1.30.1

A simple two-chip circuit shown in Figure 1.30.1 can detect and flag the presence of a noise spike or a glitch in a logic circuit. As shown, the circuit consists of a dual monostable multivibrator 74221, whose halves A and B are configured as rising-edge triggered and falling-edge triggered monostables. The time-periods of the one-shots (given by $T = RC \ln 2$) are made equal, i.e. $R_1 C_1 \ln 2 = R_2 C_2 \ln 2 = T$. The time-period $T$ is selected such that any logic pulse of duration less than $T$ is classified as a glitch or noise spike in the logic circuit. The one-shot A is triggered at the rising edges of the input waveform and the one-shot B is triggered at the falling edges (see Figure 1.30.2). The $\overline{Q}$ outputs of one-shots A and B are input to a NOR gate. The output of the NOR gate forms the Set input of a flip-flop formed by a pair of cross-coupled NOR gates from the same chip (7402). The fourth NOR gate of the quad NOR package is configured as an inverter and its output forms the Reset input for the cross-coupled NOR flip-flop.

As seen from the waveforms in Figure 1.30.2, normal logic transitions of time duration larger than $T$ do not evoke any response from the glitch-detector circuit. If a noise-spike or glitch (of duration less than $T$) appears at the input of the glitch-detector, the Set input of the cross-coupled NOR flip-flop goes HIGH for a time duration equal to the difference between the time-period $T$ of the one-shots and the duration of the glitch. This is sufficient to set the NOR flip-flop, driving its Q output HIGH. Since the $\overline{Q}$ output of the flip-flop is now low, the LED glows indicating the presence of a glitch or noise spike. Alternatively, the Q and $\overline{Q}$ outputs of the flip-flop can be used to flag an error condition. Pressing the Reset switch makes the output of the NOR gate configured as an inverter HIGH and this resets the flip-flop.

◆ ◆

## 1.31 Phase lead–lag indicator

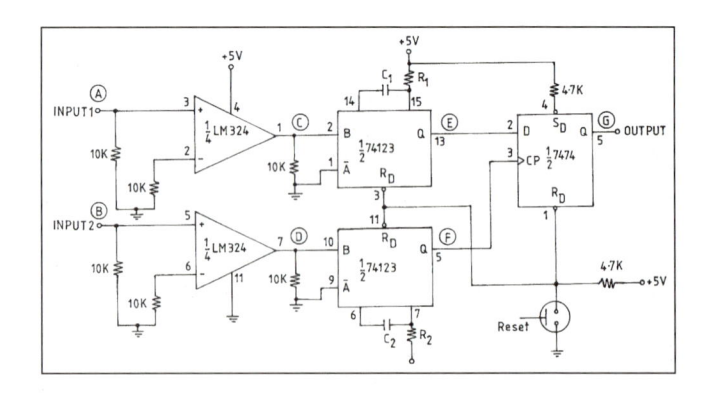

**Figure 1.31.1** Phase lead–lag indicator

Differentiation of phase of analog signals is frequently required in circuit design. As an example consider the sensing of the direction of rotation of a motor shaft. To sense direction of rotation, generally a pair of optical encoders such as slotted optical sensors are placed along the periphery of a disc mounted on the shaft; the disc has transparent and opaque patterns. We can sense direction of rotation by finding the phase difference between the outputs of the sensors.

The circuit shown in Figure 1.31.1 can be used as a phase indicator, i.e. it gives indication about the phase lead or lag of a signal with respect to a reference. The reference input and the input whose phase is to be sensed are connected to a pair of zero-crossing detectors built using quad op-amp LM 324. The outputs of the zero-crossing detectors are used to trigger a pair of retriggerable monostable

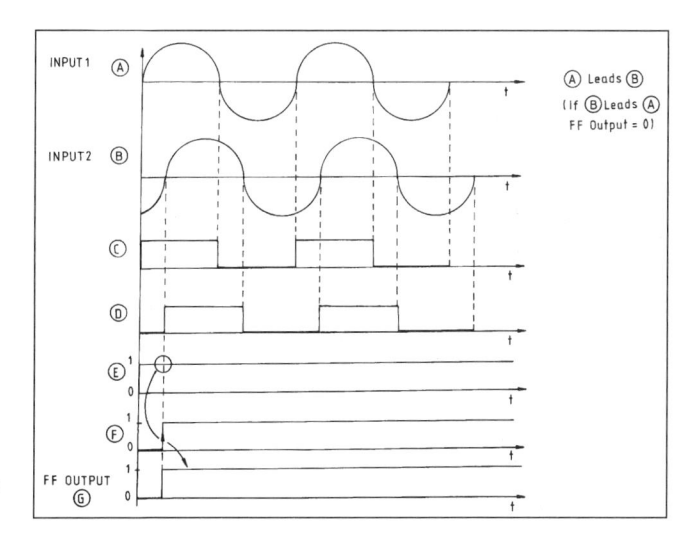

**Figure 1.31.2** Waveforms of circuit in Figure 1.31.1

multivibrators using 74123. By knowing the frequencies and hence the time-periods of the input analog signals to the zero-crossing detectors, the *RC* values of the monostables can be chosen such that they are rendered retriggerable for the frequencies of the inputs. The output pulse width of the 74123 is given by

$$t_w = 0.28R/\{C[1 + (0.7/R)]\} \text{ for } C > 1000\,\text{pF}$$

For values of $t_w$ when $C \leq 1000\,\text{pF}$, reference may be made to the nomograph in the datasheet of 74123 given in any TTL databook.

The outputs of the one-shots are connected to the D input and the clock input of an arbitration D flip-flop (1/2 7474) which extracts the difference between the frequencies of the two inputs to decide which signal leads. If the input 'A' leads input 'B' in phase, the output of the one-shot 'B' clocks a logic 1 into the Q output of the D flip-flop and if the input 'B' leads the input 'A', the output of one-shot 'B' clocks a logic 0 into the Q output of the D flip-flop. Thus the Q output of the D flip-flop is logic HIGH if signal 'A' leads signal 'B' and logic LOW if signal 'B' leads signal 'A'. This is clear from the waveforms in Figure 1.31.2.

Thus the circuit acts as a simple phase differentiator for periodic analog signals.

◆  ◆

## 1.32 Digital synchronizer

A flip-flop can be used for synchronization of digital logic signals. In circuit design the problem of interfacing asynchronous signals with synchronous, i.e. clocked systems, is frequently encountered. The circuit shown in Figure 1.32.1 shows a simple digital synchronizer using a D flip-flop. The signal to be

40 Digital circuits

synchronized is connected to the D input of the flip-flop and the system clock is logic-inverted and connected to the clock input of the flip-flop. The logic inversion of the clock is done in order to convert the positive-edge triggered flip-flop into a negative-edge triggered one (recall that the D flip-flop 7474 used in this design is a positive-edge triggered flip-flop) – the reason for this will be explained later.

As the waveforms in Figure 1.32.2 show, the flip-flop's Q output changes state at the falling edges of the clock. Assuming a pulse input at the D input of the flip-flop, not in sync with the system clock, you will observe that the flip-flop converts the asynchronous input into a synchronous one, i.e. in sync with the system clock.

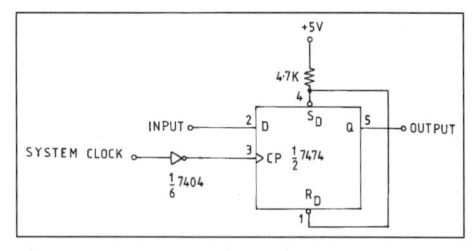

**Figure 1.32.1** Digital synchronizer

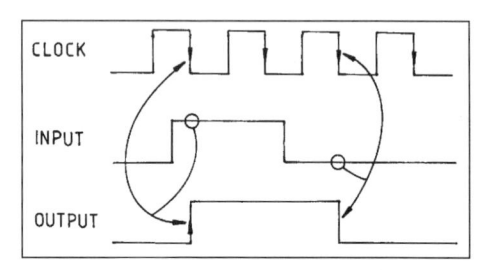

**Figure 1.32.2** Waveforms of circuit in Figure 1.32.1

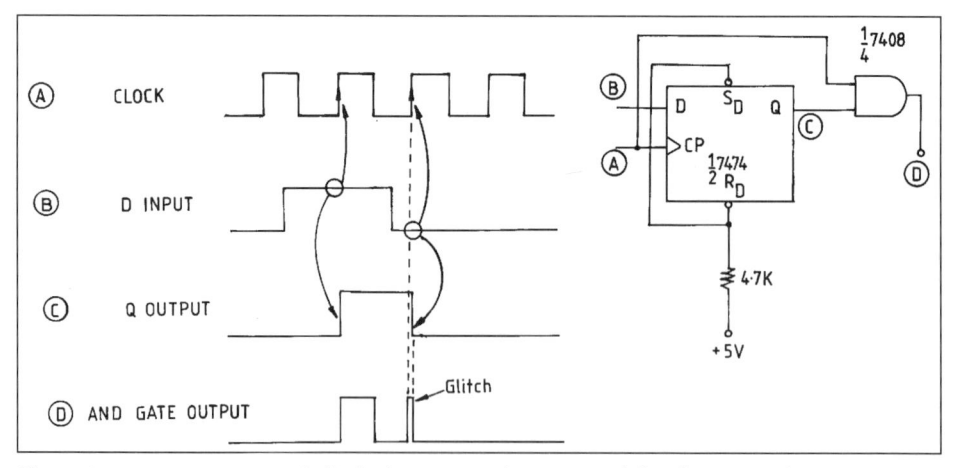

**Figure 1.32.3** Appearance of glitch if positive-edge triggered flip-flop is used

Now assume that a positive-edge triggered flip-flop is used instead of a negative-edge triggered one. As seen from the waveforms in Figure 1.32.3, a short pulse, also called a glitch, caused by the propagation delay of the digital logic used appears at the output of subsequent logic stages, such as gates connected to the Q output of the flip-flop. The glitch arises because the flip-flop changes state at the rising edges of the clock and there is a short time delay between the time taken by the flip-flop to change state and the instant at which the subsequent logic stages change state, due to the propagation delay of the digital logic. For the

Digital circuits 41

waveforms in Figure 1.32.3, it is assumed that the Q output of the flip-flop and the system clock are connected to a 2-input AND gate. You will observe from Figure 1.32.3 that a glitch appears at the AND gate output if a positive-edge triggered flip-flop is used. Glitches can cause erroneous operation of logic circuits and should be avoided.

◆ ◆ ◆

## 1.33 Electronic alternate-action switch

**Figure 1.33.1** Alternate-action switch

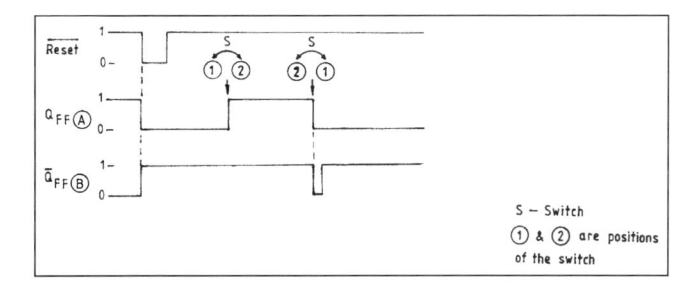

**Figure 1.33.2** Waveforms of circuit in Figure 1.33.1

An alternate-action switch is frequently required in instrumentation applications for front-panel use. This type of switch toggles alternately between its two states for successive switching actions. Although mechanical alternate-action switches are available, they tend to be rather expensive and complex in design. The circuit shown in Figure 1.33.1 achieves this type of switching action using a simple mechanical SPDT switch and two ICs – a dual D flip-flop (7474) and a quad NAND gate (7400). The operation of the circuit is as follows.

On power-up assume that a reset pulse is generated by the system in which this switch is used. The power-on reset signal is given as one of the inputs to NAND gate D, whose other input is connected to the $\overline{Q}$ output of the flip-flop B. Irrespective of the state of the $\overline{Q}$ output of flip-flop B, the output of gate D goes HIGH on power-up and this, in turn, makes the output of gate C LOW since it is configured as an inverter. The LOW output of gate C clears both the flip-flops A and B.

42  Digital circuits

The cross-coupled NAND gates A and B form a simple debouncer flip-flop for the SPDT switch. Now, assume that the SPDT switch contact is in position 1. In this state the Q output of the debouncer flip-flop (output of gate A) is HIGH and its $\overline{Q}$ output (output of gate B) is LOW. If the switch contact is now thrown to position 2, the debouncer flip-flop is reset and its Q output goes LOW (and $\overline{Q}$ output goes HIGH). This LOW-to-HIGH transition of the $\overline{Q}$ output of the debouncer flip-flop clocks flip-flop A transferring the logic 1 available at its D input to its Q output. Since the Q output of flip-flop A is connected to the D input of flip-flop B, the D input of flip-flop B is now HIGH. If the switch contact is now thrown to position 1, the debouncer flip-flop is set and its Q output goes HIGH. Since the Q output of the debouncer flip-flop is connected to the clock input of flip-flop B, the LOW-to-HIGH transition of the Q output of the debouncer flip-flop transfers the logic 1 at the D input of flip-flop B to its Q output; therefore, the Q output of flip-flop B goes high and its $\overline{Q}$ output goes LOW. The output of gate C goes LOW and this clears both the flip-flops making their Q outputs LOW.

The waveforms in Figure 1.33.2 illustrate the sequence of operations. As seen from the waveforms, for each action of the SPDT switch, the Q output of flip-flop A toggles between 1 and 0. This output can be used to turn a relay driver transistor ON and OFF, obtaining an alternating switching action. This circuit, therefore, effectively converts the mechanical action of a SPDT switch into an alternating action of an output device such as a relay.

## 1.34 Programmable undervoltage/overvoltage detector

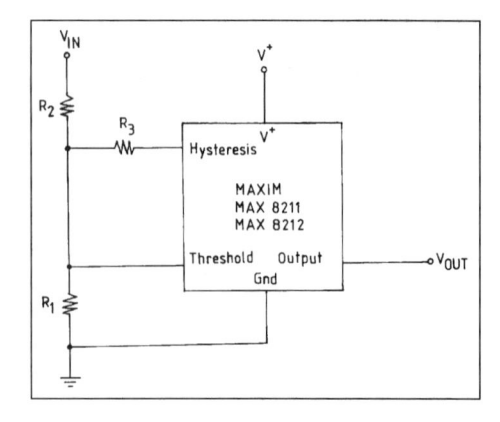

**Figure 1.34.1** Basic overvoltage/undervoltage circuit

Maxim's MAX 8211 and 8212 are CMOS micropower voltage detectors. Each of them contains a comparator, a 1.15 V bandgap reference and an open drain N-channel output driver. Two external resistors are used in conjunction with the internal reference to set the trip voltage to the desired level. A hysteresis output is also provided allowing the user to apply positive feedback for noise-free output switching. The MAX 8211 output N-channel turns ON when the voltage applied to the THRESHOLD pin is less than the internal reference (1.15 V) and the MAX

8212 output turns ON when the voltage applied to the THRESHOLD pin is greater than the internal reference.

Figure 1.34.1 shows the basic circuit for both undervoltage detection (MAX 8211) and overvoltage detection (MAX 8212). For applications where no hysteresis is needed, $R_3$ should be omitted. The ratio of $R_1$ to $R_2$ is then chosen such that for the desired trip voltage at $V_{in}$, 1.15 V is applied to the THRESHOLD pin.

## Design equations

To ensure noise-free output switching, hysteresis is frequently used in voltage detectors. For both the MAX 8211 and MAX 8212, the HYSTERESIS output is ON for THRESHOLD voltages greater than 1.15 V. Resistor $R_3$ controls the amount of current (positive feedback) supplied from the HYSTERESIS output to the mid-point of the resistor divider and hence the magnitude of the hysteresis, or deadband.

Resistor values for the basic overvoltage/undervoltage circuit are calculated as follows:

1. Choose a value for resistor $R_1$. Typical values are in the 10k to 10M range.
2. Calculate $R_2$ for the desired upper trip point $V_U$ using the formula

$$R_2 = R_1 \times (V_U - V_{TH})/V_{TH}$$
$$= R_1 \times (V_U - 1.15)/1.15$$

3. Calculate $R_3$ for the desired amount of hysteresis, where $V_L$ is the lower trip point:

$$R_3 = R_2 \times (V^+ - V_{TH})/(V_U - V_L)$$
$$= R_2 \times (V^+ - 1.15)/(V_U - V_L)$$

or if $V^+ = V_{IN}$

$$R_3 = R_2 \times (V_L - V_{TH})/(V_U - V_L)$$
$$= R_2 \times (V_L - 1.15)/(V_U - V_L)$$

*Courtesy of Maxim Integrated Products, Inc., Sunnyvale, CA*

# 2 Interface circuits

## 2.1 Logic interfacing techniques

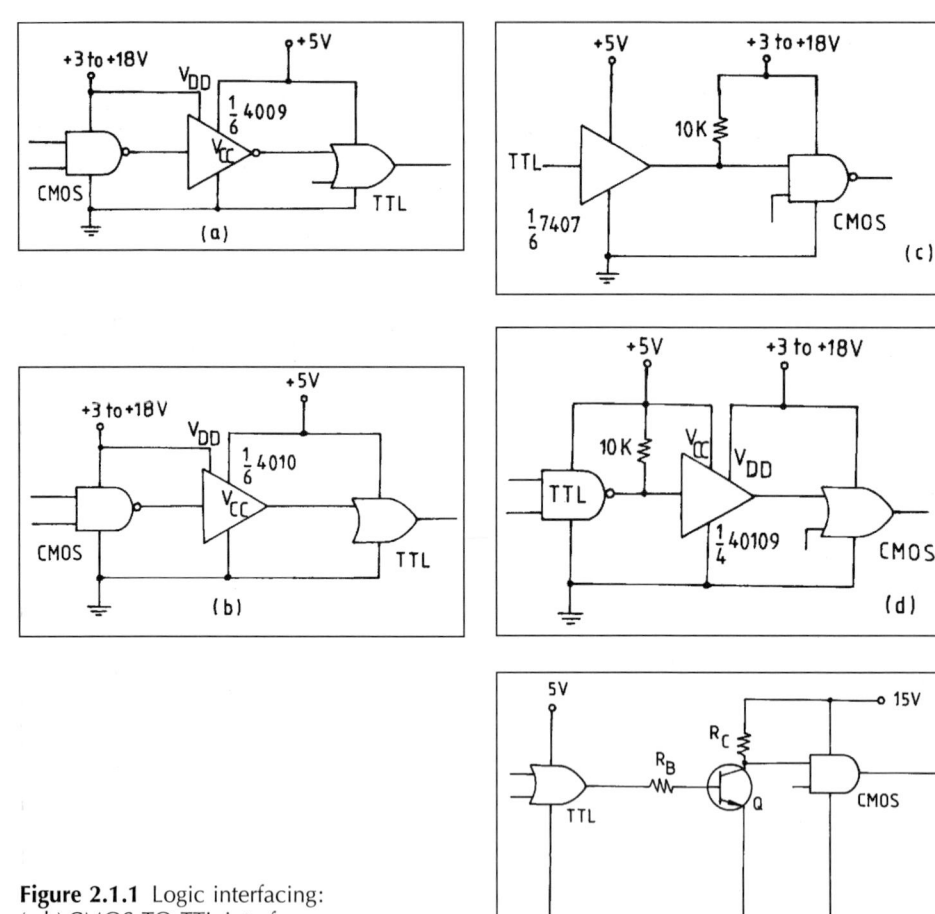

**Figure 2.1.1** Logic interfacing:
(a,b) CMOS-TO-TTL interface;
(c,d) TTL-TO-CMOS interface; (e) totem-pole TTL to 15 V CMOS interface (inverting)

The commonly used primary digital logic families in circuit design are TTL, ECL and CMOS. Variations of these primary logic families such as LSTTL, STTL, FTTL, HCMOS, etc. are also used. Since the operating supply requirements, threshold voltage levels, input and output voltage and current levels, are all different for

**Figure 2.1.1 continued** (f) totem-pole TTL to 15 V CMOS interface (non-inverting); (g) 10125 ECL-to-TTL translator; (h) 10124 TTL-to-ECL translator; (i) ECL-to-TTL translator; (j) TTL-to-ECL translator

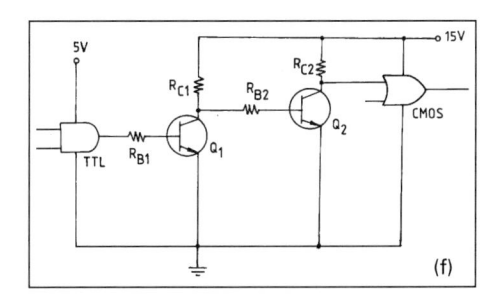

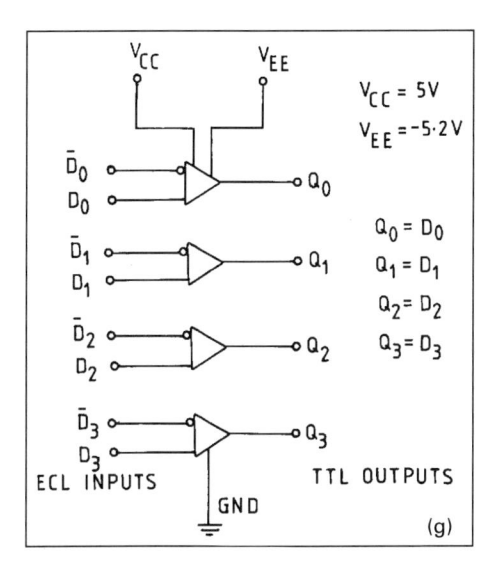

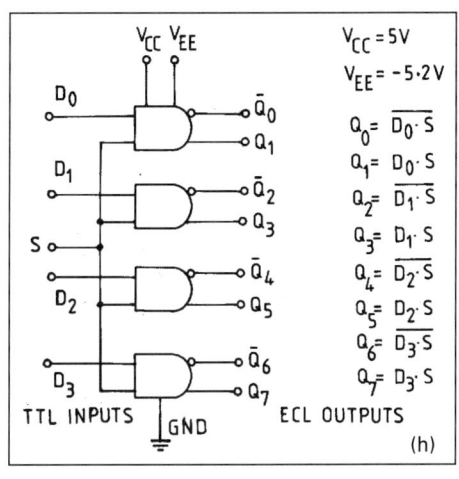

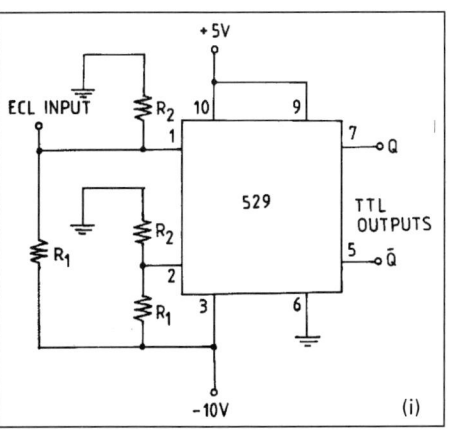

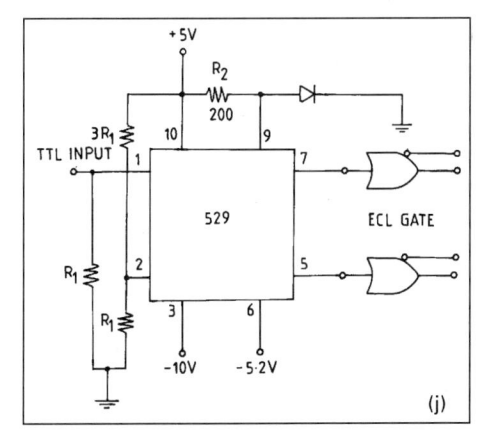

these logic families, interfacing circuits are required when connecting the output of one type of logic to the input of another type of logic.

The circuits shown in Figure 2.1.1(a) to (j) show the techniques which can be employed for interfacing the various types of logic.

## CMOS-to-TTL interface

Figure 2.1.1(a) shows a CMOS gate operating at 3 V to 18 V interfaced to a TTL gate operating at the standard 5 V, using a 4009 voltage level converter. The 4009 is an inverting type voltage level converter capable of driving up to two TTL/DTL loads operating at 3 V to 6 V. If inversion of logic signal during interface is not desired, a 4010 buffer can be used as shown in Figure 2.1.1(b). 4049 and 4050 which are inverting and non-inverting converter and buffer, respectively, can also be used instead of 4009 and 4010.

## TTL-to-CMOS interface

Figure 2.1.1.(c) shows a technique for interfacing TTL operating at 5 V to CMOS operating at 3 V to 18 V. The output of an open-collector TTL gate can be interfaced to CMOS by using a pull-up resistor to the CMOS supply, $V_{DD}$, as shown.

If TTL is to be interfaced to CMOS operating at 5 V, a pull-up resistor can be used; if the CMOS is operating at a higher voltage than 5 V and TTL is required to drive it, a level-shifter such as 40109 can be used with a pull-up resistor connected to $V_{CC}$ at the output of the TTL gate, as shown in Figure 2.1.1(d). To interface totem-pole TTL output to CMOS operating at a higher voltage than 5 V, a simple transistor inverter can be used as shown in Figure 2.1.1(e); if inversion of signal at the interface is not desirable, a pair of inverters can be used as shown in Figure 2.1.1(f).

## ECL-to-TTL interface

Emitter-coupled logic is very popular due to its speed. Systems are often built around standard TTL logic with those portions requiring higher speed being implemented with emitter-coupled logic. As soon as such a decision is made the problem of interfacing TTL-to-ECL logic levels is encountered.

The standard logic output swings of ECL are −0.8 V to −1.8 V at room temperature. Converting these signals to TTL levels can be accomplished by using the quad ECL-to-TTL translator, 10125, as shown in Figure 2.1.1(g). The input and output levels are respectively, ECL 10K and TTL Schottky. This device features a peak common-mode rejection voltage of ±1 V. The device provides a separate reference bias voltage output ($V_{BB}$) to be used in case of single-ended input busing. ECL-to-TTL interfacing can also be done using a high speed comparator NE 529, as shown in Figure 2.1.1(i). Figure 2.1.1(i) shows that the power supplies have been shifted in order to shift the common-mode range more negative. This ensures that the common-mode range is not exceeded by the logic inputs. Since ECL is extremely fast, the NE 529 is usually selected because of its superior speed so that a minimum of time is lost in transition.

## TTL-to-ECL interface

Sometimes it may be required to interface TTL to ECL. This can be accomplished using the quad TTL-to-ECL translator, 10124, as shown in Figure 2.1.1(h). 10124 has individual Data and a common Select TTL-compatible input on each gate. When the Select is in the LOW state, all ECL non-inverting outputs are in a LOW state and inverting outputs are in a HIGH state. TTL-to-ECL translation can also be done by using a NE 529 high-speed comparator as shown in Figure 2.1.1(j). In Figure 2.1.1(j), $R_2$ and the diode raise the gate supply voltage, and therefore, the NE 529 output voltage by 0.7 V, sufficient to guarantee fast switching of the translator. Resistive pull-up from the NE 529 output to $V_{CC}$ can also be used with the gate supply grounded. This method is dependent upon $RC$ time constants of distributed capacitance and is, therefore, much slower.

*Courtesy of Philips, The Netherlands*

◆ ◆

## 2.2 RS-232 line driver and receiver

**Figure 2.2.1** Typical line driver-receiver application

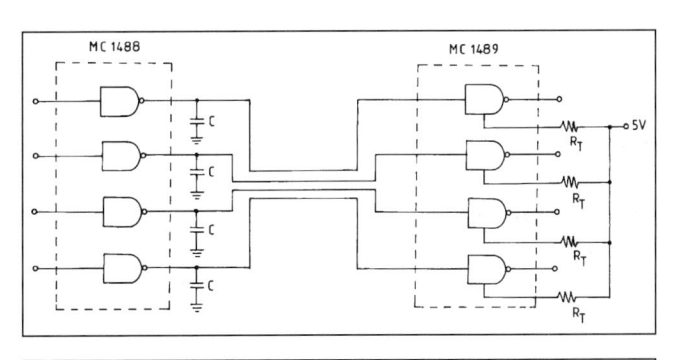

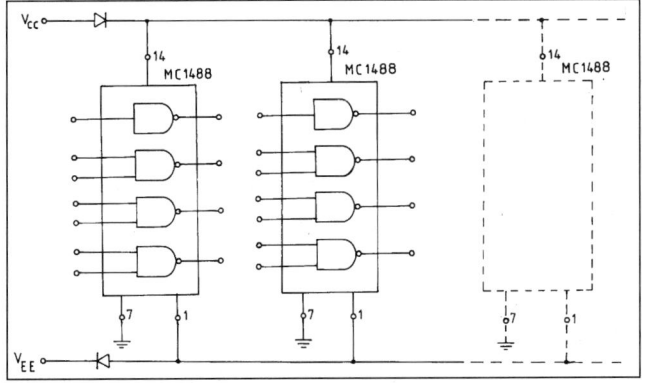

**Figure 2.2.2** Use of protection diodes

Many types of line drivers and receivers are available today. Each device has been designed to meet specific needs. For instance, the device may be extremely wideband or be intended for use in party-line systems. Some of the devices include built-in hysteresis in the receiver while others do not.

The Electronic Industries Association (EIA) has produced a number of specifications dealing with the transmission of data between data terminal and communications equipment. One of these is EIA Standard RS-232C, which delineates much information about signal levels and hardware configurations in data systems.

MC 1488 and MC 1489 are popularly used line driver and receiver respectively, which meet the RS-232C specification;.

Standard RS-232C defines the voltage level as being from 5 V to 15 V with positive voltage representing a logic 0. The MC 1488 meets these requirements when loaded with resistors from 3 kohms to 7 kohms. Output slew rates are limited by RS-232C to 30 V/µs. To accomplish this specification, the MC 1488 is loaded at its output by capacitance as shown by the typical hook-up diagram of Figure 2.2.1. For the standard 30 V/µs, a capacitance of 400 pF is selected.

The short-circuit current charges the capacitance in accordance with the relationship

$$C = I_{SC}\Delta T / \Delta V$$

where $C$ is the required capacitor, $I_{SC}$ is the short-circuit current value and $\Delta V / \Delta T$ is the slew rate.

Using the worst-case output short-circuit current of 12 mA in the above equation, calculations result in a required capacitor of 400 pF connected to each output to limit the output slew rate to 30 V/µs in accordance with the EIA standard.

The EIA standard also states that output shorts to any other conductor of the cable must not damage the driver. Thus, the MC 1488 is designed such that the output will withstand shorts to other conductors indefinitely even if these conductors are at worst-case voltage levels. In addition to output protection, the MC 1488 includes a 300 ohm resistor to ensure that the output impedance of the driver will be at least 300 ohms even if the power supply is turned off. In cases where power supply malfunction produces a low impedance to ground, the 300 ohm resistors are shortened to ground also. Output shorts then can cause excessive power dissipation. To prevent this, series diodes are included in both supply lines as shown in Figure 2.2.2.

The companion receiver, MC 1489, is also designed to meet RS-232C specifications for receivers. It must detect a voltage from ± 3 V to ± 25 V as logic signals but cannot generate an input differential voltage of greater than 2 V, should its inputs become open-circuited. Noise and spurious signals are rejected by incorporating positive feedback internally to produce hysteresis. Featured also in the receiver is an external response node so that the threshold may be externally varied to fit the application.

The design of the MC 1488 and MC 1489 makes them very versatile with many other possible applications.

*Courtesy of Philips, The Netherlands*

## 2.3 Optically isolated RS-232 interface

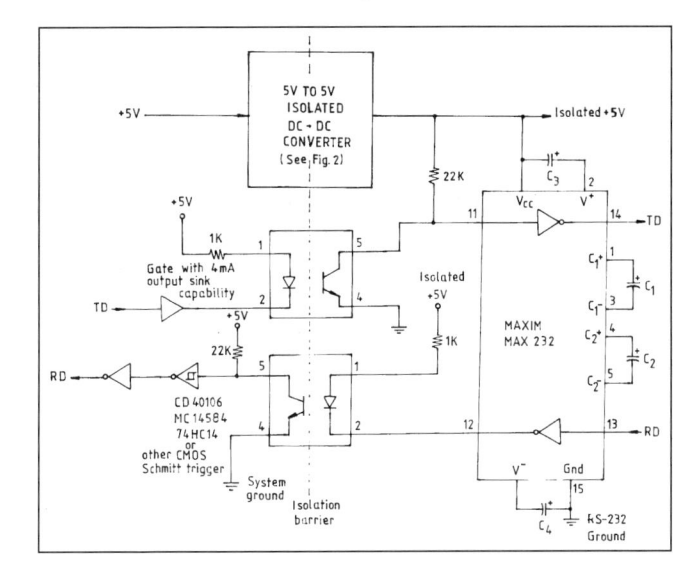

**Figure 2.3.1** Optically isolated RS-232 interface

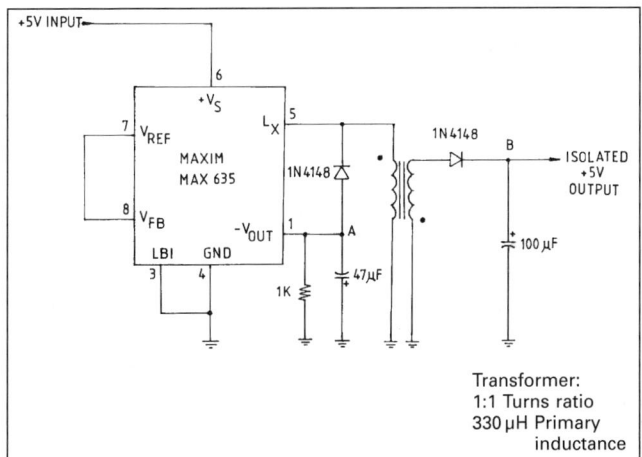

**Figure 2.3.2** +5 V isolated power supply

The most common serial interface between electronic equipment is the RS-232 interface. This serial interface has been found to be particularly useful for the interface between units made by different manufacturers since the voltage levels are defined by the EIA Standard RS-232-C and CCITT recommendation V.28. The RS-232 specification also contains signal circuit definitions and connector pin assignments, while CCITT circuit definitions are contained in a separate document, Recommendation V.24.

RS-232 and V.28 specifications require a common ground connection between the two units communicating via the RS-232/V.28 interface. In some cases, there may be large differences in ground potential between the two units, and in other cases it may be desired to avoid ground loop currents by isolating the two grounds. In other cases a computer or control system must be protected against

50  Interface circuits

accidental connection of the RS-232/V.28 signal lines to 100/220 V AC power lines. Figure 2.3.1 shows a circuit with this isolation. The power for the MAX 233 RS-232 transmitter/receiver is generated by a MAX 635 DC–DC converter (Figure 2.3.2). When the MAX 635 regulates point 'A' to –5 V, the isolated output at point 'B' will be semi-regulated to +5 V. The two opto-couplers maintain isolation between the system ground and the RS-232 ground while transferring the data across the isolation barrier. While this circuit will not withstand the 110 V AC between the RS-232 ground and either the receiver or transmitter lines, the voltage difference between the two grounds is only limited by the opto-coupler and DC–DC converter transformer breakdown ratings.

*Courtesy of Maxim Integrated Products, Inc., Sunnyvale, CA*

## 2.4 Low-power 5 V RS232 driver/receiver

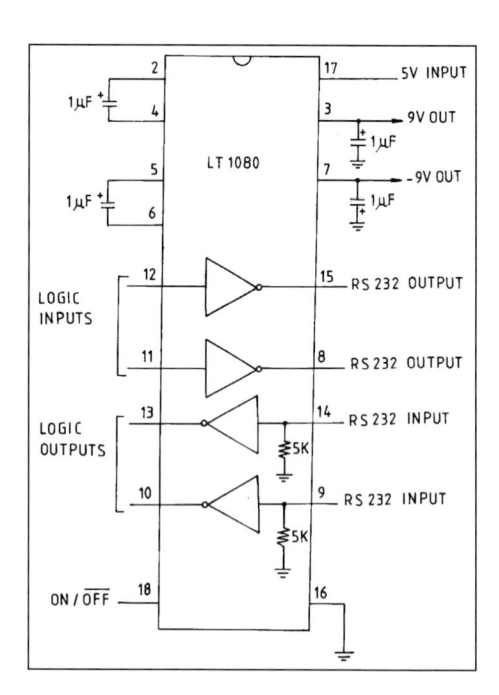

**Figure 2.4.1** Low-power 5 V RS 232 driver/receiver

The LT 1080 and LT 1081 are the only dual RS 232 driver/receiver with charge pump to guarantee absolutely no latch up. These interface optimized devices provide a realistic balance between CMOS levels of power dissipation and real-world requirements for ruggedness. The driver outputs are fully protected against overload and can be shorted to ±30 V. Unlike CMOS, the advanced architecture of LT 1080/LT 1081 does not load the signal line when 'shut down' or when power is off. Both the receiver and RS 232 outputs are put into a high impedance state. An advanced output stage allows driving higher capacitive

Interface circuits 51

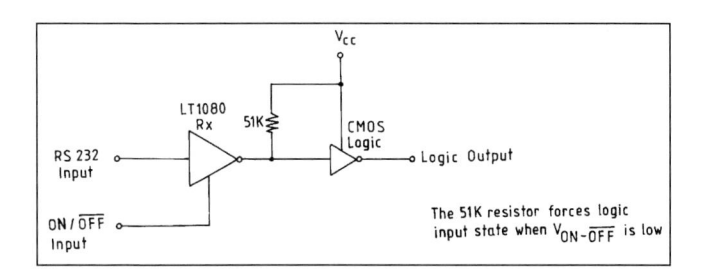

**Figure 2.4.2** Driving CMOS logic from LT 1080

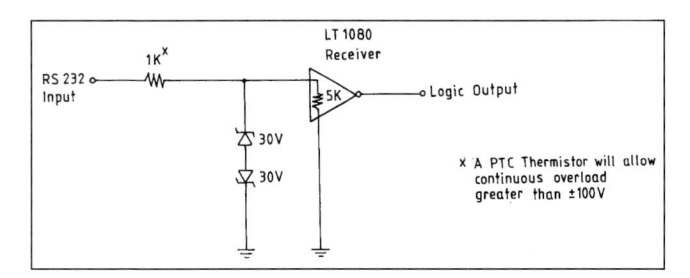

The 51K resistor forces logic input state when $V_{ON-\overline{OFF}}$ is low

**Figure 2.4.3** Protecting receiver against input overloads

x A PTC Thermistor will allow continuous overload greater than ±100 V

loads at higher speeds with exceptional ruggedness against ESD. The salient features of these devices are:

- Improved speed – operates over 65K Baud.
- Improved protection – outputs can be forced to ± 30 V without damage.
- Three-state outputs are high impedance when off.
- Needs only 1 µF capacitors (a version which uses 0.1 µF capacitors is also available).

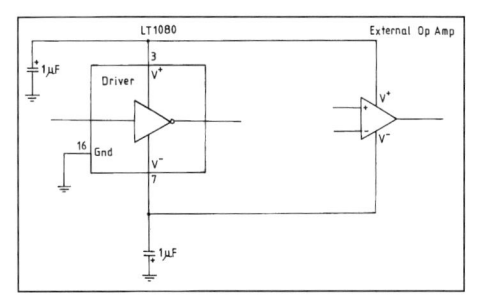

**Figure 2.4.4** Powering external devices from LT 1080

The application circuit for the LT 1080 as driver and receiver is shown in Figure 2.4.1.

The driver output stage of the LT 1080 offers significantly improved protection over older bipolar and CMOS designs. In addition to current limiting, the driver output can be externally forced to ± 30 V with no damage or excessive current flow, and will not disrupt the supplies. Some drivers have diodes connected between the outputs and the supplies, so externally applied voltages can cause excessive supply voltage to develop.

Placing the LT 1080 in the SHUTDOWN mode (pin 18 low) puts both the driver and receiver outputs in a high-impedance state. This allows data line sharing and transceiver applications.

The SHUTDOWN mode also drops input supply current ($V_{CC}$; pin 17) to zero for power-conscious systems.

When driving CMOS logic from a receiver that will be used in the

52  Interface circuits

can be placed from the logic input to $V_{CC}$ to force a definite logic level when the receiver output is in a high-impedance state (Figure 2.4.2).

To protect against receiver input overloads in excess of ±30 V, a voltage clamp can be placed on the data line and still maintain RS 232 compatibility (Figure 2.4.3).

The generated driver supplies ($V^+$ and $V^-$) may be used to power external circuitry such as other RS 232 drivers or op-amps (Figure 2.4.4). They should be loaded with care, since excessive loading can cause the generated supply voltages to drop causing the RS 232 driver output voltages to fall below RS 232 requirements. Up to about 10 mA can be drawn from the driver supplies without loading.

*Courtesy of Linear Technology Corporation, Milpitas, CA*

◆ ◆

## 2.5 Programmable micropower level translator/receiver/driver

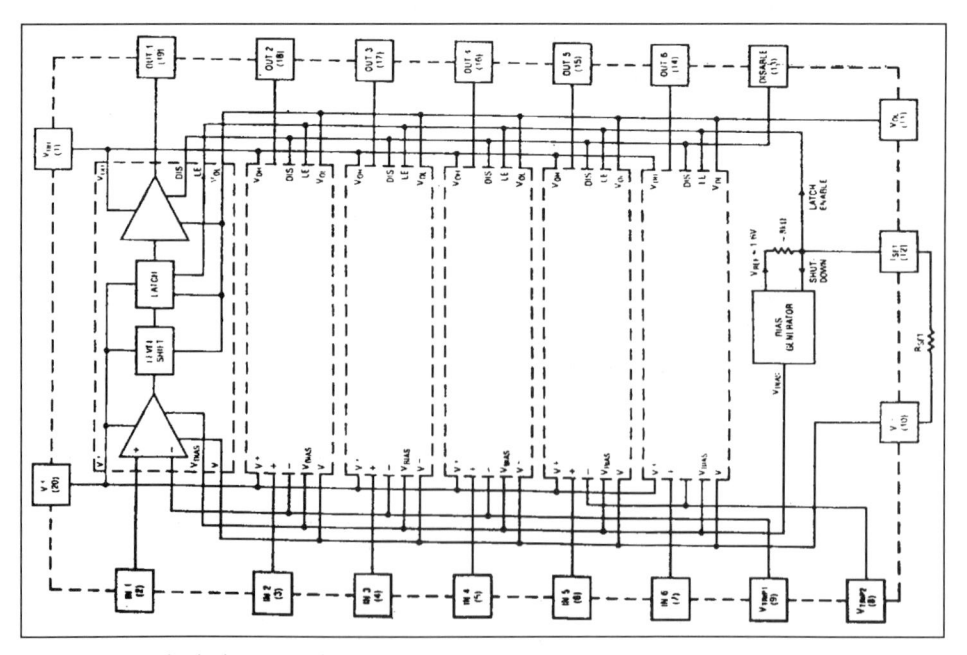

**Figure 2.5.1** Block diagram of LTC 1045

The LTC 1045 is a hex level translator/receiver/driver from Linear Technology Corporation. It consists of six voltage translators and associated control circuitry. Each translator has a linear comparator input stage with the positive input brought out separately. The negative inputs of the first four comparators are tied in common to $V_{TRIP1}$ and the negative inputs of the last two comparators are tied

Interface circuits   53

in common to $V_{TRIP1}$ and the negative inputs of the last two comparators are tied in common to $V_{TRIP2}$. With these inputs, the switching point of the comparators can be set anywhere within the common-mode range of $V^-$ to $V^+$ −2 V. To improve noise immunity, each comparator has a small built-in hysteresis. Hysteresis varies with bias current from 7 mV at low bias current to 20 mV at high bias current.

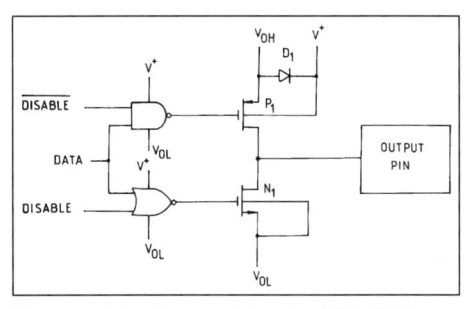

**Figure 2.5.2** Output driver in LTC 1045

## Setting the bias current

Unlike CMOS logic, any linear CMOS circuit must draw some quiescent current. The bias generator (Figure 2.5.1) allows the quiescent current of the comparators to be varied. Bias current is programmed with an external resistor. As the bias current is decreased, the LTC 1045 slows down.

## Shutting power off and latching the outputs

In addition to setting the bias current, the $I_{SET}$ pin shuts power completely off and latches the translator outputs. To do this, the $I_{SET}$ pin must be forced to

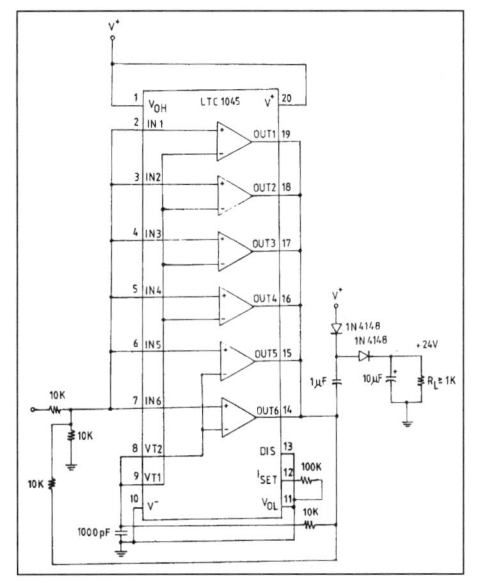

**Figure 2.5.3** LTC 1045 applications (a) 24 V relay supply from +12 V/+15 V supply

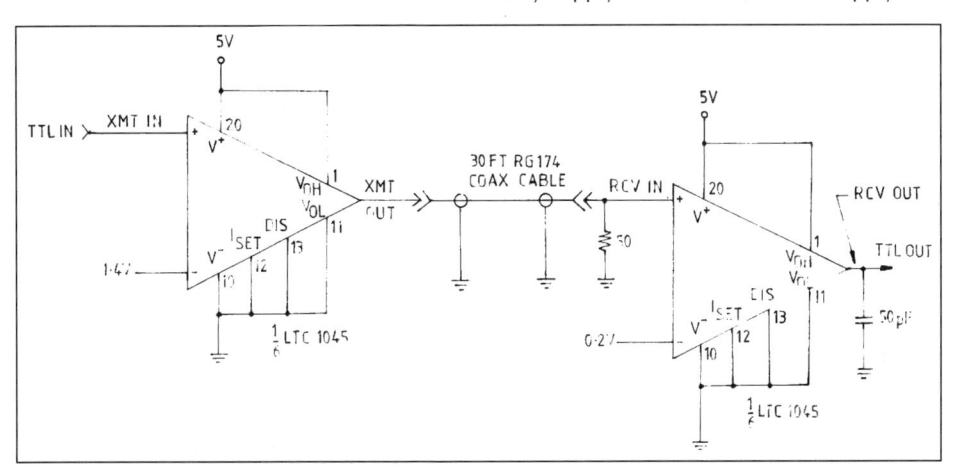

**Figure 2.5.3** (b) coaxial cable driver/receiver

54  Interface circuits

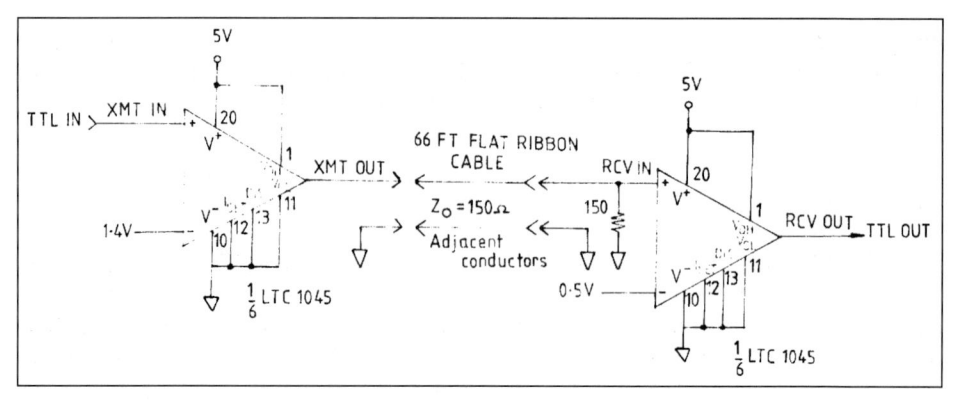

**Figure 2.5.3** (c) flat ribbon cable driver/ receiver

$V^+ - 0.5\,\text{V}$. As shown in Figure 2.5.2, a CMOS gate or a TTL gate with resistor pull-up does this quite nicely. Even though power is turned off to the linear circuitry, the CMOS output logic is powered and maintains the output state. With no DC load on the output, power dissipation, for all practical purposes, is zero.

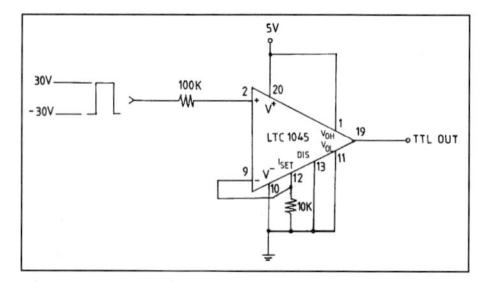

**Figure 2.5.3** (d) RS 232 receiver

Latching the output is fast – typically 80 ns from the rising edge of $I_{SET}$. Going from the latched to flow-through state is much slower – typically 1.5 μs from the falling edge of $I_{SET}$. This time is set by the comparator's power up time. During the power up time, the output can assume false states. To avoid problems, the output should not be considered valid until 2 μs to 5 μs after the falling edge of $I_{SET}$.

### *Putting the outputs in HI-Z state*

A DISABLE input sets the six outputs to a high impedance state. This allows the LTC 1045 to be interfaced to a data bus. When DISABLE = '1' the outputs are high impedance and when DISABLE = '0' they are active. With TTL supplies, $V^+$ = 4.5 V to 5.5 V and $V^-$ = GND, the DISABLE input is TTL compatible.

### *Power supplies*

There are four power supplies on the LTC 1045: $V^+$, $V^-$, $V_{OL}$ and $V_{OH}$. They can be connected almost arbitrarily, but there are a few restrictions. A minimum differential must exist between $V^+$ and $V^-$ and $V_{OH}$ and $V_{OL}$. The $V^+$ to $V^-$ differential must be at least 4.5 V and the $V_{OH}$ to $V_{OL}$ differential must be at least 3.0 V. Another restriction is caused by the internal parasitic diode $D_1$. Because of this diode, $V_{OH}$ must not be greater than $V^+$. Lastly the maximum voltage

Interface circuits 55

absolute maximum. For example, if $V^+ = 5\,V$, $V^-$ or $V_{OL}$ should be no more negative than $-10\,V$. Note that $V_{OL}$ should not be more negative than $-10\,V$ even if the $V_{OH}$ to $V_{OL}$ differential does not exceed the 15 V maximum. In this case the $V^+$ to $V_{OL}$ differential sets the limit.

## Input voltage

The LTC 1045 has no upper clamp diodes as do conventional CMOS circuits. This allows the inputs to exceed the $V^+$ supply. The inputs will break down approximately 30 V above the $V^-$ supply. If the input current is limited with 10k, the input voltage can be driven to at least ±50 V with no adverse effects for any combination of allowed power supply voltages. Output levels will be correct even under these conditions (i.e. if the input voltage is above the trip point, the output will be high and if it is below, the output will be low).

## Output drive

Output drive characteristics of the LTC 1045 will vary with the power supply voltages that are chosen. Output impedance is affected by $V^+$, $V_{OH}$ and $V_{OL}$. $V^-$ has no effect on output impedance. In general output impedance is minimized if $V^+$ to $V_{OH}$ is minimized and $V_{OH}$ to $V_{OL}$ is maximized.

Figure 2.5.3(a)–(d) shows some applications of the LTC 1045.

*Courtesy of Linear Technology Corporation, Milpitas, CA*

◆ ◆ ◆

## 2.6 RS 232 and RS 423 line driver

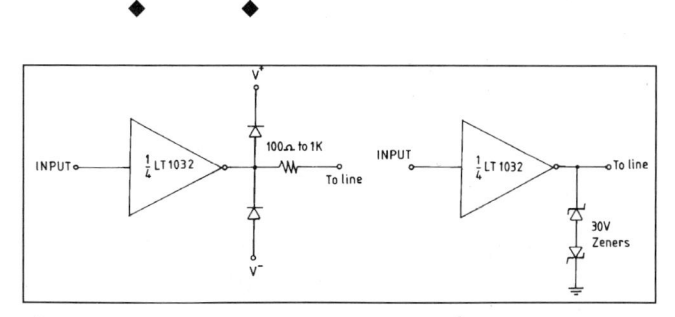

**Figure 2.6.1** Protecting against more than ±30 V output overload

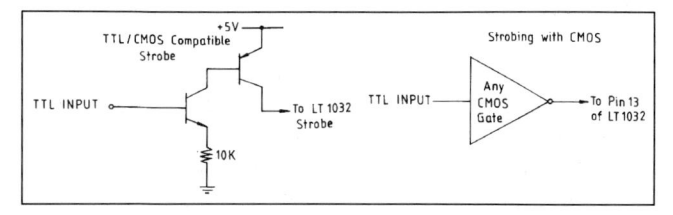

**Figure 2.6.2** Methods of strobing LT 1032 using TTL/CMOS

56  Interface circuits

**Figure 2.6.3** Slew rate adjustment method (about 4 V/μS change)

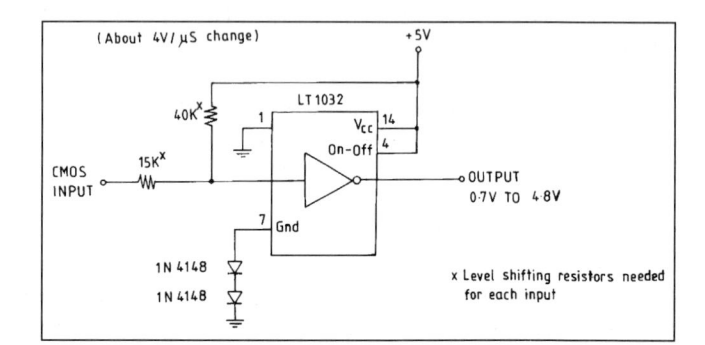

**Figure 2.6.4** Operating from a single 5 V supply

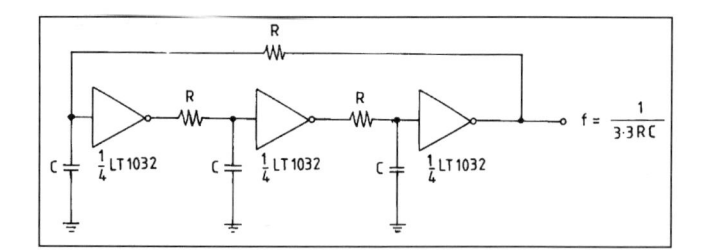

**Figure 2.6.5** (a) Phase shift oscillator

$$f = \frac{1}{3 \cdot 3 RC}$$

The LT 1032 from Linear Technology Corporation is an RS 232 and RS 423 line driver that operates over a ±5 V to ±15 V range on low supply current and can be shut down to zero supply current. Outputs are fully protected from externally applied voltages of ±30 V by both current and thermal limiting. Since the output swings to within 200 mV of the positive supply

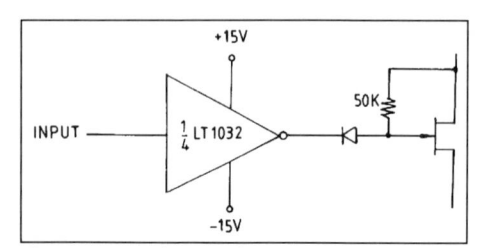

**Figure 2.6.5** (b) FET driver

and 600 mV of the negative supply, power supply needs are minimized. Also included is a strobe pin to force all outputs low independent of input or shutdown conditions. Further, slew rate can be adjusted with a resistor connected to the supply.

A major advantage of the LT 1032 is the high-impedance output state when off or powered down. The LT 1032 is exceptionally easy to use when compared to

Interface circuits 57

older drivers. Operating supply voltage can be as low as ±3 V or as high as ±15 V. Input levels are referred to ground.

The logic inputs are internally set at TTL levels. Outputs are valid for input voltages from 1 V above $V^-$ to 25 V. Driving the logic inputs to $V^-$ turns off the output stage. The 'on–off' control completely turns off all supply current of the LT 1032. The levels required to drive the device on or off are set by internal emitter–base voltages. Since the current into the 'on–off' pin is so low, TTL or CMOS drivers have no problem controlling the device.

The strobe pin is not fully logic compatible. The impedance of the strobe pin is about 2k to ground. Driving the strobe pin positive forces the output stages LOW – even if the device is shut off. Under worst-case conditions, 3 V minimum at 2 mA are needed for driving the strobe pin to ensure strobing.

The response pin can be used to make some adjustment in slew rate. A resistor can be connected between the response pin and the power supplies to drive 50 μA to 100 μA into the pin. The response pin is a low impedance point operating at about 0.75 V above ground. For supply voltage up to ±6 V, current is turned off when the device is turned off. For higher supply voltages, a Zener should be connected in series with the resistor to limit the voltage applied to the response pin to 6 V. Also, for temperatures above 100°C, using the response pin is not recommended. The leakage current into the response pin at high temperatures is excessive.

Outputs are well protected against shorts or externally applied voltage. Tested limits are ±30 V, but the device can withstand external voltages up to the breakdown of the transistors (typically about 50 V). The LT 1032 is usually immune to ESD up to 2500 V on the outputs with no damage (limit of LTC tester).

Figure 2.6.1 shows the method of protecting the LT 1032 against overloads of more than ±30 V. Figure 2.6.2 shows the method of strobing the LT 1032 using TTL and CMOS signals. Figure 2.6.3 shows a method of slew rate adjustment. Figure 2.6.4 shows the method of operating the LT 1032 from a single +5 V supply. Figure 2.6.5 circuits list some applications of the LT 1032.

*Courtesy of Linear Technology Corporation, Milpitas, CA*

◆ ◆

# Monolithic relay driver

Relays are generally driven by discrete transistors and this arrangement requires a free-wheel diode to be connected across the relay coil for providing a path for the decaying current through the coil when it is switched off so that the transistor is not damaged by collector–emitter breakdown. In addition, this arrangement requires more board space, and if in any card design several relays are used, the total board space occupied is very large. We can economize on PCB real estate, do

58  Interface circuits

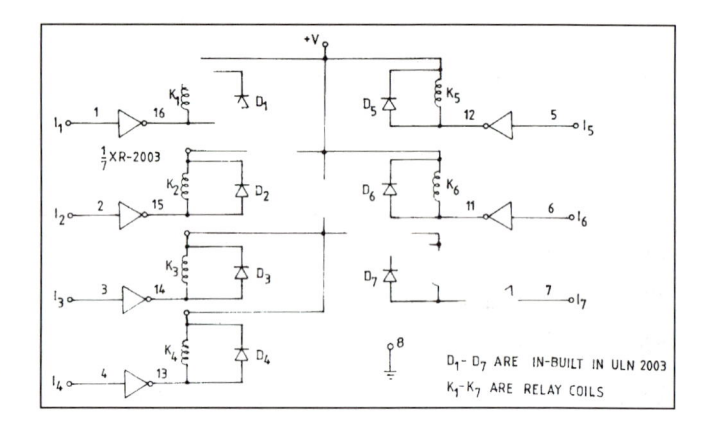

**Figure 2.7.1** Single-chip multiple relay driver

$D_1$-$D_7$ ARE IN-BUILT IN ULN 2003
$K_1$-$K_7$ ARE RELAY COILS

away with the free-wheel diodes and discrete transistors, and have better reliability of operation, using ULN 2003 which is a monolithic transistor array of seven silicon NPN high voltage, high current Darlington transistors. All the units in the monolithic package feature open-collector outputs and integral suppression (i.e. freewheel) diodes for driving inductive loads such as relays or small DC motors. There is a series base resistor to each Darlington pair, and this allows direct operation with TTL or CMOS 5 V supply voltage. The collector current rating of each Darlington pair in the array is 500 mA. However, outputs may be paralleled for higher load current capability. Since there are seven Darlington pairs in each package, there is a saving in board space compared to the discrete driver arrangement for multiple relay driving applications. The scheme for driving multiple relays using a single ULN 2003 is shown in Figure 2.7.1 and the detailed schematic of each driver in the array is shown in Figure 2.7.2.

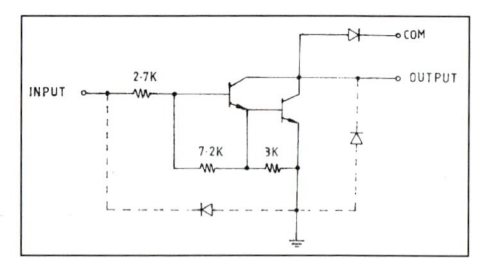

**Figure 2.7.2** Each driver in the array

*Courtesy of Philips, The Netherlands*

# 3 Timer circuits

## 3.1 Astable multivibrator using 555 timer

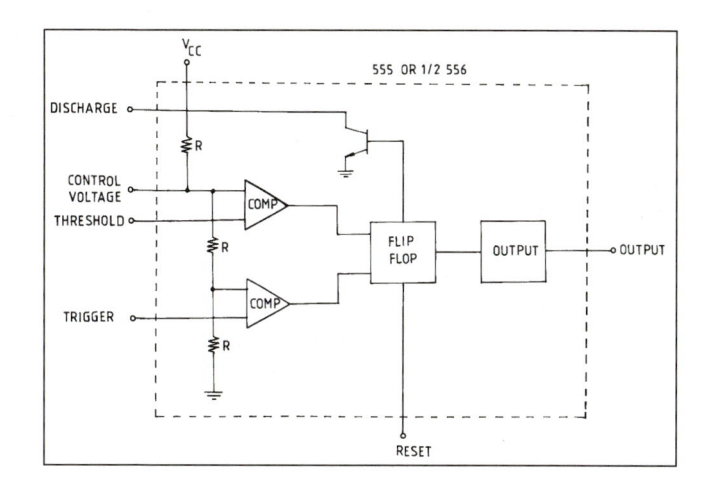

**Figure 3.1.1** 555/556 timer functional block diagram

In mid-1972 Signetics introduced the 555 timer, a unique functional building block that has enjoyed unprecedented popularity. The timer's success can be attributed to several inherent characteristics, foremost of which are versatility, stability and low cost. There can be no doubt that the 555 timer has altered the course of the electronics industry with an impact not unlike that of the IC operational amplifier.

The simplicity of the timer, in conjunction with its ability to produce long time delays in a variety of applications, has lured many designers from mechanical timers, op-amps and various discrete circuits into the ever increasing ranks of timer users.

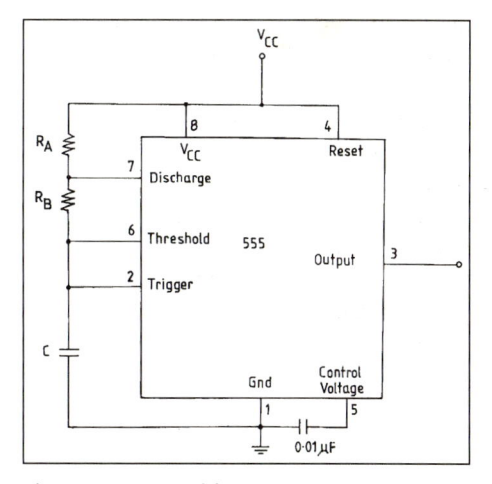

**Figure 3.1.2** Astable using 555 timer

### Description

The 555 timer consists of two voltage comparators, a bistable flip-flop, a discharge transistor and a resistor divider network. To understand the basic concept of the timer, first examine the timer in block form (shown in Figure 3.1.1).

The resistive divider network is used to set the comparator levels. Since all three resistors are of equal value, the threshold comparator is referenced internally at 2/3 of supply voltage level and the trigger comparator is referenced at 1/3 of supply voltage. The outputs of the comparator are tied to the bistable flip-flop. When the trigger voltage is moved below 1/3 of the supply voltage, the comparator changes state and sets the flip-flop, driving the output to a HIGH state. The threshold pin normally monitors the capacitor voltage of the $RC$ timing network. When the capacitor voltage exceeds 2/3 of the supply, the threshold comparator resets the flip-flop, which in turn drives the output to a low state. When the output is in a low state, the discharge transistor is ON, thereby discharging the external timing capacitor. Once the capacitor is discharged, the timer will await another trigger pulse, the timing cycle having been completed.

One of the simplest and most widely used modes of the timer is the astable or free-running mode of operation. The typical connection diagram of the 555 timer operating in the astable mode is shown in Figure 3.1.2. In this mode of operation, the trigger is tied to the threshold pin. At power-up, the capacitor is discharged, holding the trigger low. This triggers the timer, which establishes the capacitor charge path through $R_A$ and $R_B$. When the capacitor voltage reaches the threshold level of $(2/3)\,V_{CC}$, where $V_{CC}$ is the supply voltage, the output drops low and the discharge transistor turns ON. The timing capacitor now discharges through $R_B$. When the capacitor voltage drops to $(1/3)\,V_{CC}$, the trigger comparator trips, automatically retriggering the timer, creating an oscillator whose frequency is given by:

$$f = 1.44/[(R_A + 2R_B)C]$$

Selecting the ratios of $R_A$ and $R_B$ varies the duty cycle accordingly. The duty cycle is given by

$$D = R_B/(R_A + 2R_B)$$

For reliable operation a minimum value of 3k for $R_B$ is recommended to ensure that oscillation begins.

*Courtesy of Philips, The Netherlands*

◆   ◆

## 3.2  Low-cost appliance timer

You can make a low-cost timer which will energize an electrical appliance such as a radio, TV, heater, etc. for a fixed pre-settable time interval and then switch it off, using the circuit shown in Figure 3.2.1. In this circuit the 555 timer is configured as a power-on monostable. On power-up, the one-shot starts timing out for a duration $T = 1.1\,RC$. Since the threshold and trigger inputs of the 555 timer are connected together and also to $R$ and $C$, the circuit will trigger on the application

# Timer circuits

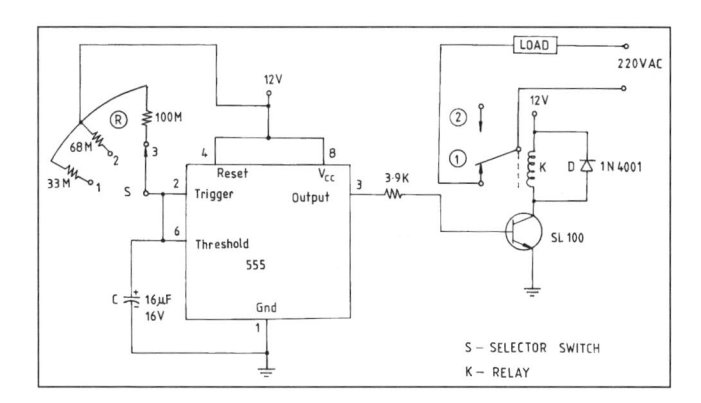

**Figure 3.2.1** A low-cost appliance timer

of power and start a timing cycle. At the instant of application of power, capacitor C is discharged and starts charging towards $V_{CC}$ through resistor R. When the voltage across the capacitor reaches a value (2/3) $V_{CC}$, the threshold of the comparator in the 555 timer is reached and it changes state. This completes the timing cycle. Only one pulse is obtained at the timer output for each application of power to the circuit. The output of the timer is used to switch a transistor which, in turn, controls the operation of a relay. The appliance can be connected as shown in Figure 3.2.1. Diode D is a free-wheel path for the decaying current when an inductive load such as a relay is turned off. The time constant of the circuit is given by $T = 1.1\ RC$. By suitably changing R and C, many different timings can be obtained. Large values of electrolytic capacitors should be avoided to prevent leakage current problems. A selector switch can be used to select one of the many resistors that can be connected for different timing requirements. In Figure 3.2.1, positions 1, 2 and 3 of the selector switch give 10 minutes, 20 minutes and half an hour timings, approximately. Such a timing device would be very useful, for example, to turn off an appliance after a pre-set time, and for similar applications.

## 3.3 Sequential timer

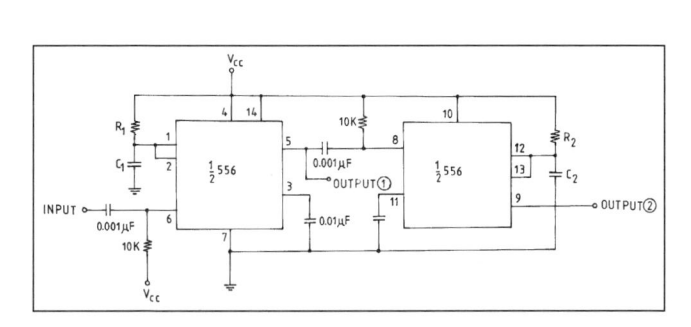

**Figure 3.3.1** Sequential timer

The circuit shown in Figure 3.3.1 uses a dual timer 556 whose both halves are configured as monostables to make a sequential timer. In this kind of timer the output of the first timer is connected as the trigger input to the second timer. An input pulse as shown in the waveforms of Figure 3.3.2 triggers the first monostable to produce an output pulse of time duration $T_1$. When the first one-shot output pulse is timing out, its negative going edge triggers the second one-shot which produces an output pulse of time duration $T_2$. This logic can be extended to design a sequential timer having more than two timers. This configuration is generally used to generate test sequences, to activate devices in a sequence, and for similar applications. The output pulse widths $T_1$, $T_2$, etc. can all be different and chosen according to the requirements of a particular design.

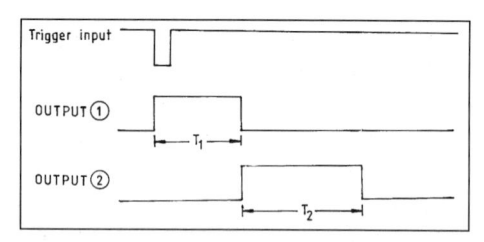

**Figure 3.3.2** Waveforms of circuit in Figure 3.3.1

## 3.4 Long duration timer using quad timer 558

**Figure 3.4.1** Long duration timer using 558

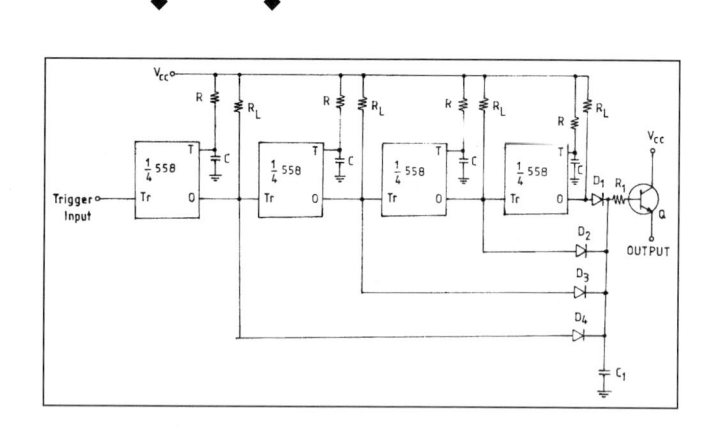

NE/SE is a quad timer. Using this device, four entirely independent timing functions can be achieved using a timing resistor and a capacitor for each section. All the four sections of the device may be used together, in tandem, for sequential timing applications up to several hours. No coupling capacitors are required when connecting the output of one timer section to the input of the next timer section. The NE/SE 558 structure is open-collector which requires a pull-up resistor to $V_{CC}$ and is capable of sinking 100 mA per unit, so long as the power dissipation and junction temperature rating of the die and package are not exceeded. The output is normally low and is switched high when triggered.

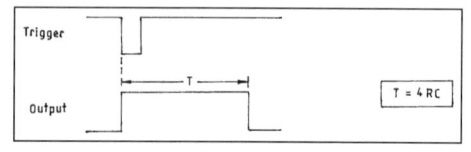

**Figure 3.4.2** Waveforms of circuit in Figure 3.4.1

$T = 4RC$

Timer circuits 63

The circuit shown in Figure 3.4.1 is a long duration timer configured using NE/SE 558. It consists of four monostable multivibrators connected in tandem, i.e. the output of the first one-shot is used to trigger the next one-shot. The output pulse width of each one-shot is given by $RC$, where $R$ and $C$ are the timing elements.

A trigger input applied to the first one-shot triggers it and it produces an output pulse of width $RC$. This output pulse charges the capacitor $C_1$ and switches ON the transistor Q giving a high output. The negative-going transition at the end of the timing duration of the first one-shot triggers the second one-shot, which now produces an output pulse of duration $RC$ and this output charges the capacitor $C_1$, keeping the transistor Q ON. This process repeats for the next monostable in the sequence. The diodes $D_1$–$D_4$ prevent the discharge of capacitor $C_1$ when the output of each one-shot goes low at the end of the timing duration $RC$. As the waveforms in Figure 3.4.2 show, a long duration pulse of width $4RC$ is available at the output of transistor Q. By choosing suitable values of the timing components $R$ and $C$, various time-periods can be obtained. This circuit can be used for very long duration timing applications.

*Courtesy of Philips, The Netherlands*

◆ ◆ ◆

## 3.5 CMOS precision programmable 0–99 seconds/minutes timer

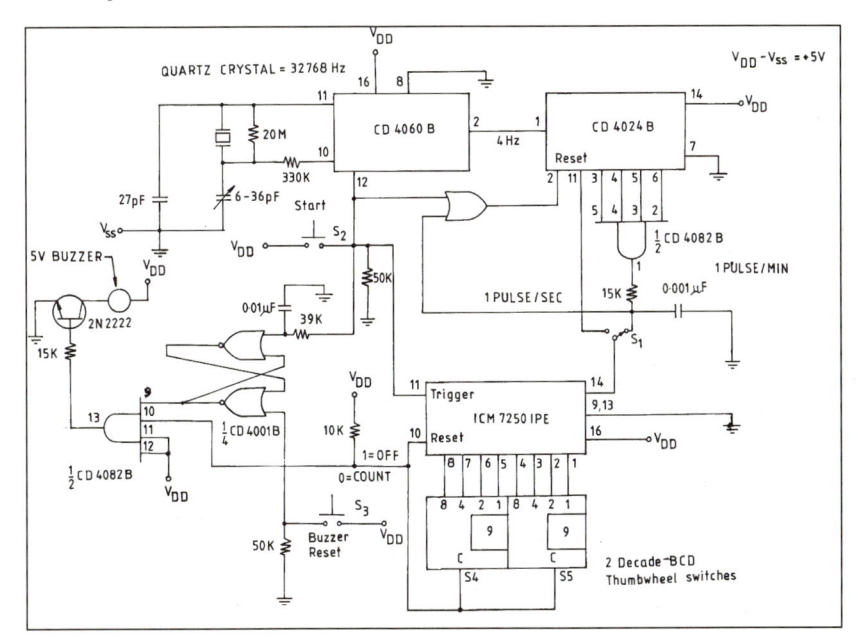

**Figure 3.5.1** 0–99 seconds/minutes timer

The ICM 7250 is a CMOS timer/counter, from Harris Semiconductor, which together with the ICM 7555/6 (CMOS versions of the NE/SE 555/6) provides a complete line of RC oscillators/timers/counters offering lower supply currents, wider supply voltage ranges, higher operating frequencies, lower component counts and a wider range of timing components. Each device consists of a counter section, control circuitry, and an RC oscillator requiring an external resistor and capacitor. The ICM 7250 is optimized for decimal counting or timing. An example of application of the ICM 7250 is a laboratory timer to alert personnel of the expiration of a preselected interval of time.

When connected as shown in Figure 3.5.1, the timer can accurately measure preselected time intervals of 0–99 seconds or 0–99 minutes. A 5 V buzzer alerts the operator when the preselected time interval is over. The circuit operates as follows.

The time base is first selected with $S_1$ (seconds or minutes); then units 0–99 are selected on the two thumbwheel switches $S_4$ and $S_5$. Finally, switch $S_2$ is depressed to start the timer. Simultaneously the quartz crystal controlled divider circuits are reset, the ICM 7250 is triggered and counting begins. The ICM 7250 counts until the pre-programmed value is reached, whereupon it is reset, pin 10 of the CD 4082B is enabled and the buzzer is turned ON. Pressing $S_3$ turns the buzzer OFF.

*Courtesy of Harris Semiconductor, Melbourne, FL*

◆ ◆ ◆

## 3.6 Retriggerable monostable using 555 timer

**Figure 3.6.1**
Retriggerable monostable using 555 timer

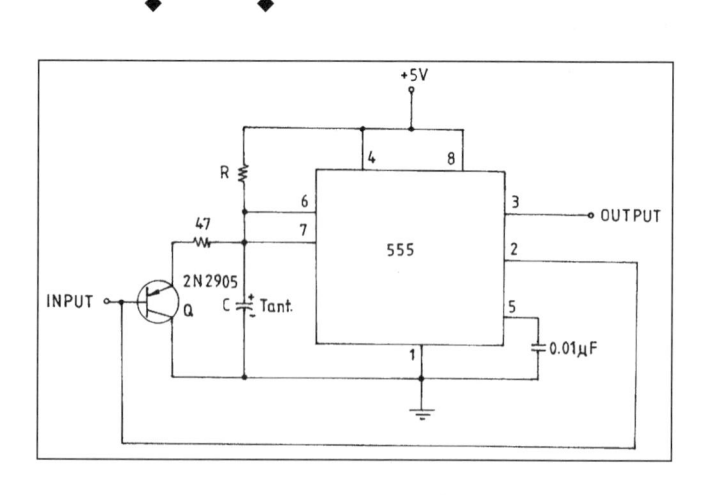

During normal operation a monostable multivibrator using a 555 timer is non-retriggerable, i.e. if trigger pulses arrive before completion of a timing cycle they are ignored and the one-shot times out. In other words, a 555 timer has no retriggerable mode of monostable operation on its own. However, in some applications a retriggerable monostable having the long timing cycle and other advantages of a 555 timer is required. This can be achieved using a 555 timer by

Timer circuits 65

connecting a single external transistor across the capacitor as shown in Figure 3.6.1. For each input pulse appearing at the input, the transistor conducts and shorts the capacitor returning its voltage to zero during the off periods of the pulse. As a result, the capacitor voltage never crosses the $(2/3)\,V_{CC}$ threshold of the internal comparator in the 555 and the one-shot output remains high as long as triggering pulses are received. The output is high provided the time-interval between pulses is less than $1.1\,RC$ which is the timing period of the monostable. Since the range of values of $R$ and $C$ that can be chosen for a 555 monostable is quite wide, the circuit can be programmed to operate for a wide range of pulse rates.

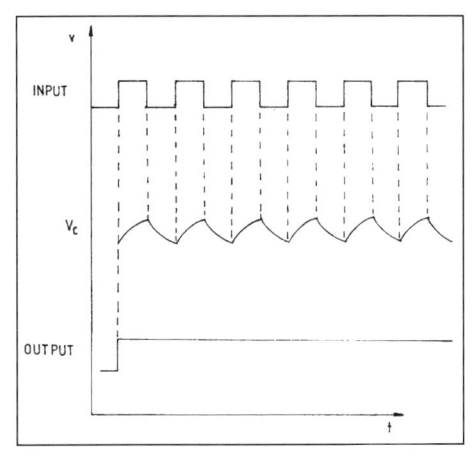

**Figure 3.6.2** Waveforms of circuit in Figure 3.6.1

The waveforms for this configuration are shown in Figure 3.6.2.

## 3.7 Low-power monostable using 555 timer

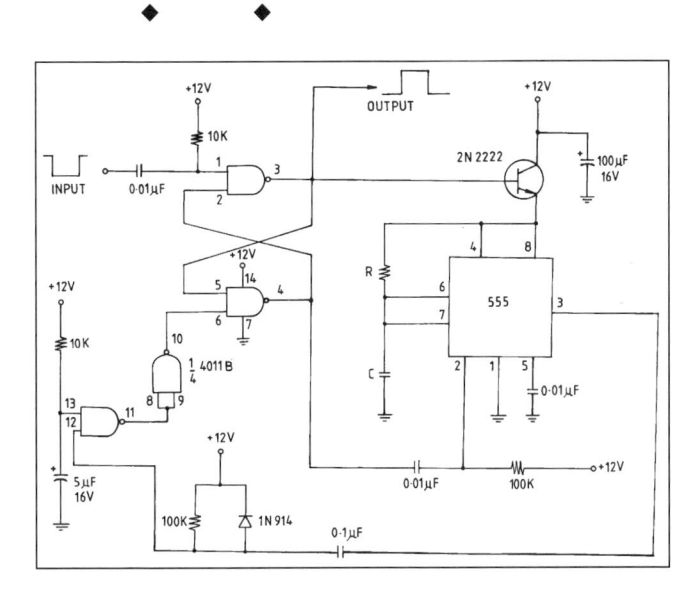

**Figure 3.7.1** Low-power monostable

In battery-operated equipment where load current is a significant factor, the circuit in Figure 3.7.1 can deliver 555 monostable operation at low standby power. This circuit interfaces directly with CMOS 4000 series and 74L00 series. During the monostable time, the current drawn is 4.5 mA for $T = 1.1RC$. The rest of the time the current drawn is less than 50 µA.

*Courtesy of Philips, The Netherlands*

# 4 Op-amp circuits

## 4.1 Integrator

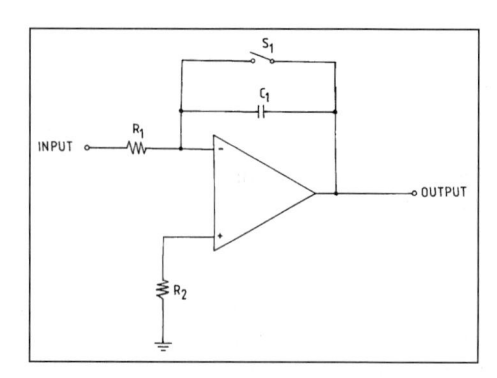

**Figure 4.1.1** Integrator

Integration can be performed with a variation of the inverting amplifier by replacing the feedback resistor with a capacitance (Figure 4.1.1). The transfer function is defined by

$$V_{OUT} = (-1/RC) \int_0^t V_{IN}\, dt$$

The gain of the circuit falls at 6 dB per octave over the range in which strays and leakages are small.

Since the gain at DC is very high, a method for resetting initial conditions is necessary. Switch $S_1$ removes the charge on the capacitor. A relay or FET may be used in the practical circuit. Bias and offset currents and offset voltage of the switch should be low in such an application.

*Courtesy of Philips, The Netherlands*

◆ ◆

## 4.2 Differentiator

The differentiator of Figure 4.2.1(a) is another variation of the inverting amplifier. The gain increases at 6 dB per octave until it intersects the amplifier open-loop gain, and then decreases because of the amplifier bandwidth. This characteristic can lead to instability and high frequency noise sensitivity. A more practical

Op-amp circuits  67

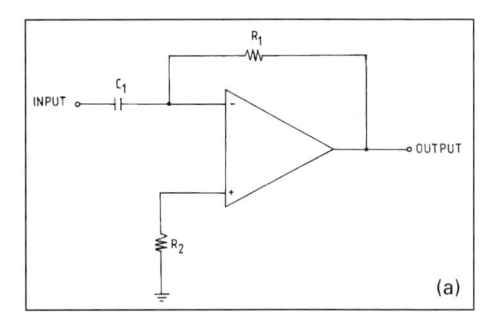

**Figure 4.2.1** (a) Differentiator; (b) practical differentiator

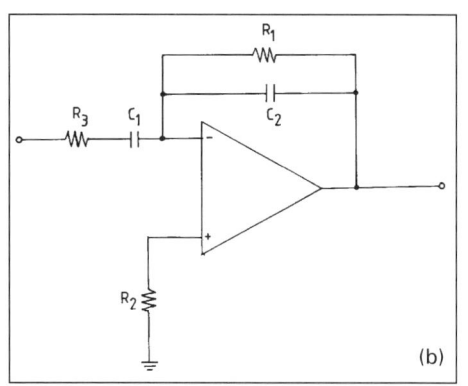

circuit is shown in Figure 4.2.1(b). The gain has been reduced by $R_3$ and the high frequency gain reduced by $C_2$, allowing better phase control and less high frequency noise. Compensation should be for unity gain.

*Courtesy of Philips, The Netherlands*

◆ ◆

## 4.3 Voltage follower

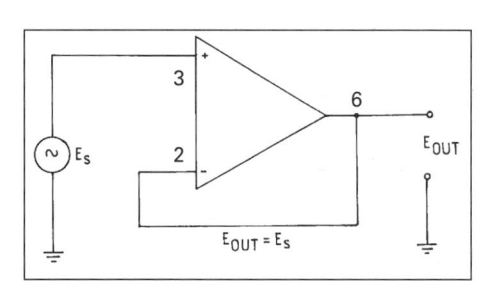

**Figure 4.3.1** Voltage follower

Perhaps the most often used and simplest circuit is that of a voltage follower. The circuit of Figure 4.3.1 illustrates the simplicity.

Applying the zero differential input theorem of op-amps, the voltages at pins 2 and 3 are equal, and since pins 2 and 6 are tied together, their voltage is equal; hence $E_{OUT} = E_{IN}$. Trivial to analyse, the circuit, nevertheless, does illustrate the power of the zero differential voltage theorem. Because the input impedance is multiplied and the output impedance divided by the loop gain, the voltage follower is extremely useful for buffering voltage sources and for impedance transformation.

The basic configuration of Figure 4.3.1 has a gain of 1 with an extremely high input impedance. Setting the feedback resistor equal to the source impedance will cancel the effects of bias current, if desired.

However, for most applications, a direct connection from output to input will suffice. Errors arise from offset voltage, common-mode rejection ratio and gain. The circuit can be used with any op-amp with the required unity gain compensation, if it is required.

*Courtesy of Philips, The Netherlands*

## 4.4 Simulated inductor

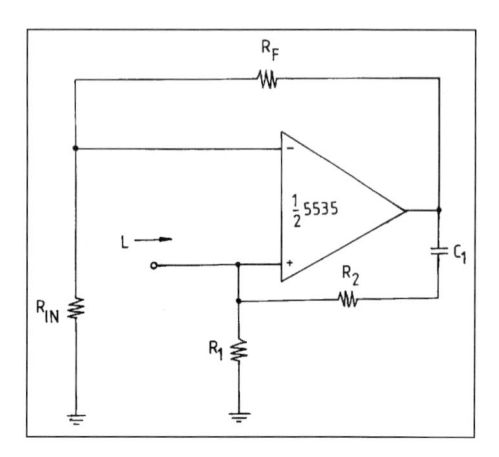

**Figure 4.4.1** Virtual inductor

With a constant current excitation, the voltage dropped across an inductance increases with frequency. Thus, an active device whose output increases with frequency can be characterized as an inductance. The circuit of Figure 4.4.1 yields such a response with the effective inductance being equal to

$$L = R_1 R_2 C$$

The Q of this inductance depends upon $R_1$ being equal to $R_2$. At the same time, however, the positive and negative feedback paths of the amplifier are equal leading to the distinct possibility of instability at high frequencies. $R_1$ should, therefore, always be slightly smaller than $R_2$ to ensure stable operation.

*Courtesy of Philips, The Netherlands*

## 4.5 Op-amp power booster

For most applications, the available power from op-amps is sufficient. There are times when more power handling capability is necessary. A simple power booster capable of driving moderate loads is shown in Figure 4.5.1. The circuit as shown

Op-amp circuits 69

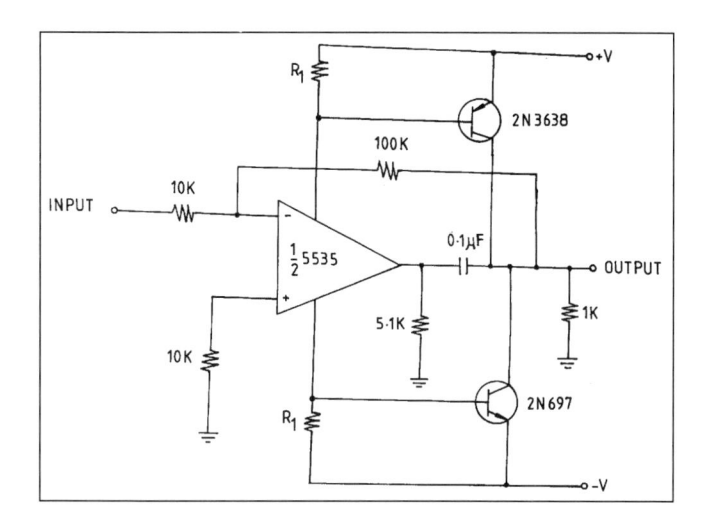

**Figure 4.5.1** Op-amp power booster

uses an NE 5535 high slew rate dual op-amp. Other amplifiers may be substituted only if $R_1$ values are changed because of the $I_{CC}$ current required by the amplifier. $R_1$ should be calculated from the expression

$$R_1 = 600\,\text{mV}/I_{CC}$$

*Courtesy of Philips, The Netherlands*

## 4.6 High current booster

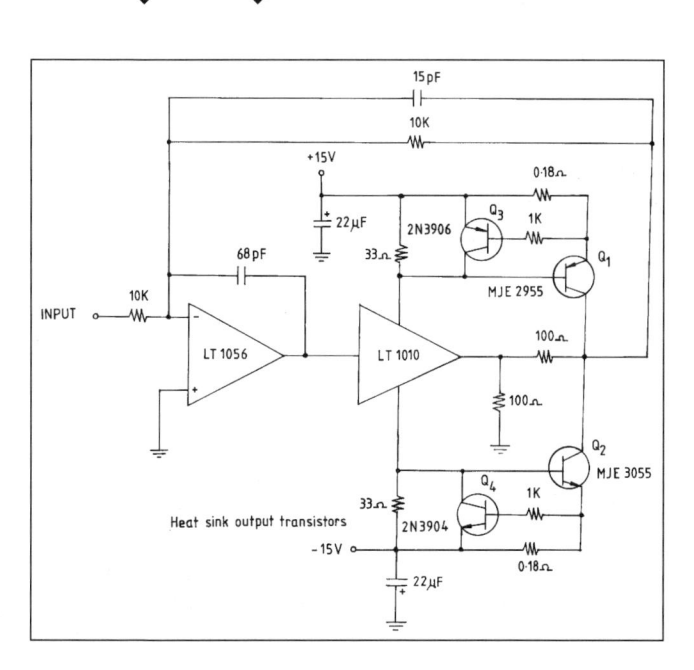

**Figure 4.6.1** High current booster

## 70 Op-amp circuits

The circuit shown in Figure 4.6.1 uses a discrete stage to obtain 3 A output current capacity. The configuration shown provides a clean, quick way to increase the output power of LT 1010. It is useful for high current loads such as linear actuator coils in disk drives.

The 33 ohm resistors sense the LT 1010's supply current, with the grounded 100 ohm resistor supplying a load for the LT 1010. The voltage drop across the 33 ohm resistors biases $Q_1$ and $Q_2$. Another 100 $\Omega$ resistor closes a local feedback loop, stabilizing the output stage. Feedback to the LT 1056 control amplifier is via the 10k value. $Q_3$ and $Q_4$, sensing across the 0.18 ohm units, furnish current limiting at about 3.3 A.

*Courtesy of Linear Technology Corporation, Milpitas, CA*

◆ ◆

## 4.7 Window comparator

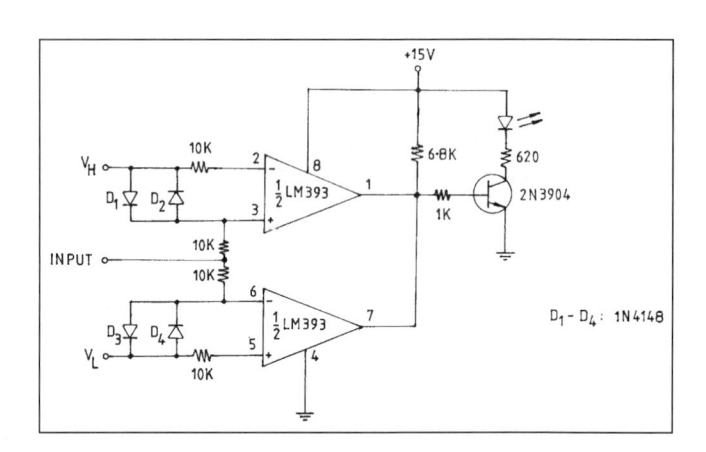

**Figure 4.7.1** Dual-limit comparator

A window comparator is a frequently used basic building block in circuit design. A window comparator also called a dual-limit comparator, as its name implies, gives a specific output if the voltage input is within a certain defined 'window' consisting of a high-voltage limit and a low-voltage limit. The Figure 4.7.1 circuit shows a simple window comparator. If the input voltage is greater than the upper-limit voltage $V_H$ or less than the lower-limit voltage $V_L$, the output of the window comparator is HIGH; if the input voltage is within the window defined by $V_1$ and $V_H$, the output of the window comparator is LOW. The output of the window comparator drives a transistor with an LED connected to its collector. The LED glows if a voltage greater than $V_H$ or less than $V_L$ appears at the window comparator input; for an input voltage within the window, the LED is OFF. Diodes $D_1$–$D_4$ are input protection clamping diodes to protect the comparator inputs from high voltages exceeding the common mode voltage.

Window comparators have wide applications in instrumentation.

◆ ◆

## 4.8 Voltage comparator with hysteresis

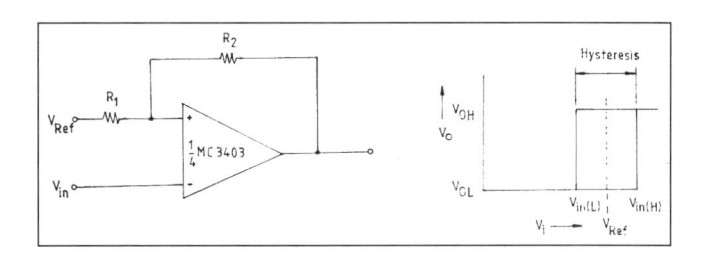

**Figure 4.8.1** Comparator with hysteresis

Voltage comparators are high gain differential input–logic output devices. They are specifically designed for open-loop operation with a minimum of delay time. Although variations of the comparator are used in a host of applications, all uses depend upon the basic transfer function. Device operation is simply a change of output voltage dependent upon whether the signal input is above or below the threshold input.

Many similarities exist between operational amplifiers and the amplifier section of voltage comparators. In fact, op-amps can be used to implement the comparator function at low frequencies. In a voltage comparator, when the input exceeds the reference voltage, the output switches either positive or negative, depending on how the inputs are connected.

Normally saturated high or low, the amplifiers used in voltage comparators are seldom held in their threshold region. They possess high gain–bandwidth products and are not compensated to preserve switching speed. Therefore, if the compared voltages remain at or near the threshold for long periods of time, the comparator may oscillate or respond to noise pulses. For instance, this is a common problem with successive approximation A/D converters where the differential voltage seen by the comparator becomes successively smaller until noise signals cause indecision. To avoid this oscillation in the linear region, hysteresis can be employed from output to input. Figure 4.8.1 shows a comparator with hysteresis using MC 3403. MC 3403 has electrical characteristics similar to the popular µA 741 but it has certain distinct advantages over standard operational amplifier types in single-supply applications. The MC 3403 can operate at supply voltages as low as 3 V or as high as 36 V. The common-mode input range also includes the negative supply, thereby eliminating the necessity for external biasing components in many applications. The output voltage range also includes the negative power supply voltage.

In the comparator of Figure 4.8.1, hysteresis occurs because a small portion of the one level output voltage is fed back in phase and added to the input signal. This feedback aids the signal in crossing the threshold. When the signal returns to the threshold, the positive feedback must be overcome by the signal before switching can occur. The threshold 'dead zone' created by this method prevents output chatter with signals having slow and erratic zero crossings. The input threshold voltage levels are given by

$$V_{IN(L)} = [R_1/(R_1 + R_2)](V_{OL} - V_{REF}) + V_{REF}$$
$$V_{IN(H)} = [R_1/(R_1 + R_2)](V_{OH} - V_{REF}) + V_{REF}$$

and the hysteresis is given by

$$H = [R_1/(R_1 + R_2)](V_{OH} - V_{OL})$$

*Courtesy of Philips, The Netherlands*

## 4.9 Voltage-to-current converter

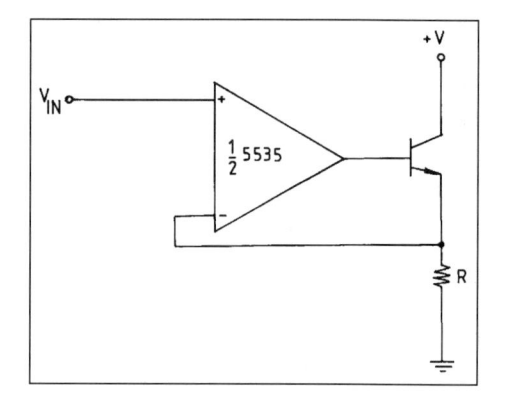

**Figure 4.9.1** Voltage-to-current converter

A simple voltage-to-current converter is shown in Figure 4.9.1. The current out is $I_{OUT} = V_{IN}/R$. For negative currents, a PNP can be used and, for better accuracy, a Darlington pair can be substituted for the transistor. With careful design, this circuit can be used to control currents of many amperes. Unity gain compensation is necessary.

The circuit in Figure 4.9.2 has a different input and will produce either polarity of output current. The main disadvantages are the error current flowing in $R_2$ and the limited current available.

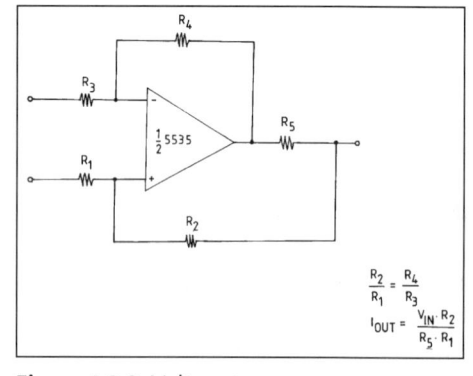

$$\frac{R_2}{R_1} = \frac{R_4}{R_3}$$
$$I_{OUT} = \frac{V_{IN} \cdot R_2}{R_5 \cdot R_1}$$

**Figure 4.9.2** Voltage-to-current converter

*Courtesy of Philips, The Netherlands*

## 4.10 Supply-frequency reject filter

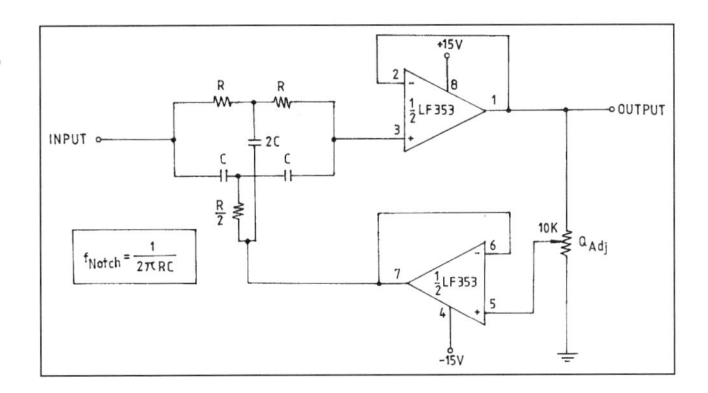

**Figure 4.10.1** Supply-frequency reject filter

The circuit shown in Figure 4.10.1 uses a dual op-amp (LF 353) and a twin-T network to configure a supply-frequency reject filter. This is essentially a notch filter which has its centre frequency at the AC supply frequency of 50 Hz or 60 Hz. This kind of filter is frequently required in instrumentation applications to eliminate AC supply signal picked up in the circuit. The frequency of the notch is given by

$$f = 1/(2\pi RC)$$

where the twin-T networks are made up of elements $R, R, 2C$ and $C, C, R/2$, as shown. The quality factor Q of the filter, which is defined as the ratio of centre frequency $f$ to the 3 dB bandwidth, is adjusted in this circuit by varying the 10k potentiometer. Being an active filter, the performance of this circuit is far superior to that of a passive twin-T network.

## 4.11 Visible voltage indicator

The circuit shown in Figure 4.11.1 allows visible indication of voltage levels using the popular quad voltage comparator LM 339. The reference voltage levels $V_{Ref1}$, $V_{Ref2}$, $V_{Ref3}$ and $V_{Ref4}$ are connected to the non-inverting inputs of the comparators. The common input voltage is connected to the inverting input terminal of all four comparators. Whenever the input voltage exceeds the reference input to a particular comparator, the comparator output goes low and the corresponding LED at the output of the comparator glows. This circuit configuration is useful for multiple voltage comparison such as sensing of transducer outputs, sensing of liquid levels and several other voltage level detection applications.

*Courtesy of Philips, The Netherlands*

74  Op-amp circuits

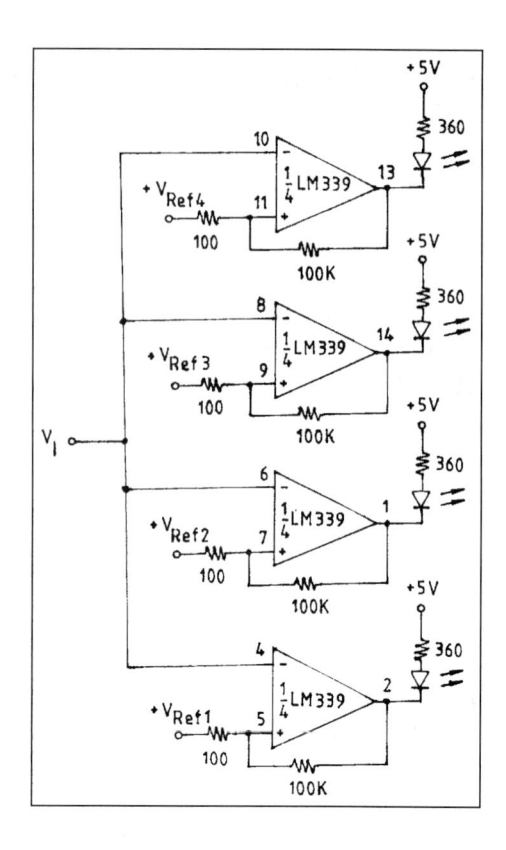

**Figure 4.11.1** Visible voltage indicator

## 4.12 Precision half-wave rectifier

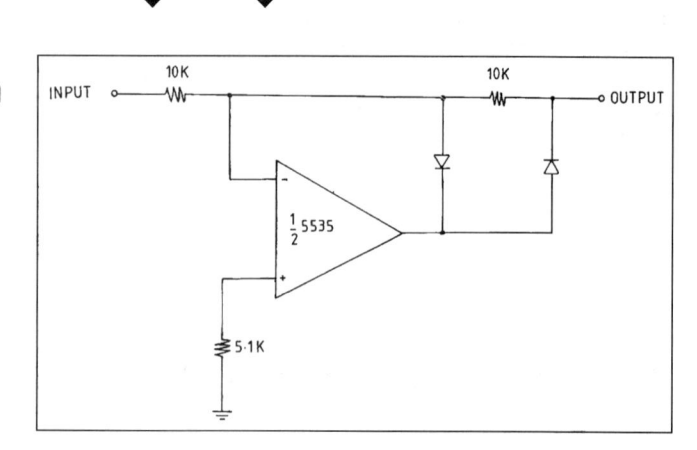

**Figure 4.12.1** Precision half-wave rectifier

Figure 4.12.1 provides a circuit for accurate half-wave rectification of the incoming signal. For positive signals, the gain is 0; for negative signals, the gain is −1. By reversing both diodes, the polarity can be inverted. This circuit provides an accurate output, but the output impedance differs for the two input polarities and buffering may be needed. The output slews through two diode drops when the input polarity reverses. The NE 5535 device will work up to 10 kHz with less than 5% distortion.

*Courtesy of Philips, The Netherlands*

## 4.13 Precision full-wave rectifier

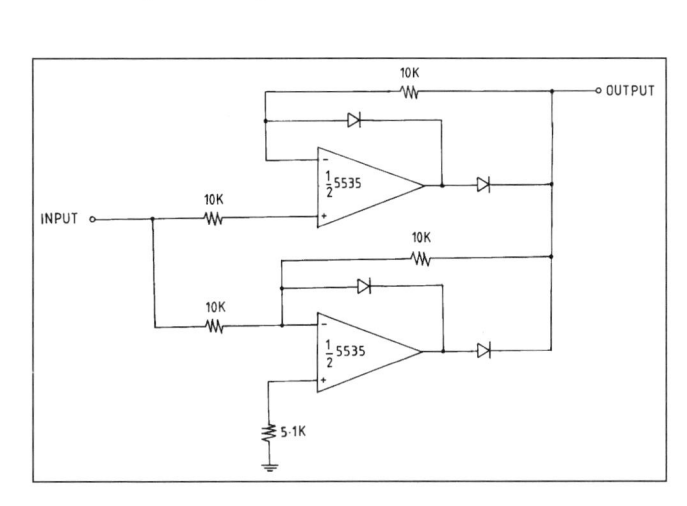

**Figure 4.13.1** Precision full-wave rectifier

The circuit in Figure 4.13.1 provides accurate full-wave rectification. The output impedance is low for both input polarities, and the errors are small at all signal levels. The output will not sink large currents except a small amount through the 10k resistors. Therefore, the load applied should be referenced to ground or a negative voltage. Reversal of all diode polarities will reverse the polarity of the output. Since the outputs of the amplifiers must slew through two diode drops when the input polarity changes, 741-type devices give 5% distortion at about 300 Hz.

*Courtesy of Philips, The Netherlands*

## 4.14 Op-amp supply splitter

You can derive bipolar supplies for an op-amp from a single power supply using the circuit shown in Figure 4.14.1 enabling it to handle bipolar signals. In the circuit, which illustrates the technique using a quad op-amp LM 324, one of the four op-amps in the quad op-amp package is used to create a pseudo-ground for bipolar operation. The two 4.7k resistors at the input to the op-amp non-inverting terminal act as a voltage splitter deriving two 5 V supplies from the single 10 V

76  Op-amp circuits

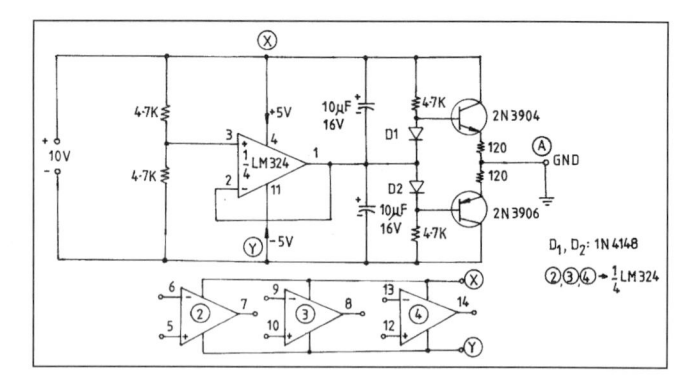

**Figure 4.14.1** Op-amp handles bipolar signals and operates on a single supply

supply. The mid-point of the resistive voltage splitter is connected to a voltage follower configured using 1/4 LM 324. The output of the op-amp now defines a new AC/DC ground – a pseudo-ground – with reference to which bipolar signals can be handled by the other three op-amps. The voltage-follower stage alone would be sufficient if the drains on the bipolar power supplies were matched and within the limits of the op-amp rating. In practice, however, unbalanced currents would flow and if it is required to handle larger currents than the op-amp permits, a matched output stage consisting of a pair of complementary transistors and switching diodes connected as shown in the circuit can be used. The common junction of the two emitter resistors now defines the pseudo-ground instead of the op-amp output. If a 30 V single supply is used instead of a 10 V supply the op-amps can be operated at ± 15 V. Thus the circuit affords a low-cost method of deriving bipolar power supplies for an op-amp using a single power supply. In general, if the single supply is $V$, the two bipolar supplies are $\pm V/2$.

## 4.15  Four-quadrant analog multiplier

The ICL 8013 is a four-quadrant analog multiplier whose output is proportional to the algebraic product of two input signals. Feedback around an internal op-amp provides level shifting and can be used to generate division and square-root functions. A simple arrangement of potentiometers may be used to trim gain accuracy, offset voltage and feedthrough performance. The high accuracy, wide bandwidth and increased versatility of the ICL 8013 make it ideal for all multiplier applications in control and instrumentation systems. Applications include RMS measuring equipment, frequency doublers, balanced modulators and demodulators, function generators and voltage-controlled amplifiers.

Some of the applications of ICL 8013 are explained in this section.

Op-amp circuits 77

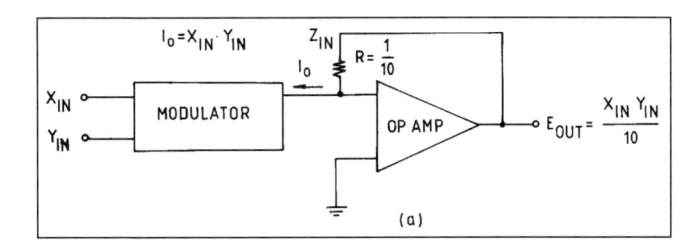

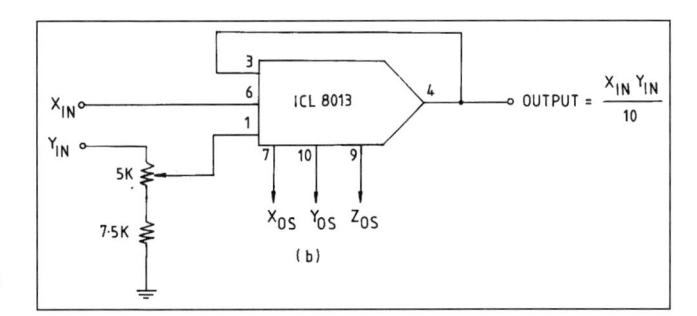

**Figure 4.15.1** (a) Multiplier block diagram; (b) circuit connection

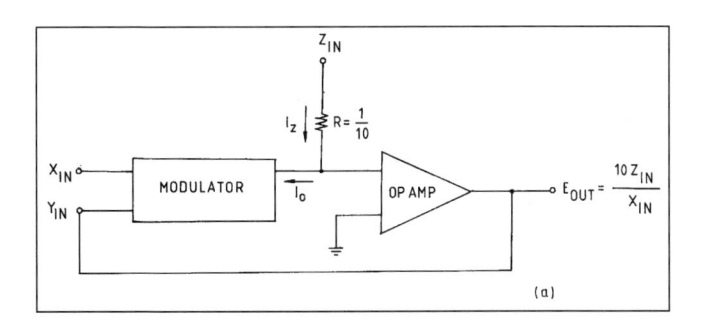

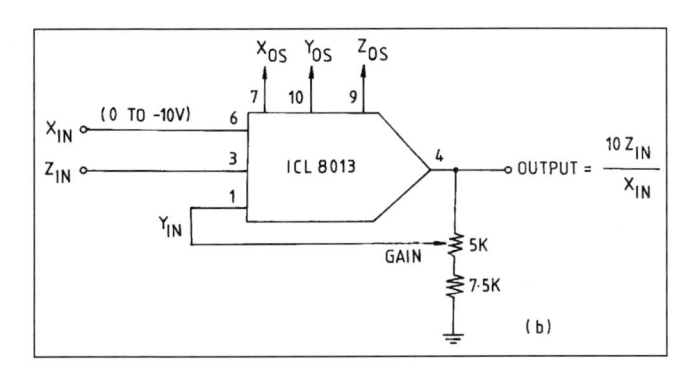

**Figure 4.15.2** (a) Divider block diagram; (b) circuit connection

78 Op-amp circuits

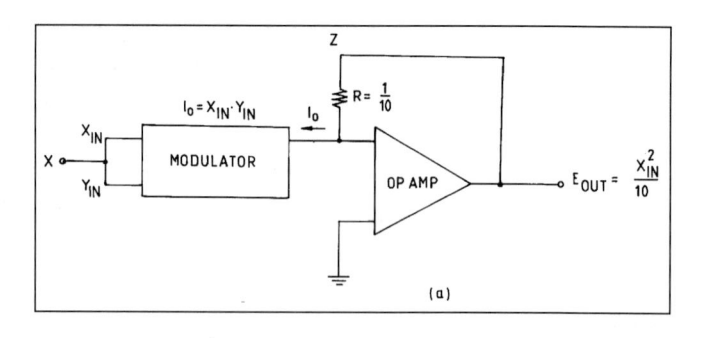

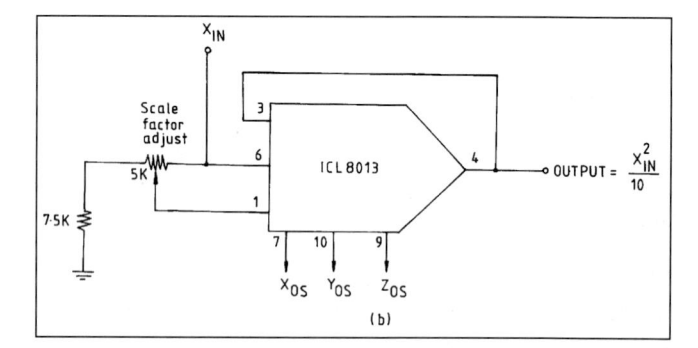

**Figure 4.15.3** (a) Squarer block diagram; (b) circuit connection

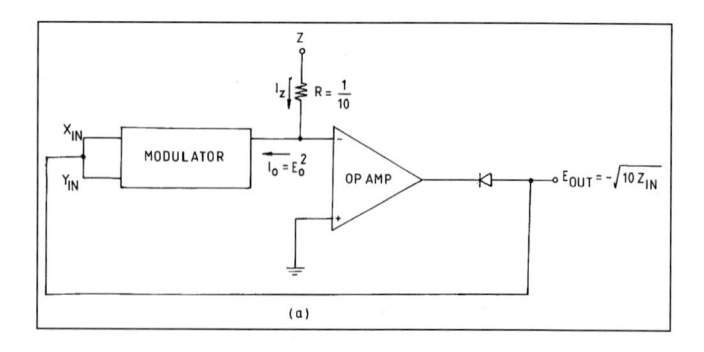

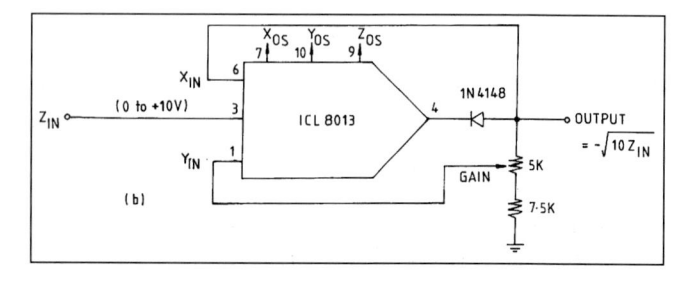

**Figure 4.15.4** (a) Square root block diagram; (b) circuit connection

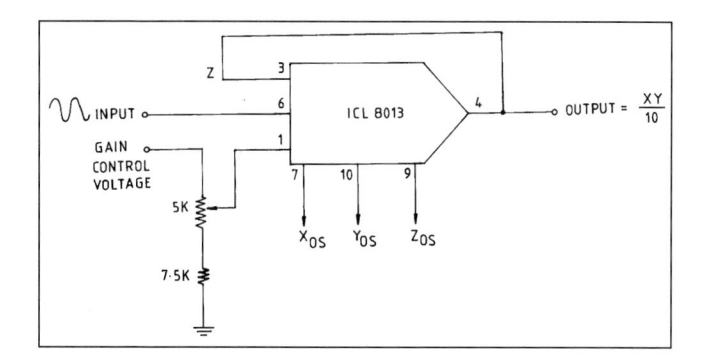

**Figure 4.15.5** Variable gain amplifier

## Multiplication

In the standard multiplier connection, the Z terminal is connected to the op-amp output as shown in Figure 4.15.1. For trimming the multiplier use the following procedure:

*Multiplier trimming procedure*
1. Set $X_{IN} = Y_{IN} = 0\,V$ and adjust $Z_{OS}$ for zero output.
2. Apply a $\pm 10\,V$ low frequency ($\leq 100\,Hz$) sweep (sine or triangle) to $Y_{IN}$ with $X_{IN} = 0\,V$, and adjust $X_{OS}$ for minimum output.
3. Apply the sweep signal of Step 2 to $X_{IN}$ with $Y_{IN} = 0\,V$ and adjust YOS for minimum output.
4. Readjust $Z_{OS}$ as in Step 1, if necessary.
5. With $X_{IN} = 10.0\,V$ DC and the sweep signal of Step 2 applied to $Y_{IN}$, adjust the gain potentiometer for output $= Y_{IN}$. This is easily accomplished with a differential scope plug-in (A + B) by inverting one signal and adjusting gain control for (output $- Y_{IN}$) = zero.

## Division

If the Z terminal is used as an input, and the output of the op-amp connected to the Y input, the device functions as a divider. Since the input to the op-amp is at virtual ground, and requires negligible bias current, the overall feedback forces the modulator output current to equal the current produced by Z.
Therefore

$$I_O = X_{IN} \cdot Y_{IN} = Z_{IN}/R = 10\, Z_{IN}$$

Since

$$Y_{IN} = E_{OUT}$$

$$E_{OUT} = 10\, Z_{IN}/X_{IN}$$

Note that when connected as a divider, the X input must be a negative voltage to maintain overall negative feedback.

80  Op-amp circuits

*Divider trimming procedure*
1. Set trimming potentiometers at mid-scale by adjusting voltage on pins 7, 9 and 10 ($X_{OS}$, $Y_{OS}$, $Z_{OS}$) for zero volts.
2. With $Z_{IN} = 0\,V$, trim $Z_{OS}$ to hold the output constant, as $X_{IN}$ is varied from $-10\,V$ to $-1\,V$.
3. With $Z_{IN} = 0\,V$ and $X_{IN} = -10.0\,V$ adjust $Y_{OS}$ for zero output voltage.
4. With $Z_{IN} = X_{IN}$ (and/or $Z_{IN} = -X_{IN}$) adjust $X_{OS}$ for minimum worst-case variation of output, as $X_{IN}$ is varied from $-10\,V$ to $-1\,V$.
5. Repeat Steps 2 and 3 if Step 4 required a large initial adjustment.
6. With $Z_{IN} = X_{IN}$ (and/or $Z_{IN} = -X_{IN}$) adjust the gain control until the output is the closest average around $+10.0\,V$ ($-10\,V$ for $Z_{IN} = -X_{IN}$) as $X_{IN}$ is varied from $-10\,V$ to $-3\,V$.

Figure 4.15.2 shows the circuit connection of the divider.

## Squaring

The squaring function is achieved by simply multiplying with the two inputs tied together. The squaring circuit may also be used as the basis for a frequency doubler using the relationship $\cos^2\omega t + (1/2)(\cos 2\omega t + 1)$.

Figure 4.15.3 shows the circuit connection of the squaring circuit.

## Square root

Tying the X and Y inputs together and using overall feedback from the op-amp results in the square root function. The output of the modulator is again forced to equal the current produced by the Z input:

$$I_O = X_{IN} \cdot Y_{IN} = (-E_{OUT})^2 = 10\, Z_{IN}$$
$$E_{OUT} = -\sqrt{10\, Z_{IN}}$$

The output is a negative voltage which maintains overall negative feedback. A diode in series with the op-amp output prevents the latch-up that would otherwise occur for negative input voltages.

*Square root trimming procedure*
1. Connect the ICL 8013 in the divider configuration.
2. Adjust $Z_{OS}$, $Y_{OS}$, $X_{OS}$ and gain using Steps 1 through 6 of divider trimming procedure.
3. Convert to the square root configuration by connecting $X_{IN}$ to the output and inserting a diode between pin 4 and the output node.
4. With $Z_{IN} = 0\,V$ adjust $Z_{OS}$ for zero output voltage.

Figure 4.15.4 shows the circuit connection of the square root circuit.

## Variable gain amplifier

The variable gain amplifier using ICL 8013 is a multiplier with the input signal applied at the X input and the control voltage applied at the Y input.

Figure 4.15.5 shows the connection diagram for variable gain amplifier using ICL 8013.

*Courtesy of Harris Semiconductor, Melbourne, FL*

## 4.16 Programmable positive and negative voltage references

**Figure 4.16.1** Programmable precision positive voltage reference

The circuit shown in Figure 4.16.1 can be used to derive a programmable positive reference voltage, which is often required in circuit design. The simple circuit provides a constant current for the reference Zener diode, adjustable by varying $R_1$, independent of changes in power supply voltage. The output voltage of the op-amp is given by

$$V_o = V_Z(1 + R_2/R_3)$$

and the Zener current by

$$I_Z = (V_O - V_Z)/R_1$$

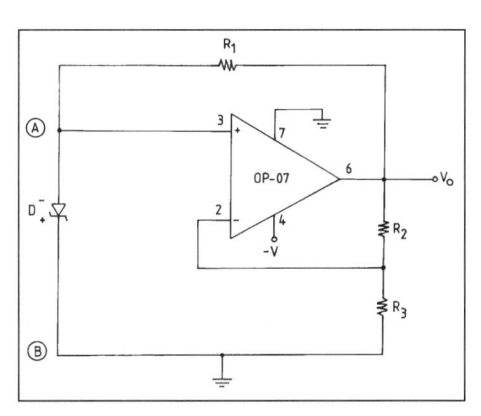

**Figure 4.16.2** Programmable precision negative voltage reference

Therefore, the output voltage of the op-amp depends on the breakdown voltage of the Zener diode connected across the terminals A and B and the values of the resistors. By selecting a reference Zener diode of suitable breakdown voltage, and setting the Zener current to the optimum value as specified by the manufacturer, a reference voltage independent of supply voltage changes can be obtained.

The circuit can go into a latch-up state on the application of power if dual supply voltages are used for the op-amp, since the circuit is based on positive feedback. Under this condition, the Zener will become forward-biased and function like an ordinary diode. To overcome this problem, the negative supply terminal of the op-amp is grounded and the op-amp is connected to a single positive supply as shown in Figure 4.16.1. The precision op-amp OP-07 used in this application has certain advantages over other op-amps, such as low offset voltage, low drift with time and temperature, very low noise, low input bias current, high common mode rejection and a wide input supply voltage range.

A negative reference voltage can be obtained by reversing the polarity of connection of the Zener diode, using a negative power supply alone and connecting the positive supply terminal of the op-amp to ground as shown in Figure 4.16.2. The same equation holds for the output voltage of the op-amp.

All resistors used in each of the voltage reference circuits should be of 1% tolerance.

◆ ◆

## 4.17 Negative voltage reference using a positive voltage reference

**Figure 4.17.1** Negative voltage reference derived using a positive reference

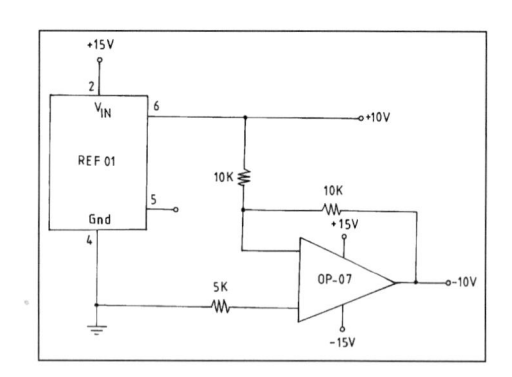

The addition of a single op-amp to the output of a positive voltage reference such as REF 01 can derive a negative reference voltage of the same value as the positive reference. The circuit shown in Figure 4.17.1 shows a method of deriving −10 V reference using a REF 01 and an OP-07 op-amp. The output drive current capability is limited by the op-amp used. In this particular case, the voltage reference used, REF 01, is pre-trimmed to within ± 0.3% of 10 V, and features an excellent temperature stability (temperature coefficient as low as 8.5 ppm/°C worst case), low current drain and low noise. These stability factors are reflected in the negative voltage reference derived.

*Courtesy of Maxim Integrated Products, Sunnyvale, CA*

# 5 Amplifier circuits

## 5.1 Inverting amplifier

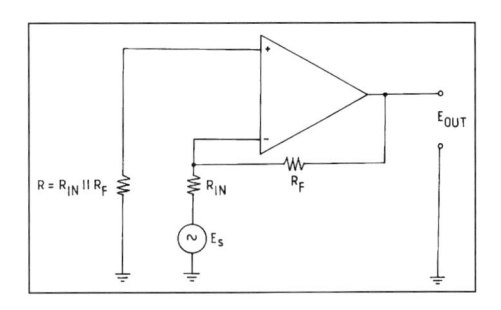

**Figure 5.1.1** Inverting amplifier

A popular application of an op-amp is in an inverting amplifier. Such an amplifier amplifies the signal applied to its inverting input and causes a phase shift of 180° of the input signal. The voltage at the inverting input is 0 and no current flows into the input. Thus the following relationships hold

$$(E_S/R_{IN}) + (E_O/R_F) = 0$$

Solving for the output $E_O$,

$$E_O = -E_S(R_F/R_{IN})$$

which gives the gain as

$$\text{Gain} = (-E_O/E_S) = (R_F/R_{IN})$$

As opposed to the non-inverting amplifier, the input impedance of the inverting amplifier is not infinite but becomes essentially equal to $R_{IN}$. This circuit has found widespread acceptance because of the ease with which input impedance and gain can be controlled to advantage.

With the inverting amplifier of Figure 5.1.1, the gain can be set to any desired value defined by $R_F$ divided by $R_{IN}$. Input impedance is defined by the value of $R_{IN}$ and R should equal the parallel combination of $R_{IN}$ and $R_F$ to cancel the effect of bias current. Offset voltage, offset current, and gain contribute most of the errors. The ground may be set anywhere within the common-mode range and any op-amp will provide satisfactory response.

*Courtesy of Philips, The Netherlands*

## 5.2 Non-inverting amplifier

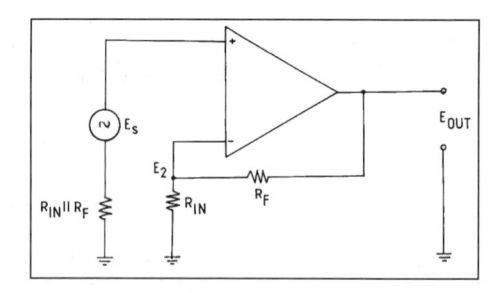

**Figure 5.2.1** Non-inverting amplifier

A popular application of an op-amp is a non-inverting amplifier, shown in Figure 5.2.1. Such an amplifier amplifies the input signal applied to its non-inverting input but does not cause any phase inversion. The voltage appearing at its inverting input is given by

$$E_2 = E_{OUT} R_{IN}/(R_F + R_{IN})$$

Since the differential voltage is zero, $E_2 = E_S$ and the output voltage becomes

$$E_{OUT} = E_S[1 + (R_F/R_{IN})]$$

which gives the value of the gain as

$$\text{Gain} = E_{OUT}/E_S = [1 + (R_F/R_{IN})]$$

It should be noted that as long as the gain of the closed loop is small compared to open-loop gain, the output will be accurate, but as the closed-loop gain approaches the open-loop value more error will be introduced.

The signal source is shown in Figure 5.2.1 in series with a resistor equal in size to the parallel combination of $R_{IN}$ and $R_F$. This is desirable because the voltage drops due to bias currents to the inputs are equal and cancel out evenly over temperature. Thus overall performance is much improved.

*Courtesy of Philips, The Netherlands*

◆  ◆

## 5.3 Summing amplifier

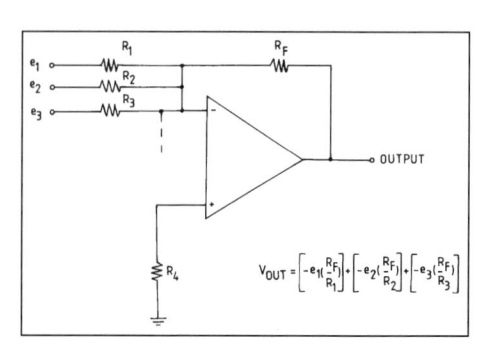

**Figure 5.3.1** Summing amplifier

The summing amplifier is a variation of the inverting amplifier. The output is the sum of the input voltages, each being weighed by $R_F/R_{IN}$ (Figure 5.3.1).

The value of $R_4$ may be chosen to cancel the effects of bias current and is selected equal to the parallel combination of $R_F$ and all the input resistors.

The output voltage is given by

$$V_{OUT} = [-e_1(R_F/R_1)] + [-e_2(R_F/R_2)] + [-e_3(R_F/R_3)]$$

where $e_1$, $e_2$ and $e_3$ are the three input voltages to the summing amplifier.

*Courtesy of Philips, The Netherlands*

## 5.4 Absolute value amplifier

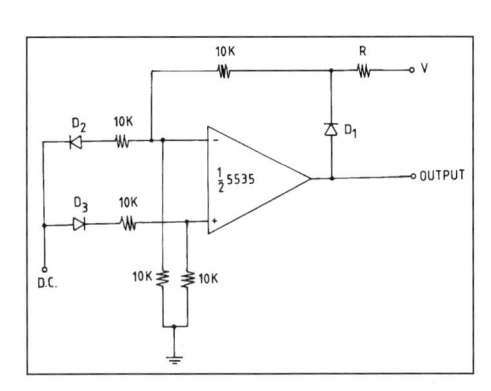

**Figure 5.4.1** Absolute value amplifier

The circuit in Figure 5.4.1 generates a positive output voltage for either polarity of input. For positive signals, it acts as a non-inverting amplifier and for negative signals, as an inverting amplifier. The accuracy is poor for input voltages under 1 V, but for less stringent applications, it can be effective.

*Courtesy of Philips, The Netherlands*

## 5.5 Differential amplifier

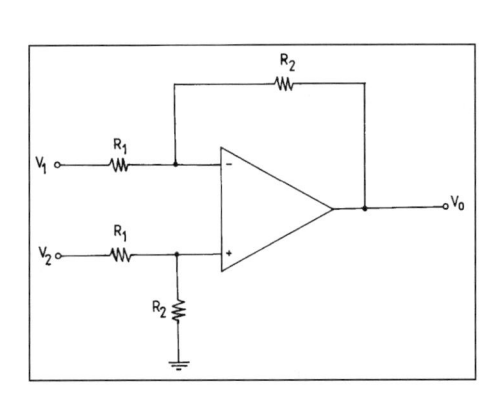

**Figure 5.5.1** Differential amplifier

## 86 Amplifier circuits

Figure 5.5.1 shows a differential amplifier using an op-amp. Ideally, a differential amplifier amplifies the signals applied to its inverting and non-inverting inputs by equal and opposite amounts. In Figure 5.5.1 the output voltage is given by

$$V_o = V_2[R_2/(R_1 + R_2)] \cdot [(R_1 + R_2)/R_1] - V_1 R_2/R_1$$
$$= (V_2 - V_1) \cdot R_2/R_1$$

The above equation shows that the output is dependent on the difference between the two input voltages. The ratio $R_2/R_1$ decides the gain of the amplifier. The ability of a differential amplifier to amplify differential input signals and reject common-mode input signals is defined in terms of the common-mode rejection ratio (CMRR), which is defined by

$$\text{CMRR} = 20 \log |\text{Differential mode gain/Common mode gain}|$$

The higher the CMRR, the better is the differential amplifier. Differential amplifiers find wide applications in instrumentation design.

## 5.6 Bridge transducer amplifier

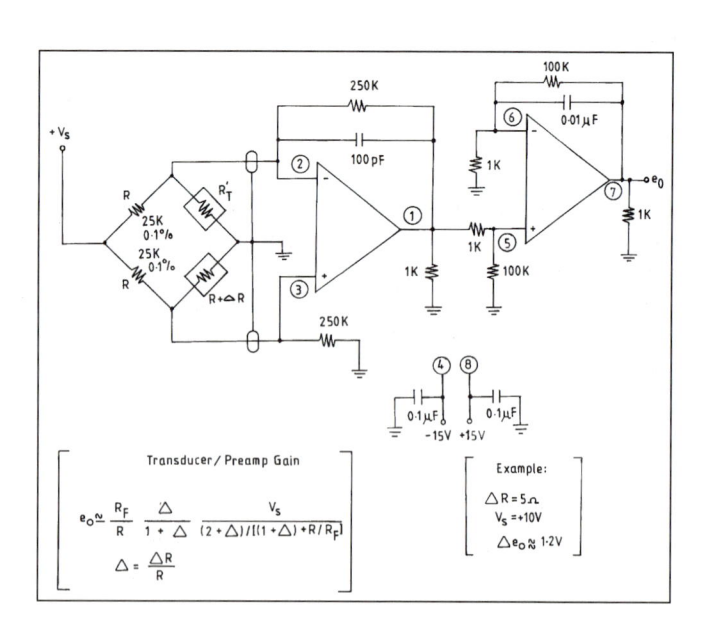

**Figure 5.6.1** Bridge transducer amplifier

In applications involving strain gauges, accelerometers and thermal sensors, a bridge transducer is often used. Frequently the sensor elements are high resistance units requiring equally high bridge resistances for good sensitivity. This type of circuit then demands an amplifier with high input impedance, low bias current and low drift. The circuit shown in Figure 5.6.1 represents a solution to these general requirements.

For $V_S = 10\,V$, the common-mode voltage is approximately $+5\,V$, well within the common-mode limits of the NE 5512.

The sensitivity of the input stage is approximately

$$R_F \cdot V_S / 2R$$

to a change in transducer resistance $\Delta R$. This gives a gain factor of $\simeq 50$ for $V_s = 10\,V$ and $R = 25\,\text{kohms}$. The second stage gain is $\times 100$ giving a total gain of $\simeq 5000$.

Noise is minimized by shielding the transducer leads and taking special care to determine a good signal ground. Common-mode noise rejection is particularly important, making matched differential impedance critical. The NE 5512 typically provides 100 dB of common-mode rejection and will considerably reduce this undesirable effect.

The following are sensitivity figures for the transducer circuits:

|       | $\Delta R$ (ohms) | $\Delta E_{OUT}$ (volts) |
|-------|-------------------|--------------------------|
| Leg 1 | 10                | −2.6                     |
|       | 5                 | −1.3                     |
| Leg 2 | 10                | + 2.4                    |
|       | 5                 | + 1.2                    |

Temperature compensation of the bridge element is accomplished by using low drift metal film resistors and also by providing a complementary non-active sensor element to thermally track the offset in the active element.

High frequency roll-off provides attenuation of unwanted noise above the pass band of the transducer. The shunt capacitors across both stage feedback resistors are for this purpose.

*Courtesy of Philips, The Netherlands*

◆ ◆

## 5.7 Monolithic logarithmic amplifier

The ICL 8048 is a monolithic logarithmic amplifier capable of handling six decades of current input, or three decades of voltage input. It is fully temperature compensated and is nominally designed to provide 1 V of output for each decade change of input. For increased flexibility, the scale factor, reference current and offset voltage are externally adjustable.

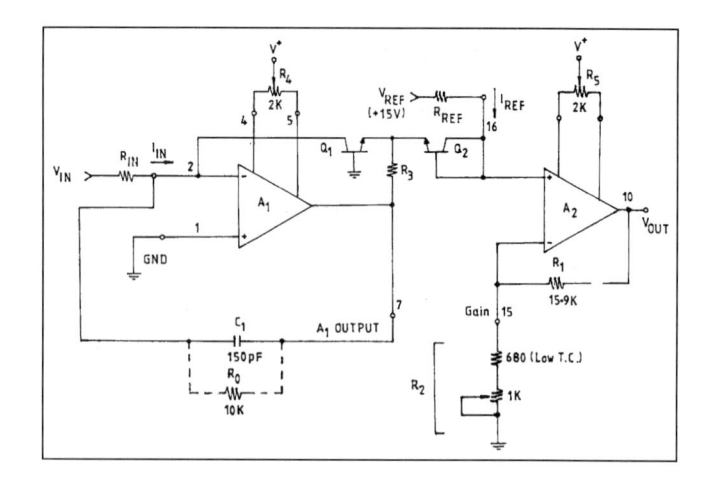

**Figure 5.7.1** ICL 8048 offset and scale factor adjustment

The ICL 8048 relies for its operation on the well-known exponential relationship between the collector current and the base–emitter voltage of a transistor:

$$I_C = I_S[e^{qV_{BE}/kT} - 1] \tag{1}$$

For base–emitter voltages greater than 100 mV, Eq. (1) becomes

$$I_C = I_S\, e^{qV_{BE}/kT} \tag{2}$$

From Eq. (2), it can be shown that for two identical transistors operating at different collector currents, the $V_{BE}$ difference ($\Delta V_{BE}$) is given by

$$\Delta V_{BE} = -2.303 \times (kT/q)\, \log_{10}(I_{C_1}/I_{C_2}) \tag{3}$$

Referring to Figure 5.7.1 it is clear that the potential at the collector of $Q_2$ is equal to the $\Delta V_{BE}$ between $Q_1$ and $Q_2$. The output voltage is $\Delta V_{BE}$ multiplied by the gain of $A_2$:

$$V_{OUT} = -2.303[R_1 + R_2]/R_2](kT/q)\log_{10}(I_{IN}/I_{REF}) \tag{4}$$

The expression $2.303 \times kT/q$ has a numerical value of 59 mV at 25°C; thus in order to generate 1 V/decade at the output, the ratio $(R_1 + R_2)/R_2$ is chosen to be 16.9. For this scale factor to hold constant as a function of temperature, the $(R_1 + R_2)/R_2$ term must have a $1/T$ characteristic to compensate for $kT/q$.

In the ICL 8048 this is achieved by making $R_1$ a thin film resistor, deposited on the monolithic chip. It has a nominal value of 15.9 k at 25°C, and its temperature coefficient is carefully designed to provide the necessary compensation. Resistor $R_2$ is external and should have a nominal value of 1k to provide 1 V/decade, and just have an adjustment range of ±20% to allow production variations in the absolute value of $R_1$.

## ICL 8048 offset and scale factor adjustment

A log amp, unlike an op-amp, cannot be offset adjusted by simply grounding the input. This is because the log of zero approaches minus infinity; reducing the input current to zero starves $Q_1$ of collector current and opens the feedback loop around $A_1$. Instead, it is necessary to zero the offset voltage of $A_1$ and $A_2$ separately, and then to adjust the scale factor. Referring to Figure 5.7.1, this is done as follows:

(1) Temporarily connect a 10k resistor ($R_O$) between pins 2 and 7. With no input voltage, adjust $R_4$ till the output of $A_1$ (pin 7) is zero. Remove $R_O$.
   Note that for a current input, this adjustment is not necessary since the offset voltage of $A_1$ does not cause any error for current-source inputs.
(2) Set $I_{IN} = I_{REF} = 1\,mA$. Adjust $R_5$ such that the output of $A_2$ (pin 10) is zero.
(3) Set $I_{IN} = 1\,\mu A$, $I_{REF} = 1\,mA$. Adjust $R_2$ for $V_{OUT} = 3\,V$ (for a 1 V/decade scale factor) or 6 V (for a 2 V/decade scale factor).

Step (3) determines the scale factor. Setting $I_{IN} = 1\,\mu A$ optimizes the scale factor adjustment over a fairly wide dynamic range, from 1 mA to 1 nA. Clearly, if the 8048 is to be used for inputs which only span the range 100 µA to 1 mA, it would be better to set $I_{IN} = 100\,\mu A$ in Step (3). Similar adjustment for other scale factors would require different $I_{IN}$ and $V_{OUT}$ values.

The scale factor adjustment procedure outlined above for the ICL 8048 is primarily directed towards setting up 1 V ($\Delta V_{OUT}$) per decade ($\Delta I_{IN}$ or $\Delta V_{IN}$) for the log amp.

This corresponds to $K = 1$ in the transfer function of the log amp:

$$V_{OUT} = -K \log_{10}(I_{IN}/I_{REF}) \qquad (5)$$

By adjusting $R_2$ in Figure 5.7.1, the scale factor $K$ in Eq. (5) can be varied. The nominal value of $R_2$ required to give a specific value of $K$ can be determined from Eq. (6) given below:

$$R_2 = \{941/(K - 0.059)\} \quad \text{ohms} \qquad (6)$$

It should be remembered that $R_1$ has a $\pm 20\%$ tolerance in absolute value, so that allowance shall be made for adjusting the nominal value of $R_2$ by $\pm 20\%$.

## Frequency compensation

Although the op-amps in the ICL 8048 are compensated for unity gain, some additional frequency compensation is required. This is because the log transistors in the feedback loop add to the loop gain. Therefore, in the ICL 8048, 150 pF should be connected between pins 2 and 7.

*Courtesy of Harris Semiconductor, Melbourne, FL*

## 5.8 Antilog amplifier

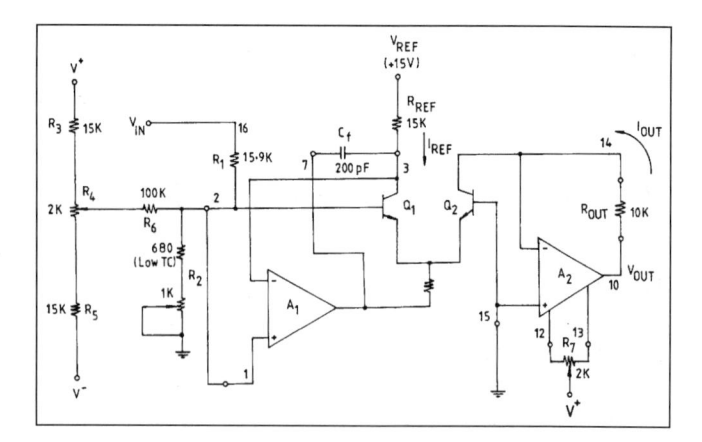

**Figure 5.8.1** ICL 8049 offset and scale factor adjustment

The ICL 8049 is a monolithic antilog amplifier – a counterpart of the log amplifier ICL 8048. It nominally generates one decade of output voltage for each 1 V change at the input.

The ICL 8049 relies on the same logarithmic properties of the transistor as the ICL 8048. The input voltage forces a specific $\Delta V_{BE}$ between $Q_1$ and $Q_2$ (Figure 5.8.1). This $V_{BE}$ difference is converted into a difference of collector currents by the transistor pair. The equation governing the behaviour of the transistor pair is derived from the equation.

$$I_C = I_S \{e^{qV_{BE}/kT} - 1\} \tag{1}$$

For base–emitter voltages greater than 100 mV, Eq. (1) becomes

$$I_C = I_S \, e^{qV_{BE}/kT} \tag{2}$$

Equation (2) yields

$$I_{C_1}/I_{C_2} = \exp[q \, \Delta V_{BE}/kT] \tag{3}$$

When numerical values for $q/kT$ are put into this equation, it is found that a $\Delta V_{BE}$ of 59 mV (at 25°C) is required to change the collector current ratio by a factor of ten. But for ease of application it is desirable that a 1 V change at the input generate a ten-fold change at the output. The required input attenuation is achieved by the network comprising $R_1$ and $R_2$. In order that scale factors other than one decade per volt may be selected, $R_2$ is external to the chip. It should have a value of 1 kohm, adjustable ± 20%, for one decade per volt. $R_1$ is a thin film resistor deposited on the monolithic chip; its temperature characteristics are chosen to compensate the temperature dependence of Eq. (3).

The overall transfer function is as follows:

$$I_{OUT}/I_{REF} = \exp\{[-R_2/(R_1 + R_2)] \times [qV_{IN}/kT]\} \tag{4}$$

Substituting $V_{OUT} = I_{OUT} \times R_{OUT}$ gives:

$$V_{OUT} = R_{OUT} \, I_{REF} \exp\{[-R_2/(R_1 + R_2)] \times [qV_{IN}/kT]\} \tag{5}$$

For voltage references, Eq. (5) becomes

$$V_{OUT} = V_{REF} \times (R_{OUT}/R_{REF}) \times \exp\{[-R_2/(R_1 + R_2)] \times qV_{IN}/kT\} \qquad (6)$$

## ICL 8049 offset and scale factor adjustment

As with the log amplifier, the antilog amplifier requires three adjustments. The first step is to null out the offset voltage of $A_2$. This is accomplished by reverse-biasing the base–emitter of $Q_2 \cdot A_2$ which then operates as a unity gain buffer with a grounded input. The second step forces $V_{IN} = 0$; the output is adjusted for $V_{OUT} = 10\,V$. The third step applies a specific input and adjusts the output to the correct voltage. This sets the scale factor. Referring to Figure 5.8.1, the exact procedure for 1 decade/volt is as follows:

1. Connect the input (pin 16) to $+15\,V$. This reverse biases the base–emitter of $Q_2$. Adjust $R_7$ for $V_{OUT} = 0\,V$. Disconnect the input from $+15\,V$.
2. Connect the input to ground. Adjust $R_4$ for $V_{OUT} = 10\,V$. Disconnect the input from ground.
3. Connect the input to a precise $2\,V$ supply and adjust $R_2$ for $V_{OUT} = 100\,mV$.

The procedure outlined above optimizes the performance over a 3-decade range at the output (i.e. $V_{OUT}$ from $10\,mV$ to $10\,V$). For a more limited range of output voltages, for example $1\,V$ to $10\,V$, it would be better to use a precise $1\,V$ supply and adjust $V_{OUT} = 1\,V$. For other scale factors and/or starting points, different values for $R_2$ and $R_{REF}$ will be needed, but the same basic procedure applies.

Although the op-amps in ICL 8049 are compensated for unity gain, some additional frequency compensation is required. This is because the log transistors in the feedback loop add to the loop gain. In the ICL 8049, 200 pF should be connected between pins 3 and 7.

*Courtesy of Harris Semiconductor, Melbourne, FL*

◆ ◆

## 5.9 Low-distortion audio amplifier

**Figure 5.9.1** Bridge audio power amplifier using LH 0101

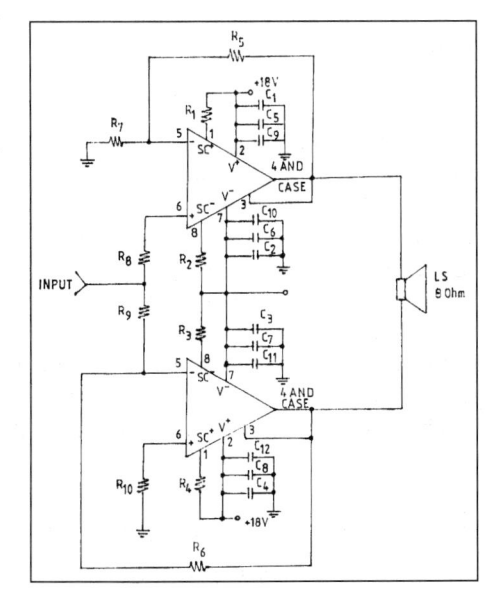

92  Amplifier circuits

**Table 5.9.1** Component list for bridge audio power amplifier

| | | |
|---|---|---|
| $R_{1-4}$ | Current limit resistors | 0.1 ohm 2 W |
| $R_5$ | Feedback resistor | 5 kohm |
| $R_6$ | Feedback resistor | 15 kohm |
| $R_{7-10}$ | Input resistors | 10 kohm |
| $C_{1-4}$ | Bypass capacitors | 47 µF, 25 V electrolytic |
| $C_{5-8}$ | Bypass capacitors | 10 µF, 25 V tantalum |
| $C_{9-12}$ | Bypass capacitors | 0.1 µF, 25 V ceramic |

The Maxim LH 0101 power operational amplifier delivers up to 5 A peak output current. It combines the ease of use and performance of a FET input op-amp with the power handling capabilities of a 5 A output stage. The output stage virtually eliminates crossover distortion while using little quiescent power. This op-amp is a wideband amplifier having a bandwidth of 300 kHz and a gain bandwidth of 5 MHz. The hermetically sealed LH 0101 is well suited to severe environments. Figure 5.9.1 shows two LH 0101s used in a bridge audio power amplifier. The bridge configuration doubles the voltage that can be delivered to the load, in this case delivering 50 V peak-to-peak to an 8 ohm speaker. This means that 40 W RMS can be delivered to the 8 ohm speaker while using only ± 18 V power supplies. The harmonic distortion is a respectable 0.1%, which should suffice for all but the most demanding applications.

Table 5.9.1 shows the component values used for this audio amplifier.

*Courtesy of Maxim Integrated Products, Inc., Sunnyvale, CA*

## 5.10 Low-noise, high-speed precision op-amp

**Figure 5.10.1** Phono preamplifier

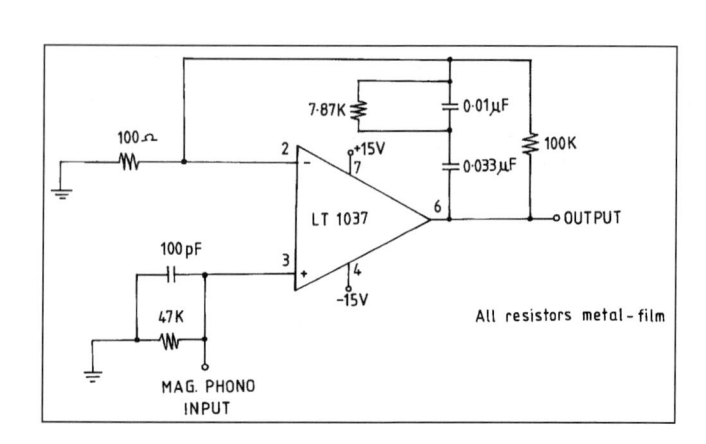

## Amplifier circuits

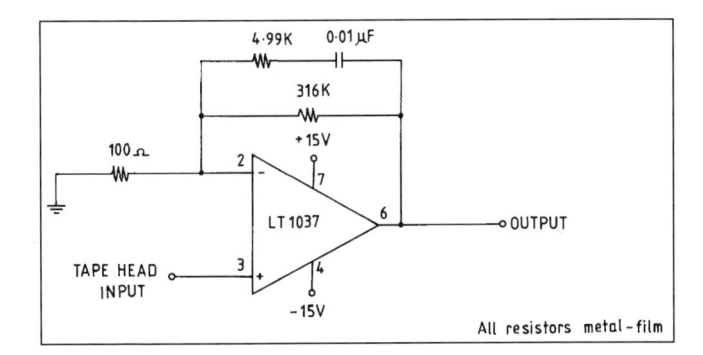

**Figure 5.10.2** Tape head amplifier

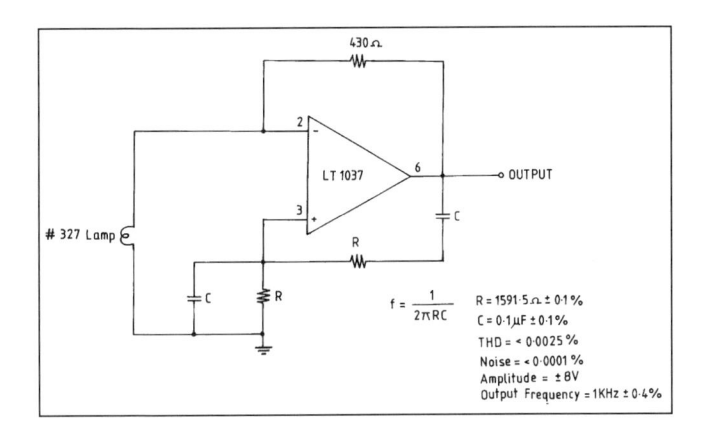

**Figure 5.10.3** Ultra-pure 1 kHz sine wave generator

$$f = \frac{1}{2\pi RC}$$

R = 1591·5 Ω ± 0·1%
C = 0·1 μF ± 0·1%
THD = < 0·0025 %
Noise = < 0·0001 %
Amplitude = ± 8V
Output Frequency = 1KHz ± 0·4%

The LT 1037 from Linear Technology Corporation is a low-noise op-amp featuring a wideband noise figure of 2.5 nV/Hz (less than the noise of a 400 ohm resistor), 1/f corner frequency of 2 Hz and 60 nV peak-to-peak 0.1 Hz to 10 Hz noise. In this device, low noise is combined with outstanding precision and speed specifications: 10 μV offset voltage, 0.2 μV/C drift, 130 dB common-mode and power supply rejection and 60 MHz gain–bandwidth product on the decompensated LT 1037, which is stable for closed-loop gains of 5 or greater. The voltage gain of LT 1037 is an extremely high 20 million driving a 2k load and 12 million driving a 600 ohm load to ± 10 V.

Figure 5.10.1 shows a phono preamplifier and Figure 5.10.2 shows a tape head preamplifier using the LT 1037. An ultra-pure 1 kHz sine wave generator based on a Wien-bridge oscillator and using the LT 1037 is shown in Figure 5.10.3.

*Courtesy of Linear Technology Corporation, Milpitas, CA*

## 5.11 Low-cost 50 W per channel audio amplifier

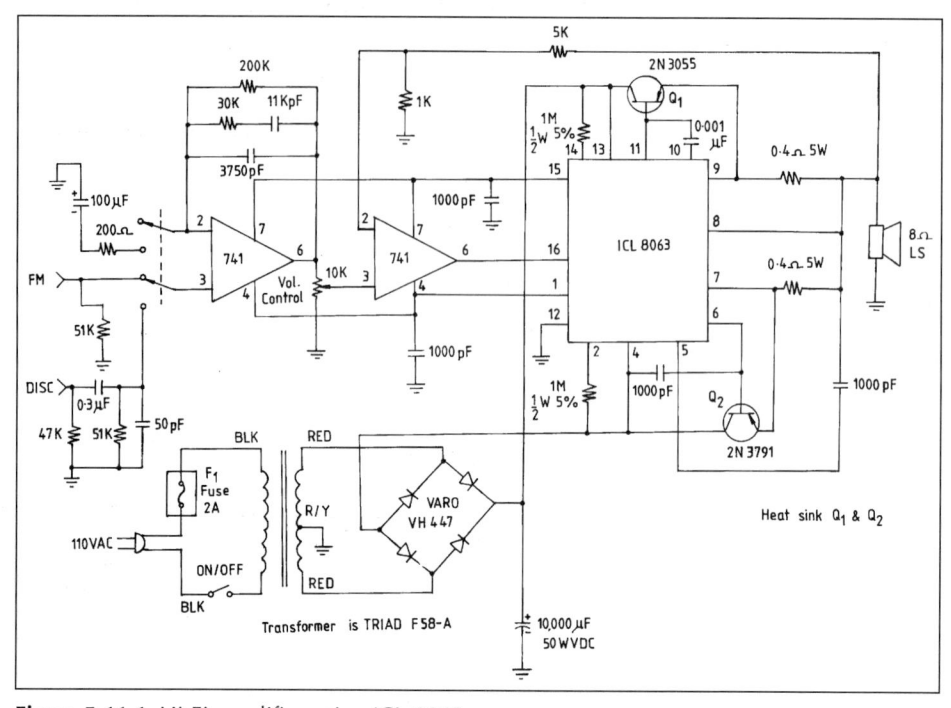

**Figure 5.11.1** Hi-Fi amplifier using ICL 8063

The circuit shown in Figure 5.11.1 can be used as a high fidelity amplifier. The circuit uses ICL 8063 power transistor driver/amplifier. A channel is here defined as all amplification between turn-table or tape output and power output.

The input 741 stage is a preamplifier with RIAA equalization for records. Following the first 741 stage is a 10k control potentiometer, whose wiper arm feeds into the power amplifier stage consisting of a second 741, the ICL 8063 and the power transistors. To achieve good listening results, selection of proper resistance values in the power amplifier stage is important. Best listening is to be found at a gain value of 6[(5k + 1k)/1k = 6]. A gain value of 3 is a practical minimum, since the first stage 741 preamp puts out only ± 10 V maximum signals, and if maximum power is necessary this value must be multiplied by 3 to obtain ± 30 V levels at the output of the power amp stage.

Each channel delivers about 56 V peak-to-peak across an 8 ohm speaker and this converts to 50 W RMS power. This is derived as follows:

$$\text{Power} = V_{rms}^2/8 \text{ ohms}; V_{rms} = 56 \text{ Vp-p}/2.82 = 20 \text{ V rms}$$

$$\text{Therefore, power} = 20^2/8 = 400/8 = 50 \text{ W rms power}$$

Distortion will be <0.1% up to about 100 Hz, and then it increases as the frequency increases, reaching about 1% at 20 kHz.

The ganged switch at the input is for either disk playing or FM, either from an FM tuner or a tape amplifier. Assuming DC coupling on the outputs, there is no need for a DC reference to ground (resistor) for FM position. To clear the signal in the FM position, place a 51k resistor to ground as shown in Figure 5.11.1 (from FM input position to ground).

*Courtesy of Harris Semiconductor, Melbourne, FL*

◆ ◆ ◆

## 5.12 Power amplifier

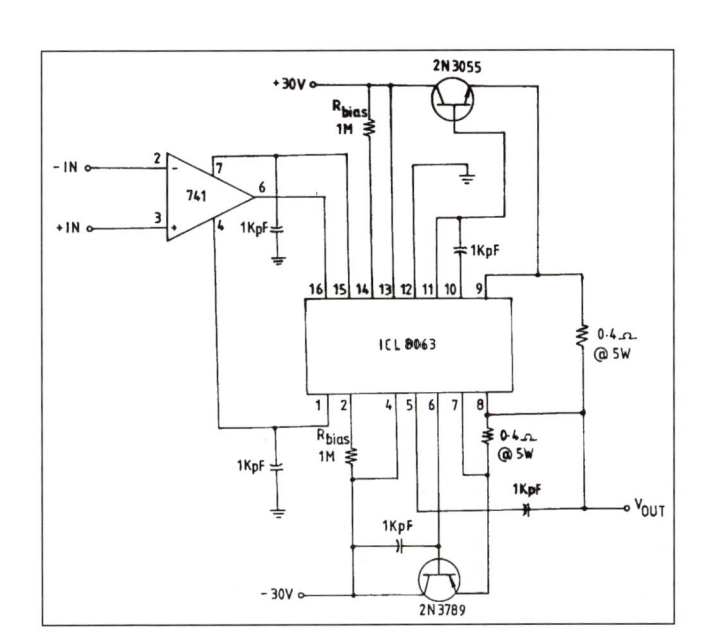

**Figure 5.12.1** Power amplifier using ICL 8063

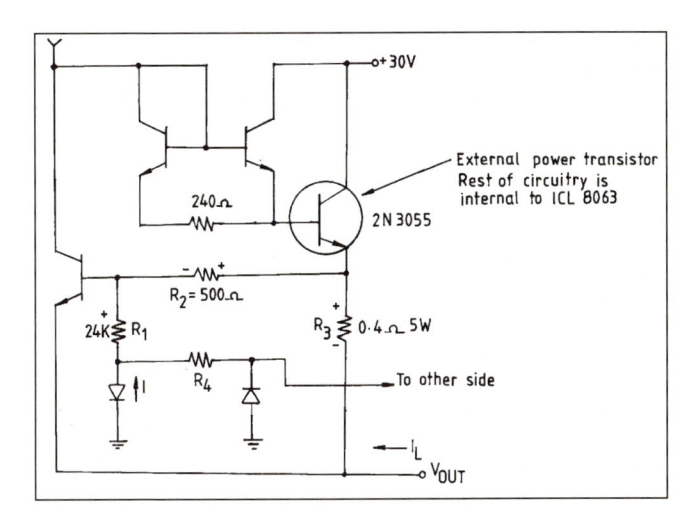

**Figure 5.12.2** Current limiting protection circuit (one side shown)

One problem faced almost every day by circuit designers is how to interface the low voltage, low current outputs of linear and digital devices to that of power transistors and Darlingtons.

For example, a low level op-amp has a typical output voltage range of ±6 V to ±12 V, and output current usually on the order of about 5 mA. A power transistor with a ±35 V supply, a collector current of 5 A, and a beta, or gain of 100 needs at least 50 mA of drive.

The ICL 8063 is a monolithic power transistor driver and power transistor amplifier circuit on the same chip. It has all the necessary safe operating area circuitry and short-circuit protection, and has on-chip ±13 V voltage regulators to eliminate the need for extra external power supplies. Designed to operate with all varieties of operational amplifiers and other functions, two external power transistors and 8 to 10 passive components, the ICL 8063 is ideal for use in such applications as linear and rotary actuator drivers, stepper motor drivers, servo motor drivers, power supplies, power DACs and electronically controlled orifices. The ICL 8063 takes the output levels (typically ±11 V) from an op-amp and boosts them to ±30 V to drive power transistors (e.g. 2N 3055 (NPN) and 2N 3789 (PNP)). The outputs of the ICL 8063 can supply up to 100 mA of base drive to the external power transistors.

The circuit shown in Figure 5.12.1 using the ICL 8063 allows the circuit designer to build a power amplifier block capable of delivering ±2 A at ±25 V (50 W) to any load, with only three additional discrete devices and eight passive components. Moreover, the circuit draws only about ±30 mA of quiescent current from either of the ±30 V power supplies. A similar design using discrete components would require anywhere from 50 to 100 components.

Slew rate is about the same as that of a 741 op-amp, approximately 1 V/μs. Input current, voltage offset, CMRR and PSRR are also the same. Use of 1000 pF compensation capacitors in the circuit, as shown, allows good stability down to unity gain non-inverting (the worst case). This circuit will drive a 1000 pF $C_L$ to ground, or in other words, the circuit can drive 30 feet of RG-58 coaxial cable for line driver applications, with no problems.

As Figure 5.12.2 indicates, setting up a current limiting (safe area) protection circuit is straightforward. The 0.4 ohm, 5 W resistors set the maximum current one can get out of the output. The equation this SOA circuit follows is, for $V_{OUT}$ positive,

$$V_{be} = I_{L_{R_3}} - \{R_2/R_1 + R_2)\}(V_{OUT} + I_{L_{R_3}} - 0.7)$$

$$= I_{L_{R_3}} - \{R_2/(R_1 + R_2)\} V_{OUT}$$

for $V_{OUT}$ negative,

$$V_{be} = I_{L_{R_3}} - \{R_2/(R_1 + R_2 + R_4)\}(V_{OUT} + I_2 R_3 + 0.7)$$

$$= I_{L_{R_3}} - \{R_2/(R_1 + R_2 + R_4)\} V_{OUT}$$

Solving these equations we obtain the following:

| $V_{OUT}$ (volts) | $I$ (µA) | $I_L$ @ 25°C (A) | $I_L$ @ 125°C (A) |
|---|---|---|---|
| 24 | 1000 | 3 | 2.4 |
| 20 | 830 | 2.8 | |
| 16 | 670 | 2.6 | |
| 12 | 500 | 2.4 | 1.8 |
| 8 | 333 | 2.1 | |
| 4 | 167 | 1.9 | |
| 0 | 0 | 1.7 | 1.1 |

As this table indicates, maximum power delivered to a load is obtained when $V_{OUT} \geq 24\,V$.

Often design requirements necessitate an unsymmetrical output current capability. In that case, instead of the 0.4 ohm resistors protecting the NPN and PNP output stages, as shown in Figure 5.12.1, simply substitute any other value. For example, if up to 3 A are required when $V_{OUT} \geq +24\,V$, and only 1 A out when $V_{OUT} \geq -24\,V$, use a 0.4 ohm resistor between pin 8 and pin 9 on the ICL 8063 and a 1 ohm, 2 W resistor between pin 7 and pin 8. Maximum output current versus $V_{OUT}$ for varying values of protection resistors are as follows:

| | Maximum output current (A) | | |
|---|---|---|---|
| $V_{OUT}$ (V) | 0.4 ohm @ 25°C | 0.68 ohm @ 25°C | 1 ohm @ 25°C |
| 24 | 3 | 1.7 | 1.2 |
| 12 | 2.4 | 1.4 | 0.9 |
| 0 | 1.7 | 1.0 | 0.7 |

The biasing resistors located between pin 13 and pin 14 and between pin 2 and pin 4 are typically 1 milliohm for $V_{SUPPLY} = \pm 30\,V$, which guarantees adequate performance in such applications as DC motor drivers, power DACs, programmable power supplies and line drivers (with $\pm 30\,V$ supplies). The table that follows shows the proper value for $R_{BIAS}$ for optimum output current capability with supply voltages between $\pm 5\,V$ and $\pm 30\,V$.

| $\pm V_{CC}$ (V) | $R_{BIAS}$ (ohms) |
|---|---|
| 30 | 1M |
| 25 | 680k |
| 20 | 500k |
| 15 | 300k |
| 10 | 150k |
| 5 | 62k |

When buying external power transistors, careful attention should be paid to beta values. For 2N 3055 and 2N 3789 transistors used in this circuit, beta should be no more than 150 max at $I_c = 20\,mA$ and $V_{CE} = 30\,V$. This beta value sets the quiescent current at less than 30 mA when not delivering power to a load.

The design in Figure 5.12.1 will tolerate a short-circuit to ground indefinitely, provided adequate heat sinking is used.

However, if $V_{OUT}$ is shunted to $\pm 30\,V$ the output transistors (2N 3055 and 2N 3789) will be destroyed, but since the safe operating area for these devices is 4 A at 30 V, the problem does not occur for $V_{SUPPLY} = \pm 15\,V$.

*Courtesy of Harris Semiconductor, Melbourne, FL.)*

## 5.13 Single-supply instrumentation amplifier

**Figure 5.13.1** Differential voltage amplification from a resistance bridge

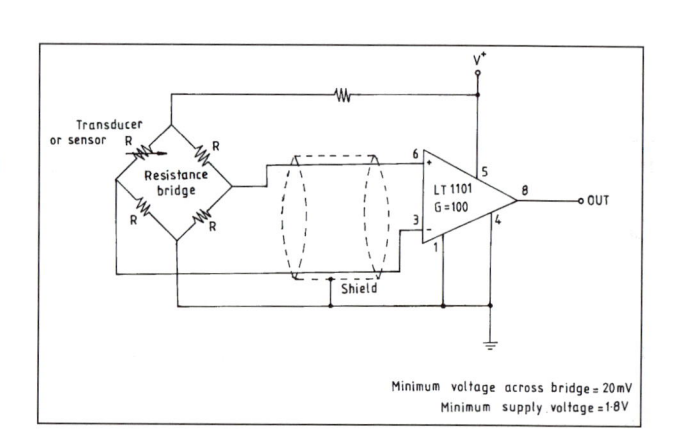

The LT 1101 from Linear Technology Corporation is the first instrumentation amplifier fully specified for single-supply operation, i.e. when the negative supply is 0 V. Both the input common-mode range and the output swing are within a few millivolts of ground.

The LT 1101 establishes the following milestones:

- It is the first micropower instrumentation amplifier.
- It is the first single-supply instrumentation amplifier.
- It is the first instrumentation amplifier to feature fixed gains of 10 and/

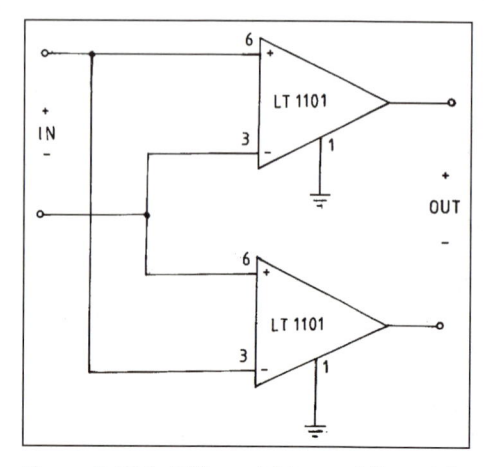

**Figure 5.13.2** Differential input-differential output instrumentation amplifier

or 100 in a low-cost space-saving 8-lead package. The LT 1101 is completely self-contained: no external gain-setting resistor is required. The LT 1101 combines its micropower operation (75 µA supply current) with a gain error of 0.008%, gain linearity of 3 ppm and gain drift of 1 ppm/°C. The output is guaranteed to drive a 2k load to ± 10 V with excellent gain accuracy. The other precision specifications of the device are also outstanding: 50 µV input offset voltage, 130 pA input offset current and low drift (0.4 µV/°C and 0.7 pA/°C). In addition, unlike other instrumentation amplifiers, there is no output offset voltage contribution to total error. The device can be operated from a single lithium cell or two Ni-Cd batteries. Battery voltage can drop as low as 1.8 V, yet the LT 1101 still maintains its gain accuracy. The output sinks current while swinging to ground – no external, power-consuming pull-down resistors are needed.

Probably the most common application for instrumentation amplifiers is amplifying a differential signal from a transducer or sensor resistance bridge. All competitive instrumentation amplifiers have a minimum required common-mode voltage which is 3 to 5 V above the negative supply. This means that the voltage across the bridge has to be 6 V to 10 V or dual supplies have to be used, i.e. micropower single battery usage is not attainable on competitive devices.

## Offset nulling

The LT 1101 is not equipped with dedicated offset null terminals. In many bridge transducer or sensor applications, calibrating the bridge simultaneously eliminates the instrumentation amplifier's offset as a source of error.

## Gains between 10 and 100

Gains between 10 and 100 can be achieved by connecting two equal resistors ($= R_x$) between pins 1 and 2 and pins 7 and 8.

$$\text{Gain} = 10 + R_x / \{R + (R_x/90)\}$$

The nominal value of $R$ is 9.2k. The usefulness of this method is limited by the fact that $R$ is not controlled to better than ± 10% absolute accuracy in production. However, on any specific unit 90$R$ can be measured between pins 1 and 2.

## Input protection

Instrumentation amplifiers are often used in harsh environments where overload conditions can occur. The LT 1101 employs PNP input transistors, consequently the differential input voltage can be ± 30 V (with ± 15 V supplies, ± 36 V with ± 18 V supplies) without an increase in input bias current. Competitive instrumentation amplifiers have NPN inputs which are protected by back-to-back diodes. When the differential input voltage exceeds ± 1.3 V on these competitive devices, input current increases to the milliampere level; more than ± 10 V differential voltage can cause permanent damage.

When the LT 1101's inputs are pulled above the positive supply, the inputs will clamp a diode voltage above the positive supply. No damage will occur if the input current is limited to 20 mA.

500 ohm resistors in series with the inputs protect the LT 1101 when the inputs are pulled as much as 10 V below the negative supply.

The circuit in Figure 5.13.1 shows a simple resistance transducer bridge interfaced to the LT 1101 to illustrate the application of LT 1101 for differential voltage amplification. Figure 5.13.2 shows a differential input – differential output instrumentation amplifier using a pair of LT 1101s.

*Courtesy of Linear Technology Corporation, Milpitas, CA)*

# 6 Waveform generators and oscillators

## 6.1 Monolithic precision waveform generator

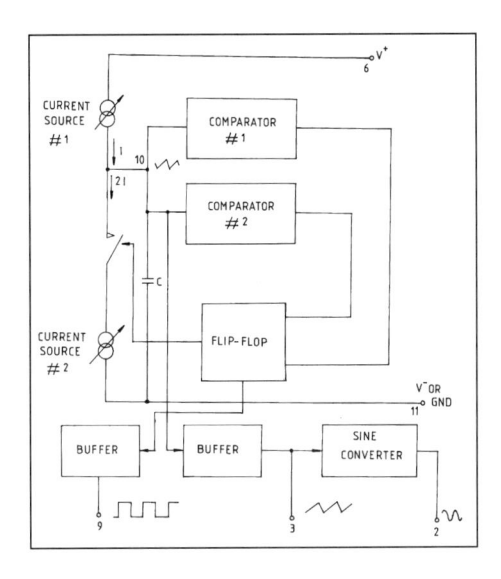

**Figure 6.1.1** Functional diagram of ICL 8038

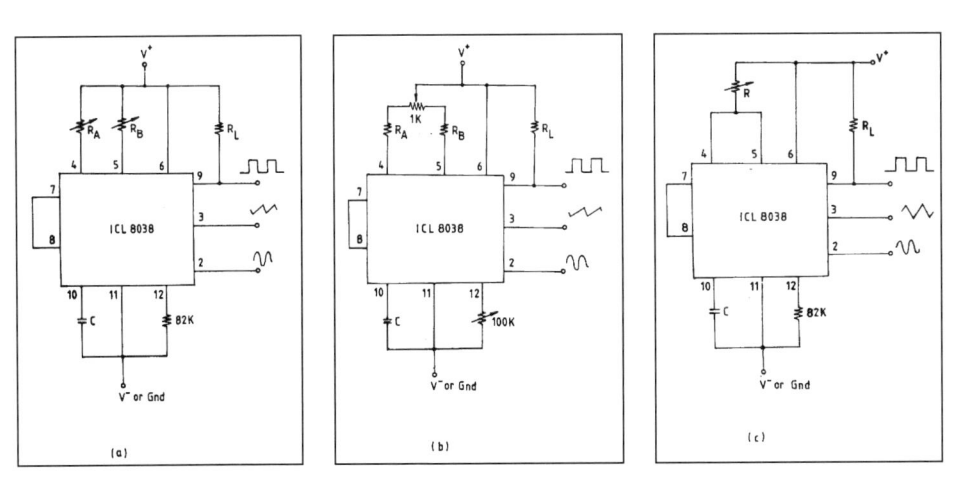

**Figure 6.1.2** Possible connections for the external timing resistors

The ICL 8038 waveform generator is a monolithic IC capable of producing high accuracy sine, square, triangular and pulse waveforms with a minimum of external components. The frequency can be selected externally from 0.001 Hz to more than 300 kHz using either resistors or capacitors, and frequency modulation and sweeping can be accomplished with an external voltage. The sine wave output has a low distortion of 1% and the triangular wave output has a high

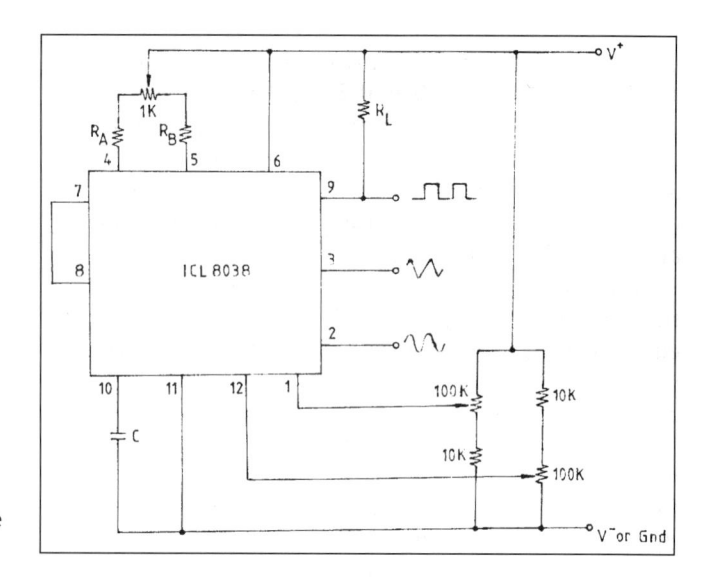

**Figure 6.1.3** Connection to achieve minimum sine wave distortion

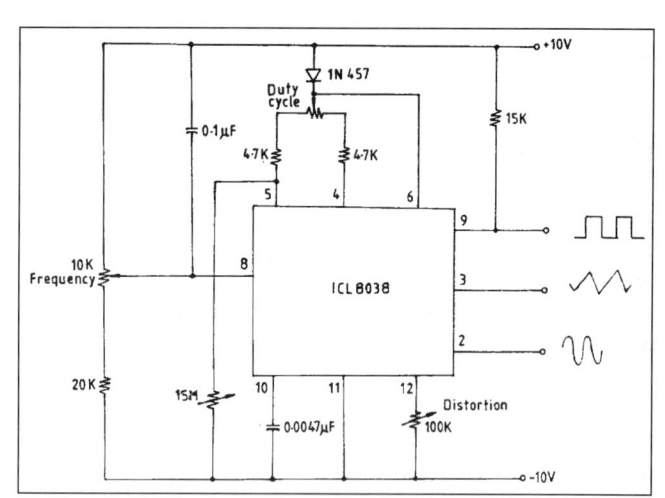

**Figure 6.1.4** Variable audio oscillator, 20 Hz to 20 kHz

linearity of 0.1%. The duty cycle of the output is variable from 2% to 98%, and the output levels can be varied from TTL level to up to 28 V. The chip is quite versatile for waveform generation applications, with only a handful of components required for configuring the required circuit.

### Device operation

An external timing capacitor $C$ is charged and discharged by two current sources (see Figure 6.1.1). Current source 2 is switched ON and OFF by a flip-flop, while current source 1 is ON continuously. Assuming that the flip-flop is in a state such that current source 2 is OFF, the capacitor is charged with a current $I$, and the voltage across the capacitor rises linearly with time. When this voltage reaches the level of comparator 1 (set at 2/3 of the supply voltage), the flip-flop is triggered, changing state, and releases current source 2. This current source normally carries a current 2$I$, thus the capacitor is recharged with a net current $I$ and the voltage across it drops linearly with time. When it has reached the level of comparator 2 (set at 1/3 of the supply voltage), the flip-flop is triggered into its original state and the cycle starts again.

Four waveforms are readily obtainable from this basic generator circuit. With the current sources set at $I$ and 2$I$, respectively, the charge and discharge times are equal. Thus a triangle waveform is created across the capacitor and the flip-flop produces a square wave. Both waveforms are fed to buffer stages and are available at pins 3 and 9.

The levels of the current sources can, however, be selected over a wide range with two external resistors. Therefore, with the two currents set at values different from $I$ and 2$I$, an asymmetrical sawtooth appears at terminal 3 and varies with a duty cycle from less than 1% to greater than 99% are available at terminal 9.

The sine wave is created by feeding the triangular wave into a non-linear network (sine-converter). This network provides a decreasing shunt-impedance as the potential of the triangle moves towards the two extremes.

## Waveform timing

The symmetry of all waveforms can be adjusted with the external timing resistors. Two possible ways to accomplish this are shown in Figure 6.1.2. Best results are obtained by keeping the timing resistors $R_A$ and $R_B$ separate (Figure 6.1.2(a)). $R_A$ controls the rising portion of the triangle and sine wave and 1 state of the square wave. A 50% duty cycle is achieved when $R_A = R_B$.

If the duty cycle is to be varied over a small range about 50% only, the connection shown in Figure 6.1.2(b) is slightly more convenient. If no adjustment of the duty cycle is desired, terminals 4 and 5 can be shorted together, as shown in Figure 6.1.2(c). This connection, however, causes an inherently larger variation of the duty cycle, frequency, etc.

With two separate timing resistors, the frequency is given by

$$f = 1/\{R_A C[1 + R_B/(2R_A - R_B)]/0.66\}$$

or, if $R_A = R_B = R$,

$$f = 0.33/RC \text{ (for Figure 6.1.2(a))}$$

If a single timing resistor is used (Figure 6.1.2(c) only), the frequency is

$$f = 0.165/RC$$

Neither time nor frequency are dependent on supply voltage, even though none of the voltages are regulated inside the IC. This is due to the fact that both currents and thresholds are direct, linear functions of the supply voltage and thus their effects cancel.

To minimize sine wave distortion the 82k resistor between pins 11 and 12 is best made variable. With this arrangement distortion of less than 1% is achievable. To reduce this even further, two potentiometers can be connected as shown in Figure 6.1.3; this configuration allows a typical reduction of sine wave distortion close to 0.5%.

### Selecting $R_A$, $R_B$ and C

For any given output frequency, there is a wide range of RC combinations that will work; however, certain constraints are placed upon the magnitude of the charging current for optimum performance. At the low end, currents of less than 1 µA are undesirable because circuit leakages will contribute significant errors at high temperatures. At higher currents ($I > 5\,\text{mA}$), transistor betas and saturation voltages will contribute increasingly larger errors. Optimum performance will, therefore, be obtained with charging currents of 10 µA to 1 mA.

The capacitor value should be chosen at the upper end of its possible range.

### Waveform output level control and power supplies

The waveform generator can be operated from a single power supply (10 to 30 V) or a dual power supply ($\pm 5\,\text{V}$ to $\pm 15\,\text{V}$). With a single power supply the average levels of the triangle and sine wave are exactly one-half of the supply voltage, while the square wave alternates between $V^+$ and ground. A split power supply has the advantage that all waveforms move symmetrically about ground.

The square wave output is not committed. A load resistor can be connected to a different power supply, as long as the applied voltage remains within the breakdown capability of the waveform generator (30 V). In this way, the square wave output can be made TTL compatible (load resistor connected to +5 V) while the waveform generator itself is powered from a much higher voltage.

Figure 6.1.4 shows a variable audio oscillator of frequency range 20 Hz to 20 kHz.

*Courtesy of Harris Semiconductor, Melbourne, FL*

◆  ◆

## 6.2 Single-supply function generator

The circuit shown in Figure 6.2.1 is a simple square wave and triangle wave generator. In the circuit of Figure 6.2.1, the first op-amp is configured as a voltage comparator with the reference voltage input set at one-half the supply voltage. The resistor $R_1$ connected in the feedback path provides positive feedback. The output of the first op-amp is fed to the inverting input of the second op-amp whose non-inverting input is connected to a reference voltage of half the supply

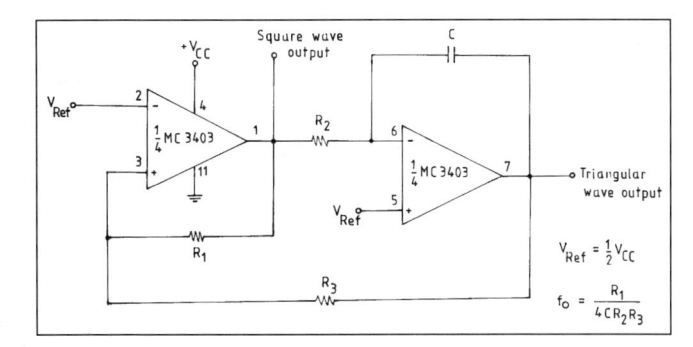

**Figure 6.2.1** Single-supply function generator

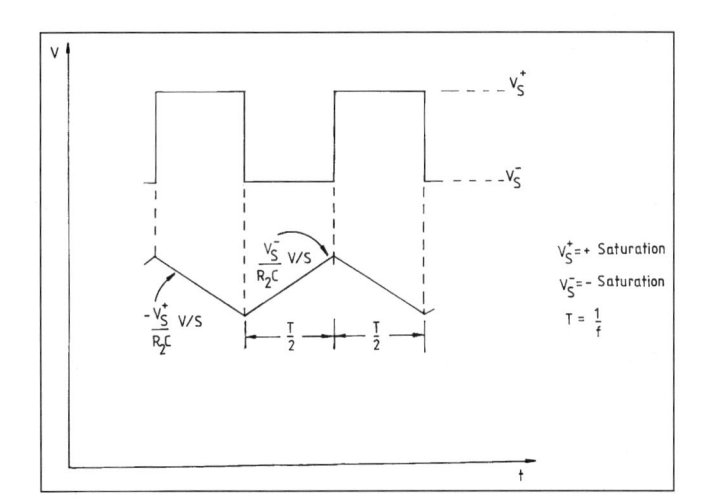

**Figure 6.2.2** Waveforms of circuit in Figure 6.2.1

voltage. The second op-amp is configured as an integrator with capacitor C in the feedback path and resistor $R_2$ as the input resistor. The output of the integrator is connected to the non-inverting input of the first op-amp through resistor $R_3$. The first op-amp gives a square wave output and the integrator gives a triangular wave output. The frequency of either waveform is given by

$$f = R_1/(4CR_2R_3)$$

The waveforms in Figure 6.2.2 illustrate the technique.

*Courtesy of Philips, The Netherlands*

## 6.3 Triangle and square wave generator

**Figure 6.3.1** Function generator using 566

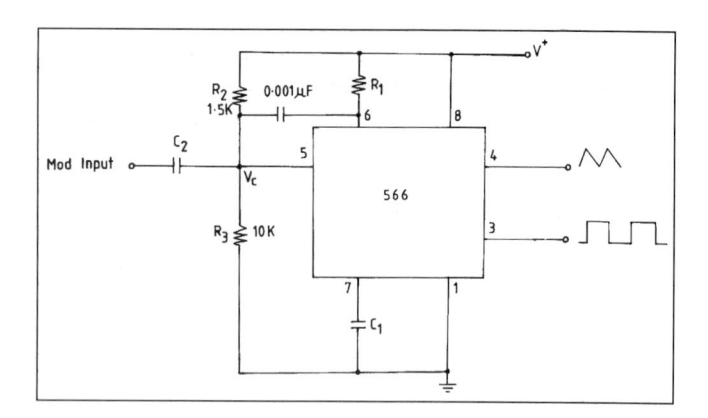

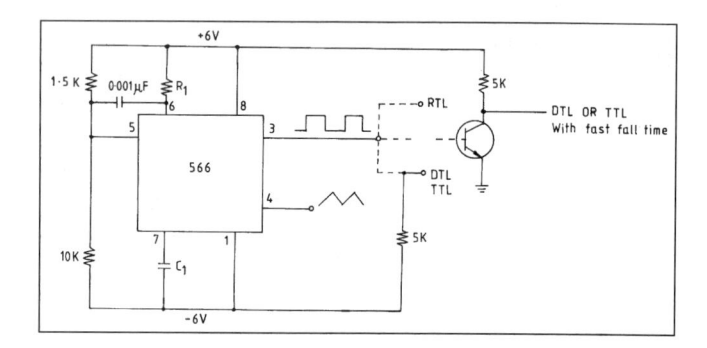

**Figure 6.3.2** Function generator using 566

The NE/SE 566 function generator is a general purpose voltage-controlled oscillator designed for highly linear frequency modulation. The circuit provides simultaneous square wave and triangle wave outputs at frequencies up to 1 MHz. A typical connection diagram is shown in Figure 6.3.1. The control terminal (pin 5) must be biased externally with a voltage $V_C$ in the range

$$(3/4)\,V^+ < V_C < V^+$$

where $V^+$ is the total supply voltage. In Figure 6.2.1 the control voltage is set by the voltage divider formed by $R_2$ and $R_3$. The modulating signal is then AC coupled with the capacitor $C_2$. The modulating signal can be direct coupled as well, if the appropriate DC bias voltage is applied to the control terminal. The frequency is given approximately by

$$f_o = \{2[(V^+) - (V_C)]/R_1 C_1 V^+\}$$

and $R_1$ should be in the range $2k < R_1 < 20k$.

A small capacitor (typically 0.001 µF) should be connected between pins 5 and 6 to eliminate possible oscillation in the control current source.

If the VCO is to be used to drive standard logic circuitry, it may be desirable to use a dual supply as shown in Figure 6.3.2. In this case the square wave output has the proper DC levels for logic circuitry. RTL can be driven directly from pin 3. For

DTL or TTL gates which require a current sink of more than 1 mA, it is usually necessary to connect a 5k resistor between pin 3 and negative supply. This increases the current sinking capability to 2 mA. The third type of interface shown uses a saturated transistor between the 566 and the logic circuitry. This scheme is used primarily for TTL circuitry which requires a fast fall time (<50 ns) and a large current sinking capability.

*Courtesy of Philips, The Netherlands*

## 6.4 Ramp generators

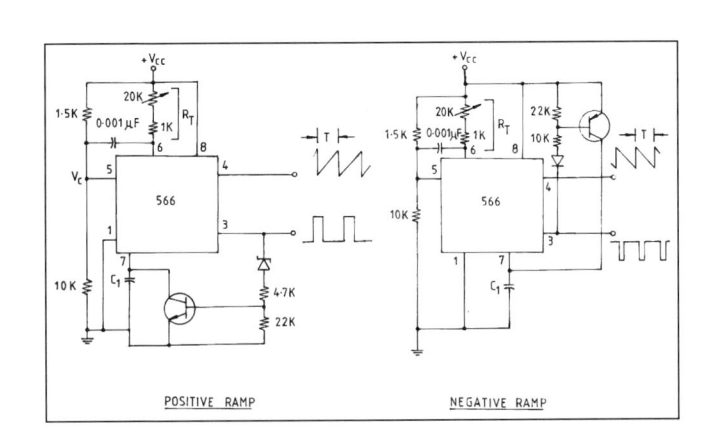

**Figure 6.4.1** Ramp generators

The oscillator portion of many of the PLLs can be used as a precision, voltage-controllable waveform generator. Figure 6.4.1 shows how the 566 can be wired as a positive or negative ramp generator. In the positive ramp generator, the external transistor driven by the pin 3 output rapidly discharges $C_1$ at the end of the charging period so that charging can resume instantaneously. The PNP transistor of the negative ramp generator likewise rapidly charges the timing capacitor $C_1$ at the end of the discharge period. Because the circuits are reset so quickly, the temperature stability of the ramp generator is excellent. The time-period is given by

$$T = 1/2f_o$$

where $f_o$ is the 566 free-running frequency in normal operation. Therefore,

$$T = 1/2f_o = \{R_T C_1 V_{CC}/5(V_{CC} - V_C)\}$$

where $V_C$ is the bias voltage at pin 5 and $R_T$ is the total resistance between pin 6 and $V_{CC}$. Note that a short pulse is available at pin 3. (Placing collector resistance in series with the external transistor collector will lengthen the pulse.)

# 6.5 Sawtooth and pulse generator

**Figure 6.5.1** (a) Positive sawtooth and pulse generator

Figure 6.5.1 shows how the pin 3 output of the 566 can be used to provide different charge and discharge currents for $C_1$ so that a sawtooth output is available at pin 4 and a pulse at pin 3. The PNP transistor should be well saturated to preserve good temperature stability. The charge and discharge times may be estimated by using the formula

$$T = \{R_T C_1 V_{CC}/5(V_{CC} - V_C)\}$$

where $R_T$ is the combined resistance between pin 6 and $V_{CC}$ for the interval considered.

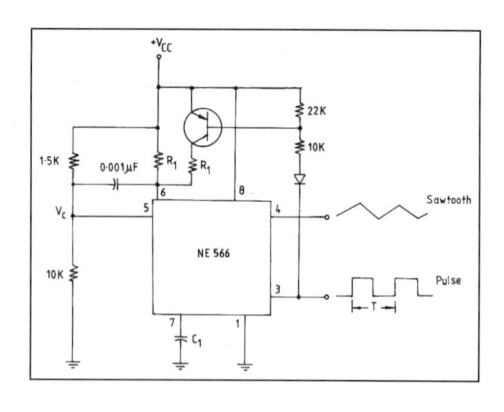

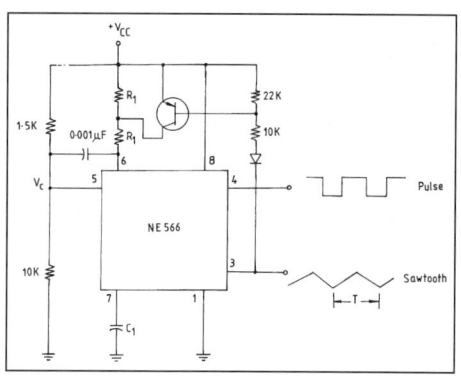

**Figure 6.5.1** (b) negative sawtooth and pulse generator

*Courtesy of Philips, The Netherlands*

# 6.6 Square wave tone-burst generator

**Figure 6.6.1** Square wave tone-burst generator

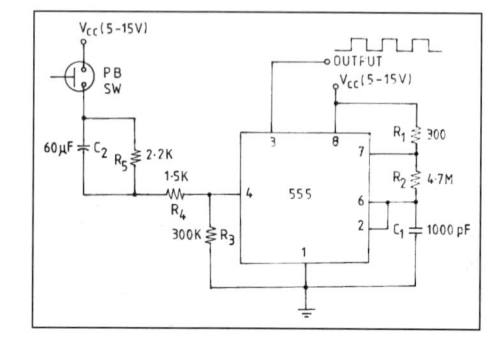

Waveform generators and oscillators 109

Depressing the pushbutton in Figure 6.6.1 provides square wave tone bursts whose duration depends on the duration for which the voltage at pin 4 exceeds a threshold. Components $R_1$, $R_2$ and $C_1$ cause the astable action of the timer IC as follows:

Since the reset terminal of the 555 timer is pulled low by the 300k resistor, when the pushbutton switch is open, there is no output at pin 3, i.e. the astable does not oscillate. When the pushbutton is pressed, the capacitor $C_2$ charges towards the supply voltage $V_{CC}$. As the voltage across the capacitor increases, the reset terminal of the 555 timer is pulled up towards $V_{CC}$ and the astable is activated producing a square wave output at pin 3. The frequency of the astable output is given by

$$f = 1.44/(R_1 + 2R_2) C_1$$

and duty cycle by

$$D = R_2/(R_1 + 2R_2)$$

Replacing the pushbutton switch by a transistor switch, a controlled ON/OFF square wave tone-burst signal can be obtained by driving the transistor ON and OFF at its base by a control signal.

*Courtesy of Philips, The Netherlands*

◆ ◆

## 6.7 Single-tone-burst generator

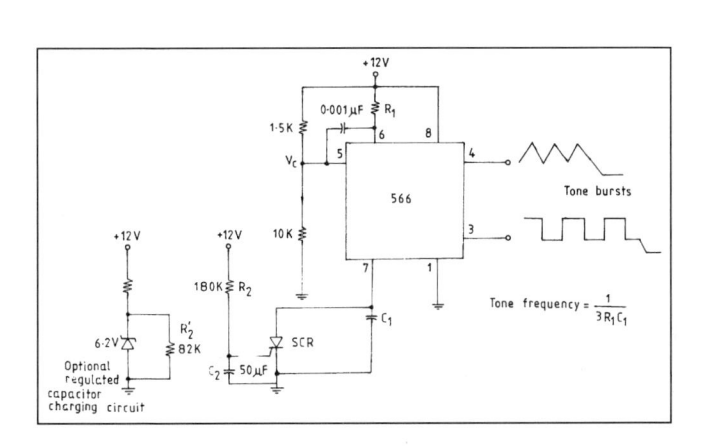

**Figure 6.7.1** Single-burst tone generator

Figure 6.7.1 shows a tone-burst generator which supplies a tone for one-half second after the power supply is activated; its intended use is a communications network alert signal. Cessation of the tone is accomplished at the SCR, which shunts the timing capacitor $C_1$ charge current when activated. The SCR is gated ON when $C_2$ charges up to the gate voltage which occurs in 0.5 seconds. Since only 70 µA is available for triggering, the SCR must be sensitive enough to trigger at this level. The triggering current can be increased, of course, by reducing $R_2$ (and increasing $C_2$ to keep the same time constant). If the tone duration must be constant under widely varying supply voltage conditions, the optional Zener diode regulator circuit can be added, along with the new value for $R_2$, $R_2' = 82$ kohm.

If the SCR is replaced by an NPN transistor, the tone can be switched ON and OFF at will at the transistor base terminal.

*Courtesy of Philips, The Netherlands*

## 6.8 Linear ramp voltage generator

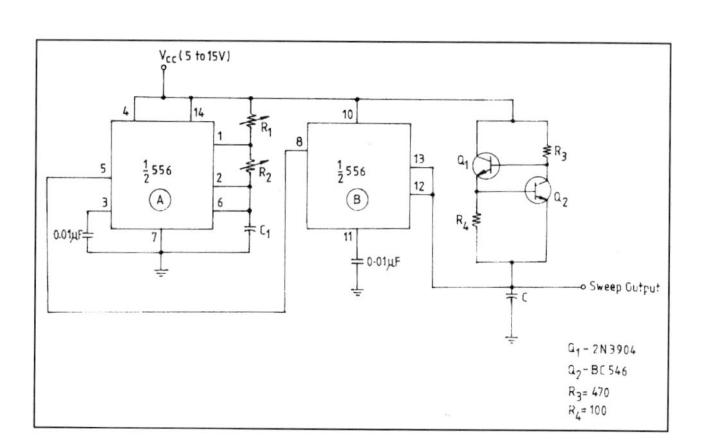

**Figure 6.8.1** Linear ramp voltage generator

The circuit shown in Figure 6.8.1 can be used as a simple ramp voltage generator. It consists of a dual timer 556 one half of which is configured as a free-running astable multivibrator and the other half as a monostable. The capacitor in the monostable is charged by a floating constant current source instead of through a resistor. The astable 'A' repeatedly triggers the one-shot 'B'.

In the floating constant current source consisting of transistors $Q_1$ and $Q_2$, and resistors $R_3$ and $R_4$, current initially flows through $R_3$, to the base of transistor $Q_1$ and through $R_4$. When the voltage across the terminals of the constant current source rises above $2V_{BE}$, the voltage drop across $R_4$ increases and at some point the minimum forward voltage drop $V_{BE}$ for transistor $Q_2$ is exceeded and it is turned ON. If the voltage across the terminals of the current source increases further, $Q_2$ conducts more heavily and diverts base current away from $Q_1$. This maintains a constant current through the floating current source and hence the charging current for the capacitor is constant. This constant current charging makes the voltage across the capacitor rise linearly with time. When the capacitor voltage exceeds $(2/3)V_{CC}$, since the capacitor is connected to the discharge and threshold terminals of the timer, the timer changes state and the capacitor C begins to discharge towards zero voltage. As the astable repeatedly triggers the one-shot, since this configuration of the timer one-shot is not retriggerable, the capacitor repeatedly charges linearly and then discharges, thereby generating a ramp voltage waveform. The capacitor charging current is given by

$$I = \{(V_{BE}/R_4) + (V_{CC} - 2V_{BE})/R_3\}$$

and the ramp time $t$ is given by

$$t = (CV/I) \text{ sec}$$

where $C$ is in farads, $V$ is in volts and $I$ is in amps.

In the above equation, $V = (2/3)V_{CC}$, $V_{CC}$ being the supply voltage, $C$ is the capacitance of the capacitor and $I$ is the charging current. $t$ is defined as the time taken by the capacitor to charge from zero voltage to $(2/3)V_{CC}$ at which point the capacitor begins to discharge.

Using these equations a ramp voltage generator to meet a specific requirement can be easily designed.

## 6.9 Staircase waveform generator

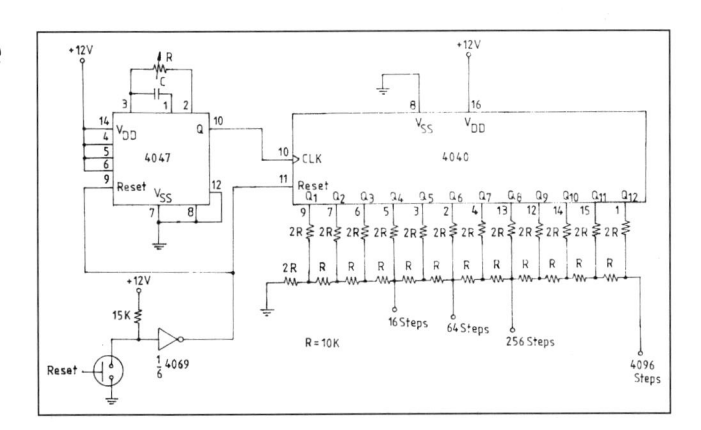

**Figure 6.9.1** Staircase waveform generator

The circuit shown in Figure 6.9.1 generates a staircase or stepped waveform of a periodic nature. The circuit consists of an astable configured using a 4047 whose output is used to clock a 12-stage binary counter, 4040. The outputs of the counter, $Q_1$–$Q_{12}$, are connected to a R–2R resistor summing network as shown. The counter is incremented on the negative-going clock transitions of each input pulse. The frequency of the 4047 astable is given by

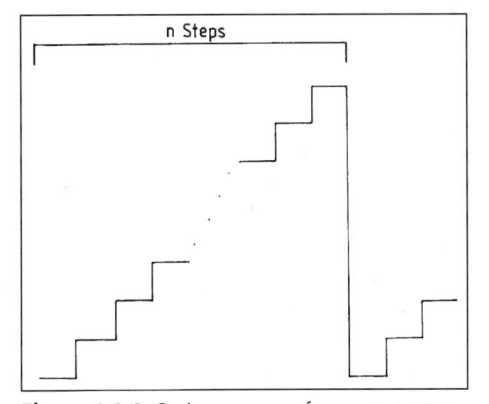

**Figure 6.9.2** Staircase waveform generator waveforms

$$f = 1/4.4RC$$

Adjustment of the potentiometer R varies the frequency. The duty cycle of the output is 50%.

The summing network of resistors connected to the outputs of the counter generates a staircase waveform. The staircase steps start at the ground reference and go up to the positive supply $V_{DD}$. The output $Q_4$ provides a waveform having 16 steps, output $Q_6$ provides a 64-step waveform and so on. After going through the steps, the waveform returns to zero voltage level and this process repeats, as shown in Figure 6.9.2. The duration of each step is related to the input clock frequency by $T = 1/f$ where T is the step duration.

This kind of a stepped waveform generator is useful in instrumentation applications such as in a transistor curve tracer.

## 6.10 Two-phase sine wave generator

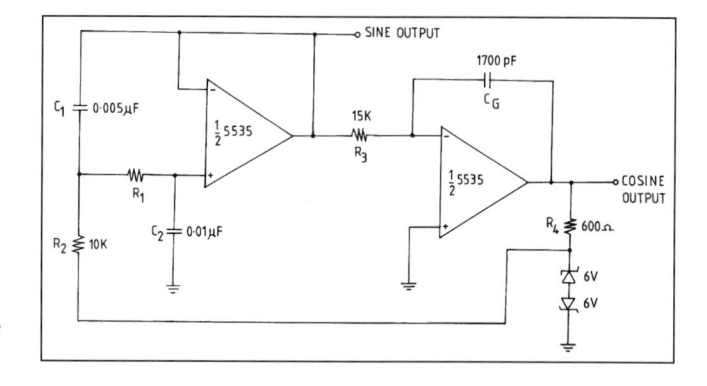

**Figure 6.10.1** Two-phase sine wave oscillator

The circuit shown in Figure 6.10.1 uses a 2-pole pass Butterworth followed by a phase-shifting single-pole stage, fed back through a voltage limiter to achieve sine and cosine outputs. The values shown using the µA 741 amplifiers give about 1.5% distortion at the sine output and about 3% distortion at the cosine output. By careful trimming of $C_G$ and/or limiting network, better distortion figures are possible. The component values shown give a frequency of oscillation of about 2 kHz. The values can be readily selected for other frequencies. The NE 5535 should be used at higher frequencies to reduce distortion due to slew limiting.

*Courtesy of Philips, The Netherlands*

## 6.11 Get ±15 V square waves from +5 V

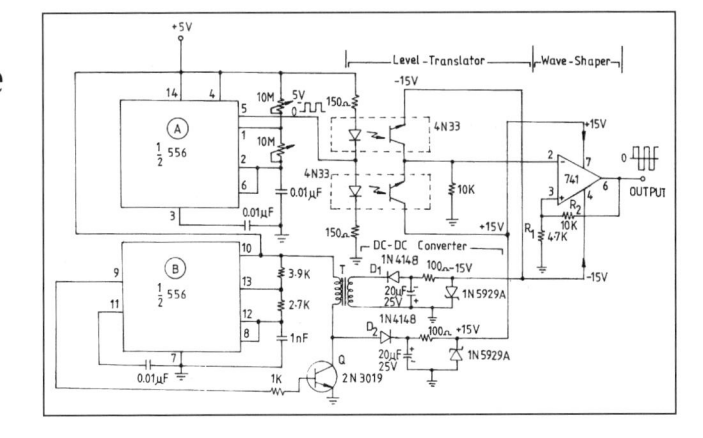

**Figure 6.11.1** Low-cost bipolar square wave generator

This low-cost circuit should prove useful when bipolar square waves are needed and only 5 V system power is available. It is much less expensive than the more obvious expedient of buying a dual-output DC–DC converter. Although this implementation generates ± 15 V square waves, slight modifications, such as using different Zeners, can be made to yield other voltages. The heart of the circuit is a 556 dual timer, half configured as a variable-frequency astable multivibrator (A) and the other as a DC–DC converter (B) (see Figure 6.11.1).

The output of the 556's 'B' half is a 100 kHz signal that drives transistor Q. Because the collector load on Q is the primary of a 1 mH pulse transformer, the switching of the collector current generates a train of voltage spikes (about four times the supply voltage, or around 20 V in this case) at the collector of Q. The spikes are rectified by diode $D_2$, filtered by the 20 μF capacitor, and regulated by a 15 V Zener diode to yield a + 15 V supply rail. In a similar fashion, the output of the transformer secondary is rectified by diode $D_1$ and used to form a −15 V rail.

With the ± 15 V rails established, they can be used to power the output transistors in a pair of 4N 33 opto-couplers, whose inputs are driven by the 'A' half of the 556. As the diagram shows, the common output of the 4N 33's is terminated by a 10k load and fed through a wave shaper to form the output of the generator. The wave shaper is a Schmitt-trigger circuit based on an op-amp.

The generator output is a bipolar pulse train with a peak-to-peak swing of about 30 V. Its frequency and duty cycle can be varied by adjusting the two 10 Mohm pots on astable multivibrator 'A'. The circuit thus serves as a variable-frequency square wave generator with a variable pulse width.

*Reprinted with permission from Electronic Design (Vol.39, No.21) November 7, 1991. Copyright 1991, Penton Publishing Inc.*

## 6.12 Programmable pulse generator

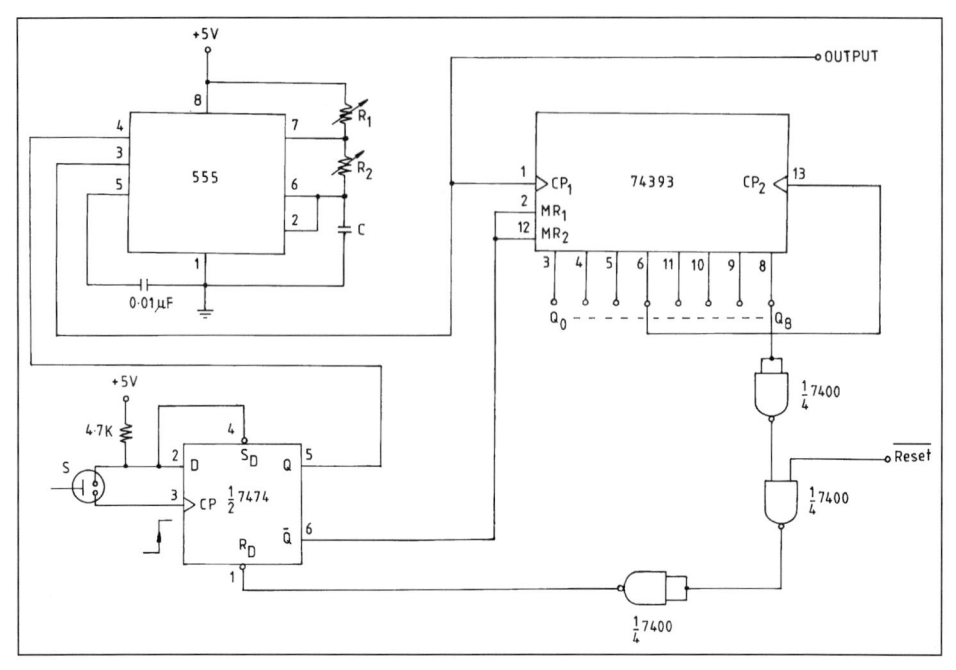

**Figure 6.12.1** Programmable pulse generator

The circuit shown in Figure 6.12.1 generates a preset number of pulses in powers of two, i.e. it is a $2^n$ pulse generator, where $n$ is programmable up to 8. The circuit consists of an astable multivibrator using a 555 timer whose output drives a dual 4-bit binary counter, 74393, configured in this circuit as an 8-bit binary counter by connecting the $Q_3$ output of the first 4-bit counter to the clock input of the second 4-bit counter. The frequency of the astable multivibrator is given by

$$f = 1.44/[(R_1 + 2R_2)C]$$

and the duty-cycle by

$$D = R_2/(R_1 + 2R_2)$$

The 74393 is triggered by a HIGH-to-LOW transition of the clock. The Reset pin of the 555 timer (pin 4) is connected to the Q output of a D flip-flop (1/2 7474), whose Reset ($R_D$) terminal is controlled by the output of the counter through an inverter.

To obtain $2^n$ pulses, the $Q_n$ output of 74393 is connected to an inverter as shown in Figure 6.12.1. On power up, the Reset from the system resets the flip-flop making its Q output low and its $\overline{Q}$ output high. The high level on the $\overline{Q}$ resets the counter 74393 and all its outputs go to a logic low state. Since the $Q_n$ output of the counter is connected to the input of the inverter, the output of the inverter goes

HIGH. Since the system reset signal is a short-duration low-level pulse and thereafter it goes high, the output of the enable AND gate goes high. With $S_D$ and $R_D$ of the flip-flop both high, pressing the pushbutton switch S sets the flip-flop and its Q output goes high. This takes the Reset pin of the 555 timer to a high level and the astable oscillates giving an output. The output of the astable clocks the binary counter; when the programmed count is reached, the $Q_n$ output of the counter goes HIGH and the output of the inverter goes LOW. This resets the flip-flop making its Q output LOW, which in turn resets the timer and the counter is disabled and the counter output terminates at the programmed number of pulses.

## 6.13 Nanoseconds pulse generator

**Figure 6.13.1** Nanoseconds pulse generator

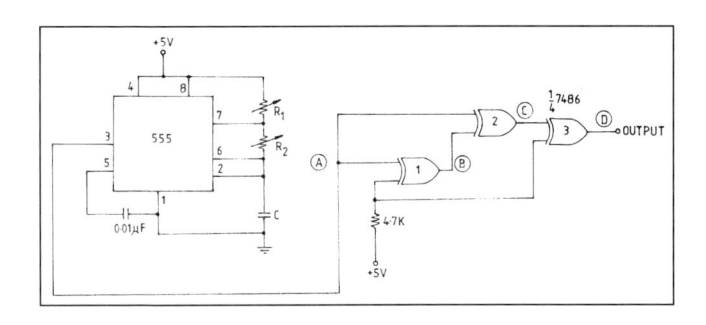

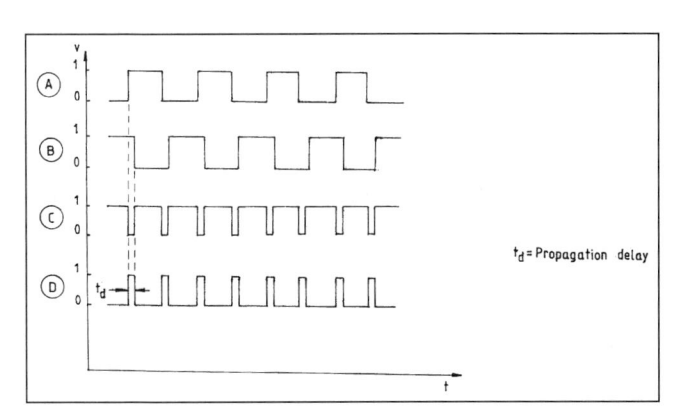

**Figure 6.13.2** Waveforms of circuit in Figure 6.13.1

The circuit shown in Figure 6.13.1 can be used to generate very narrow pulses – of nanoseconds duration – which can be used to simulate glitches and transition spikes and for synchronization purposes in logic circuits. The circuit consists of a variable frequency astable multivibrator configured using a 555 timer and three EX–OR gates configured as shown. The frequency of the astable is given by

$$f = 1.44/(R_1 + 2R_2)C$$

and duty cycle by

$$D = R_2/(R_1 + 2R_2)$$

The output of the astable multivibrator is fed to a pair of EX–OR gates. A time-delayed and inverted version of the astable output is EX–ORed with the output of the astable to obtain the narrow pulses of duration equal to the propagation delay $t_d$ of the logic used (in this case about 20 ns). The output of the EX–OR gate at point C in the circuit consists of a series of negative-going spikes of duration $t_d$ and this is inverted by the last EX–OR gate to derive the waveform at D, which is a series of positive-going spikes from a logic LOW base-line (Figure 6.13.2). The same technique can be used also with LS, CMOS or any other logic type to derive spikes of different widths corresponding to their propagation delays. By varying the frequency of the astable, the repetition rate of the spikes can be varied.

## 6.14  3-phase clock generator

**Figure 6.14.1** 3-phase clock generator

The simple 2-chip circuit shown in Figure 6.14.1 can be used to generate 3-phase clocks, i.e. clock waveforms at 120° with respect to each other, from an available clock waveform. The circuit essentially consists of a Johnson decade counter with decoded outputs – the 74HC4017 – which is clocked by the available clock. The 74HC4017 is configured as a modulo-6 counter by feeding back its $Q_6$ output to its Master Reset input. The $Q_0$–$Q_5$ outputs of the counter are ORed three at a time by means of three OR gates as shown, i.e. ($Q_0,Q_1,Q_2$), ($Q_2,Q_3,Q_4$) and ($Q_4,Q_5,Q_0$) to derive the A, B and C phases as shown. Since $Q_6$ is fed back to the Master Reset, the counter is reset at every sixth clock pulse. The clock input CP1 is tied to ground and the available clock is input to CP0; therefore, the counter advances for every LOW-to-HIGH transition of the clock. The waveforms in Figure 6.14.2 illustrate the technique.

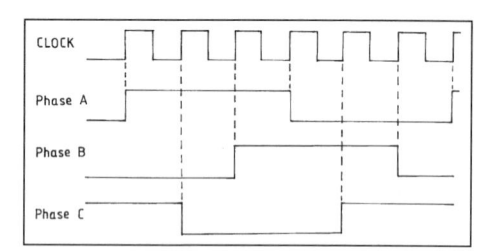

**Figure 6.14.2** Waveforms of circuit in Figure 6.14.1

# 6.15 4-phase clock generator

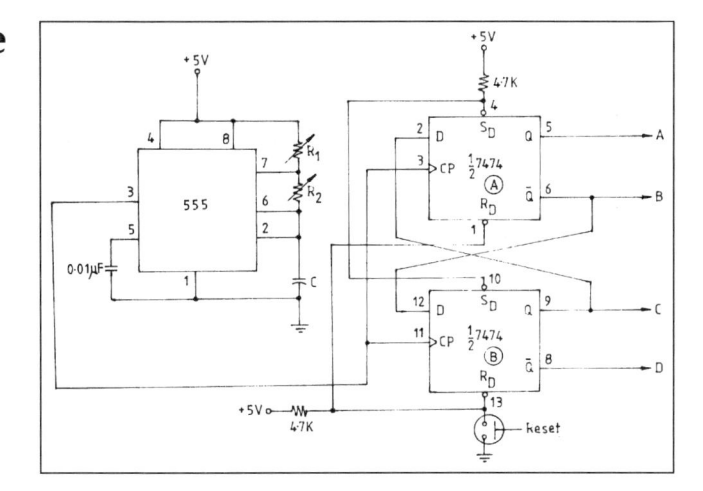

**Figure 6.15.1** 4-phase clock generator

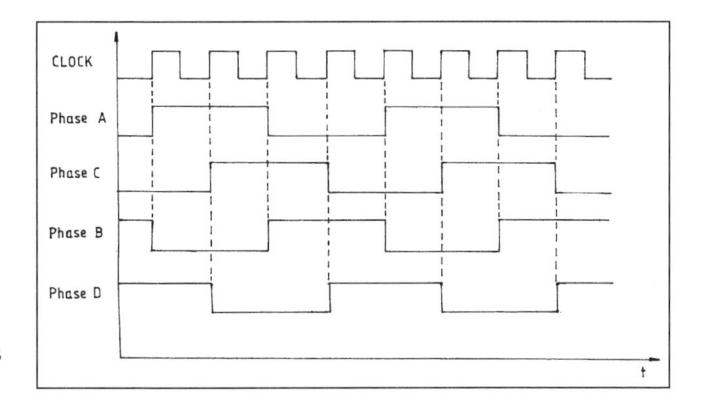

**Figure 6.15.2** Waveforms of circuit in Figure 6.15.1

The simple 2-chip circuit shown in Figure 6.15.1 can be used as a 4-phase clock generator. The circuit consists of an astable multivibrator, built using a 555 timer, synchronously clocking a pair of D flip-flops configured as shown in Figure 6.15.1. The Q output of the flip-flop B is tied to the D input of flip-flop A and the $\overline{Q}$ output of flip-flop A is tied to the D input of the flip-flop B. The Set inputs of both the flip-flops are tied to a logic HIGH level and the Reset inputs are connected to a push-button switch as shown. As the waveforms in Figure 6.15.2 show, the outputs of the flip-flops labelled A,B,C,D are in quadrature sequence. The time-period of each of these outputs is $4T$ if $T$ is the time-period of the input clock. In other words, the frequency of each of the flip-flop outputs is $f/4$ if $f$ is the input clock frequency. The frequency and duty cycle of the clock input to the D flip-flops can be varied by adjusting the potentiometers $R_1$ and $R_2$ and are given in terms of the RC components by

$$f = \frac{1.44}{(R_1 + 2R_2)C}$$

$$\text{Duty-cycle} = \frac{R_2}{(R_1 + 2R_2)}$$

Such a 4-phase clock generator is required in many applications, for example, to drive a stepper motor.

## 6.16 Digital phase-shifted clock generator

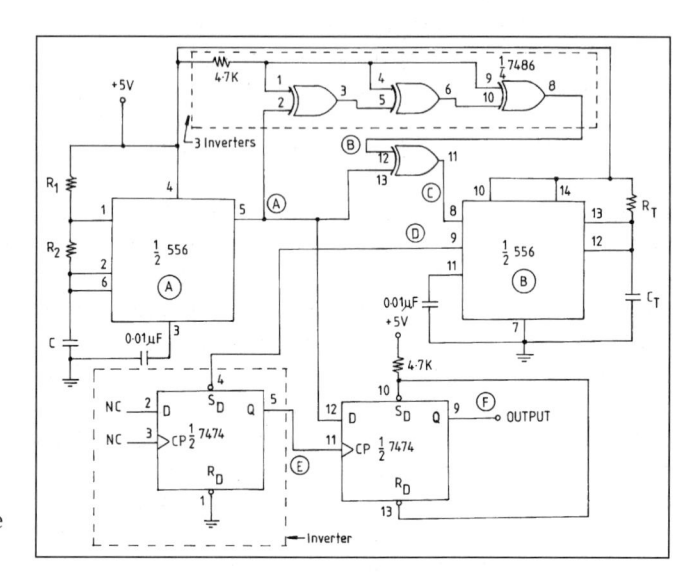

**Figure 6.16.1** Quadrature pulse train generator

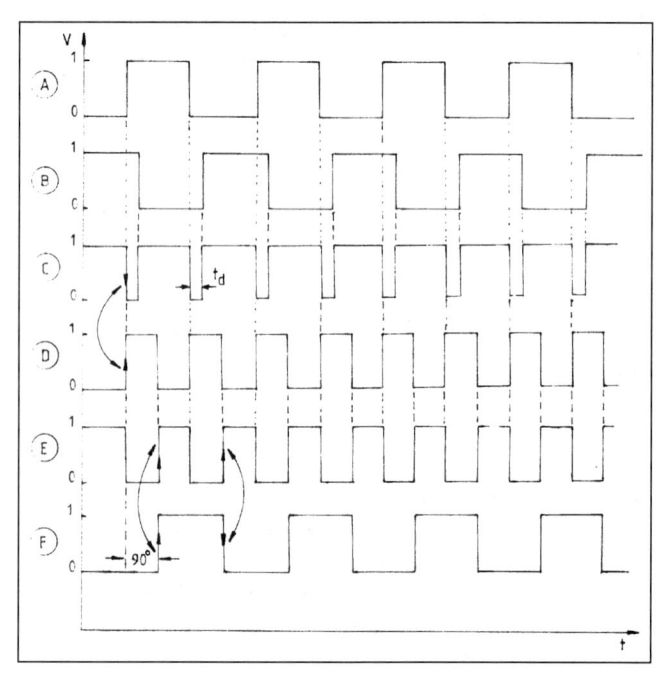

**Figure 6.16.2** Waveforms of circuit in Figure 6.16.1

The circuit shown in Figure 6.16.1 generates phase-shifted pulse trains, i.e. pulse trains having a phase difference between each other. The circuit consists of a dual timer 556, one half of which is configured as an astable multivibrator (A) and the other half as a monostable multivibrator (B). An edge-detection technique using the propagation delay of digital logic elements is used to detect the transitions of the astable waveform. The output of the astable is fed to an EX–OR gate in two forms – one is the astable output and the other is a delayed and inverted version of it. The delayed and inverted signal is derived by passing the astable output through three EX–OR gates which are configured as inverters (see Figure 6.16.1). The delayed version is the complement of the input but shifted in time by a duration $t_d$ which is equal to the combined propagation delay of the three inverters. The output of the EX–OR gate, which is fed by the true signal and the delayed and inverted signal, consists of a series of negative-going transitions of width $t_d$ (see waveforms in Figure 6.16.2). These sequential and repetitive negative-going transitions are used to trigger the monostable B. The pulse width of the output of B is given by $(1.1)R_T C_T$. The output of the one-shot is inverted using a D flip-flop configured as a logic inverter, and the other D flip-flop in the 7474 package is used as digital frequency mixer whose inputs consist of the astable output and the inverted monostable output. The inverted one-shot output waveform clocks the logic levels of the astable output waveform at its rising edges. The output of the digital frequency mixer at point F in Figure 6.16.1 is a phase-shifted version of the astable output at point A. By varying the $R_T C_T$ values of the monostable (conveniently done by making $R_T$ variable) it is possible to achieve phase shift in the range 0° to 180°. If the one-shot output pulse width $(1.1)R_T C_T$, is made equal to half the time period of the astable output, 90° phase shift between the waveforms at point A and point F can be obtained as shown in Figure 6.16.2. Since quadrature waveform generation is a frequent requirement in digital design this is shown in Figure 6.16.2.

## 6.17 Wien-bridge oscillator

The Wien-bridge oscillator is a good low-distortion sine wave generator for low to moderate frequencies. The standard Wien network is shown in Figure 6.17.1. The transfer function of this network is given by

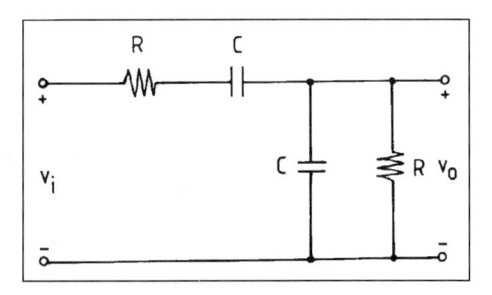

**Figure 6.17.1** Wien network

# 120 Waveform generators and oscillators

$H(s) = V_o(s)/V_i(s)$
$= (R/sC)/\{(R + 1/sC) + [(R.\,1/sC)/(R + 1/sC)]\}$
$= j\omega RC/(1 + 3j\omega RC - \omega^2 R^2 C^2)$

on simplification. $H(s)$ becomes real if $\omega RC = 1$ and is given by $H(s) = 1/3$ and the phase shift becomes zero, the gain being 3. The frequency of oscillations derived from the relation $\omega RC = 1$ is

$f = 1/2\pi RC$

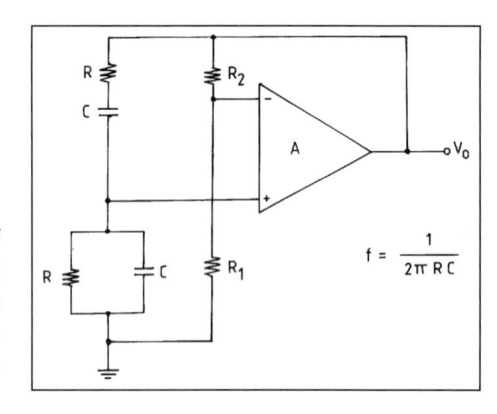

$$f = \frac{1}{2\pi RC}$$

**Figure 6.17.2** Wien-bridge oscillator

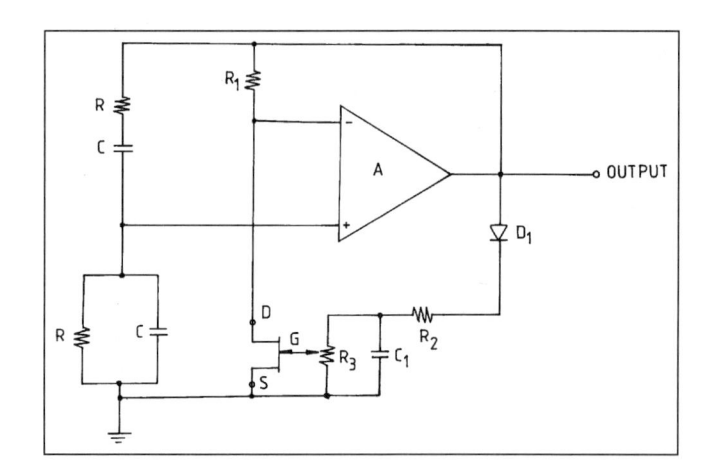

**Figure 6.17.3** Wien-bridge oscillator with a FET providing stability

The attenuation introduced by the Wien network is 3 and, therefore, for sustaining oscillations, the gain should be slightly greater than 3. An op-amp based Wien-bridge oscillator is shown in Figure 6.17.2. To achieve stability of oscillations, resistor $R_1$ is made non-linear by replacing it with a PTC resistor such as an incandescent lamp. The cold resistance of the lamp is less and increases as its temperature goes up and this, in turn, increases the negative feedback and reduces overall loop gain, thereby stabilizing the oscillator. Alternatively resistor $R_2$ can be replaced by a NTC resistor for stabilizing the oscillator. Another method to stabilize the oscillator is to use a FET in the feedback loop as a voltage-dependent resistor (VDR) by biasing the FET suitably using the rectified oscillator output to derive DC bias voltage. The FET is suitably biased so that oscillations are maintained. This technique is shown in Figure 6.17.3.

## 6.18 Wien-bridge oscillator using spare logic inverters

**Figure 6.18.1** Wien-bridge oscillator using spare inverters

A digitally controlled Wien-bridge oscillator can be configured using spare logic inverters as quasilinear amplifiers. Such oscillators are useful in applications such as keyed oscillators, FSK generation and signalling. The gains of the two 74HCU04 (unbuffered) inverters provide sufficient regenerative action to cause oscillations (Figure 6.18.1). Application of a control logic input to the input of the first inverter can control the oscillator output. If the control input is a logic 0, the oscillator gives an output and if a logic 1 signal is applied to the control input terminal, the oscillator output is disabled. The control input thus effectively switches the oscillator output ON and OFF. The frequency of oscillations is given by

$$f = 1/(2\pi RC)$$

For applications where the accuracy and stability of oscillator frequency are not critical, this simple oscillator can be used.

◆ ◆

## 6.19 RC phase shift oscillator

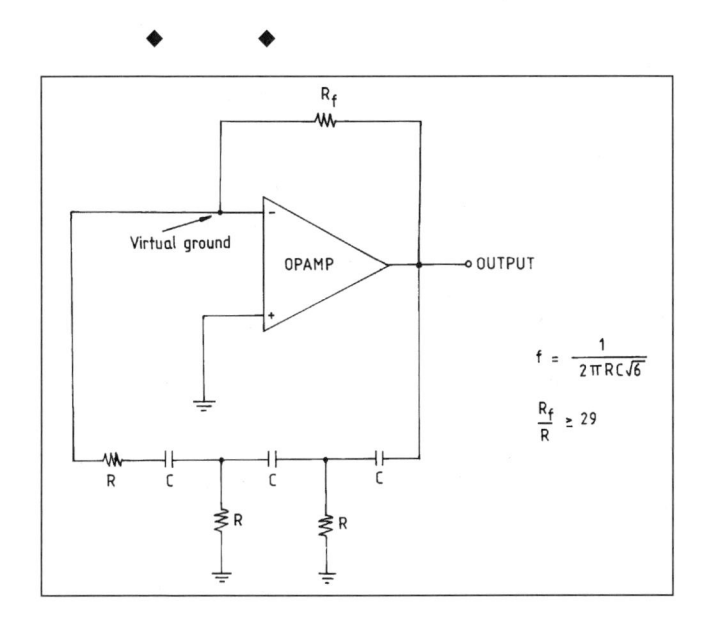

$$f = \frac{1}{2\pi RC\sqrt{6}}$$

$$\frac{R_f}{R} \geq 29$$

**Figure 6.19.1** *RC* phase shift oscillator

This is one of the popularly used *RC* oscillators. A *RC* phase shift oscillator based on an op-amp is shown in Figure 6.19.1. The op-amp introduces a phase-shift of 180° (inverting amplifier) and the phase shift introduced by the *RC* network should be 180° to satisfy the Barkhausen criterion of 360° ($2\pi fn$) phase shift through the amplifier and the feedback network. The frequency at which this occurs is the frequency of oscillation of the oscillator circuit and is given by

$$f = 1/(2\pi RC \sqrt{6})$$

At this frequency, the attenuation of the *RC* network is 29 and, therefore, the gain of the amplifier should be at least 29 in order to satisfy the loop gain requirement for oscillation.

## 6.20 Crystal oscillator

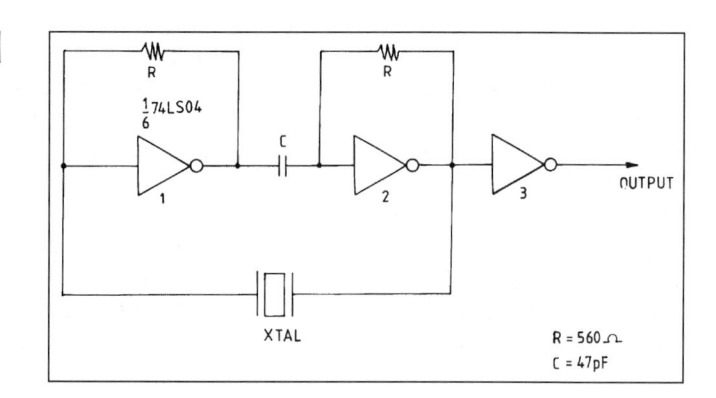

**Figure 6.20.1** Crystal oscillator

Crystal oscillators are characterized by stability of oscillations. Quartz crystals are popular for crystal controlled oscillators since they have a good stability of resonance frequency with respect to temperature variation. Crystal oscillators are invariably used in digital circuits since accuracy of timing is an important consideration. In general, any device with a reasonable gain can be used in an oscillator circuit by using a positive feedback configuration. The circuit shown in Figure 6.20.1 is a typical crystal oscillator configuration frequently used by digital circuit designers. Here the two inverters are used as quasi-linear amplifiers. The two 560 ohm resistors apply positive feedback from the output to the input of the corresponding inverter. The configuration shown is a parallel resonance oscillator. The inverters provide sufficient gain and a phase-shift near zero, which produces the oscillations. The frequency of oscillation of the oscillator circuit is the same as the resonance frequency of the crystal used.

## 6.21 Gated oscillator

**Figure 6.21.1** Gated oscillator

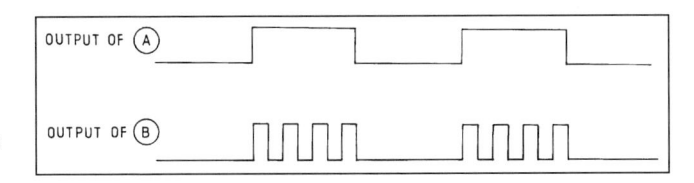

**Figure 6.21.2** Waveforms of gated oscillator

The circuit shown in Figure 6.21.1 generates a pulsed output of square waves as shown by the waveforms of Figure 6.21.2. The circuit consists of a dual timer 556 both halves of which are configured as astable multivibrators. The frequency of the astable A is lower than that of B. The output of A is connected to the reset pin of B. When the output of A is LOW, the reset pin of the astable B is pulled LOW and its output is LOW; when the output of A is HIGH, the reset pin of astable B is pulled HIGH and B oscillates to give an output. In other words, the output of A effectively gates the output of B at its frequency. The circuit, therefore, acts as a gated oscillator or pulsed oscillator, i.e. astable B oscillates for the duration set by the HIGH output level of astable A and a gated pulse train is obtained.

# 7 Phase locked loops

## 7.1 Monolithic phase locked loop

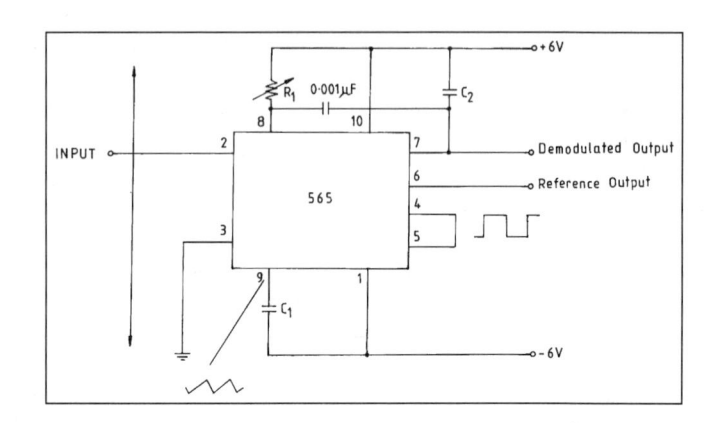

**Figure 7.1.1** PLL 565 connection diagram

The 565 is a general purpose PLL designed to operate at frequencies below 1 MHz. The circuit comprises a voltage-controlled oscillator of exceptional stability and linearity, a phase comparator, an amplifier and a low-pass filter (ref. 26, p.1298). The centre frequency of the PLL is determined by the free-running frequency of the VCO; this frequency can be adjusted externally with a resistor or a capacitor. The low-pass filter, which determines the capture characteristics of the loop, is formed by an internal resistor and an external capacitor.

The device has a highly stable centre frequency with a factor of around 200 ppm/°C, a wide operating power supply range of ±6 V to ±12 V, a highly linear demodulated output of 0.2% linearity and a bandwidth adjustable from < ±1% to > ±60%.

A simple scheme using the 565 to receive FSK signals of 1070 Hz and 1270 Hz is shown in Figure 7.1.1. FSK refers to data transmission by means of a carrier which is shifted between two preset frequencies. This frequency shift is usually accomplished by driving a VCO with the binary data signal so that the two resulting frequencies correspond to the 0 and 1 states (commonly called space and mark) of the binary data signal. In Figure 7.1.1, as the signal appears at the input, the loop locks to the input frequency and tracks between the two frequencies with a corresponding DC shift at the output (pin 7).

The design formulas are:

*Free-running frequency of VCO*

$$f_o = 1.2/4R_1C_1 \text{ in Hz}$$

Lock range

$$f_L = \pm 8f_o/V_{CC} \text{ in Hz}$$

Capture range

$$f_C = \pm 1/2\pi \sqrt{2\pi f_L/T}$$

where $T = 3.6 \times 10^3 \times C_2$

The loop filter capacitor $C_2$ is chosen to set the proper overshoot on the output and a three-stage RC ladder filter is used to remove the sum frequency components. The band edge of the ladder filter is chosen to be approximately half-way between the maximum keying rate (300 baud or bits per second or 150 Hz) and twice the input frequency (approximately 2200 Hz). The free-running frequency should be adjusted (with $R_1$ so that the DC voltage level at the output is the same as that at pin 6 of the loop. The output signal can now be made logic compatible by connecting a voltage comparator between the output and pin 6.

The input connection is typical for cases where a DC voltage is present at the source and, therefore, a direct connection is not desirable. Both input terminals are returned to ground with identical resistors (in this case, the values are chosen to achieve a 600 ohm input impedance).

*Courtesy of Philips, The Netherlands*

◆ ◆ ◆

## 7.2 Monolithic tone decoder

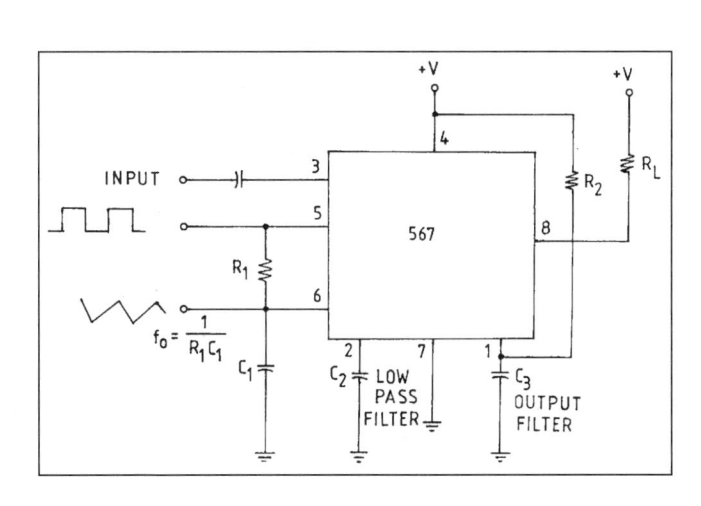

**Figure 7.2.1** Tone decoder using 567 PLL

The 567 tone and frequency decoder is a highly stable phase locked loop with synchronous AM lock detection and power output circuitry. Its primary function is to drive a load whenever a sustained frequency within its detection band is present at the self-biased input. The bandwidth centre frequency and output delay are independently determined by means of four external components. The salient features of this device are: wide frequency range (0.01 Hz to 500 kHz), high stability of centre frequency, independent controllable bandwidth (up to 14%), high out-band signal and noise rejection, and a logic compatible output with 100 mA current sinking capability.

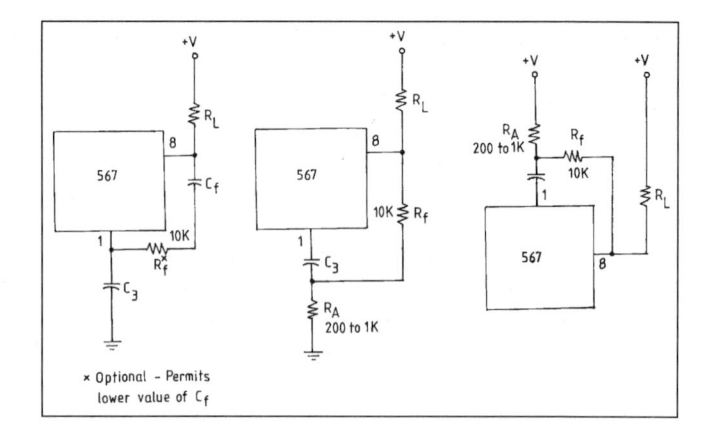

**Figure 7.2.2** Methods of chatter prevention

× Optional – Permits lower value of $C_f$

## *Phase-locked loop terminology*

*Centre frequency* ($f_o$)  The free running frequency of the current controlled oscillator (CCO) in the absence of an input signal.

*Detection bandwidth* (*BW*)  The frequency range, centred about $f_o$ within which an input signal above the threshold voltage (typically 20 mV RMS) will cause a logical zero state on the output. The detection bandwidth corresponds to the loop capture range.

*Lock range*  The largest frequency range within which an input signal above the threshold voltage will hold a logical zero state on the output.

*Detection band skew*  A measure of how well the detection band is centred about the centre frequency, $f_o$. The skew is defined as:

$$(f_{MAX} + f_{MIN} - 2f_o)/2f_o$$

where $f_{MAX}$ and $f_{MIN}$ are the frequencies corresponding to the edges of the detection band. The skew can be reduced to zero if necessary by means of an optional centring adjustment.

## *Design formulas*

$$f_o = 1/(1.1)R_1C_1$$
$$BW = 1070\sqrt{(V_i/f_oC_2)} \text{ in \% of } f_o$$
$$V_i \leq 200 \text{ mV}_{RMS}$$

where $V_i$ = input voltage ($V_{RMS}$) and $C_2$ = low-pass filter capacitor (µF).

## Selection of components for a simple tone decoder using the 567

Figure 7.2 shows a typical connection diagram for the 567. For most applications, the following three-step procedure will be sufficient for choosing the external components $R_1$, $C_1$, $C_2$ and $C_3$.

1. Select $R_1$ and $C_1$ for the desired centre frequency. For best temperature stability, $R_1$ should be between 2k and 20k, and the combined temperature coefficient of the $R_1 C_1$ product should have sufficient stability over the projected temperature range to meet the necessary requirements.
2. Select the low-pass capacitor, $C_2$, by referring to the bandwidth versus input signal amplitude graph, given in the datasheet of 567 tone decoder. If the input amplitude variation is known, the appropriate value of $f_o C_2$ necessary to give the desired bandwidth may be found. Conversely, an area of operation may be selected on this graph and the input level and $C_2$ may be adjusted accordingly. For example, constant bandwidth operation requires that input amplitude be above 200 mV rms. The bandwidth, as noted on the graph, is then controlled solely by the $f_o C_2$ product ($f_o$ (Hz), $C_2$ (µF)).
3. The value of $C_3$ is generally non-critical. $C_3$ sets the band edge of a low-pass filter which attenuates frequencies outside the detection band to eliminate spurious outputs. If $C_3$ is too small, frequencies just outside the detection band will switch the output stage on and off at the beat frequency, or the output may pulse on and off during the turn-on transient. If $C_3$ is too large, turn-on and turn-off of the output stage will be delayed until the voltage on $C_3$ passes the threshold voltage. (Such delay may be desirable to avoid spurious outputs due to transient frequencies.) A typical minimum value for $C_3$ is $2C_2$.
4. Optional resistor $R_2$ sets the threshold for the largest 'no output' input voltage. A value of 130k is used to ensure the tested limit of 10 mV$_{RMS}$ min. This resistor can be referenced to ground for increased sensitivity.

## Available outputs

The primary output is the uncommitted output transistor collector, pin 8. When an in-band input signal is present, this transistor saturates; its collector voltage being less than 1.0 V (typically 0.6 V) at full output current (100 mA). The voltage at pin 2 is the phase detector output which is a linear function of frequency over the range of 0.95 to 1.05 $f_o$ with a slope of about 20 mV per 1% of frequency deviation. The average voltage at pin 1 is, during lock, a function of the in-band input amplitude in accordance with the transfer characteristic given. Pin 5 is the controlled oscillator square wave output of magnitude $(+V - 2V_{BE}) \approx (+V - 1.4\ V)$ having a DC average of $+V/2$. A 1k load may be driven from pin 5. Pin 6 is an exponential triangle of 1 V p–p with an average DC level of $+V/2$. Only high impedance loads may be connected to pin 6 without affecting the CCO duty cycle or temperature stability.

## Operating precautions

A brief review of the following precautions will help the user achieve the high level of performance of which the 567 is capable.

1. Operation in the high input level mode (above 200 mV) will free the user from bandwidth variations due to changes in the in-band signal amplitude. The input stage is now limiting, however, so that out-band signals or high noise levels can cause an apparent bandwidth reduction as the in-band signal is suppressed. Also the limiting action will create in-band components from sub-harmonic signals at $f_o/3, f_o/5$, etc.

2. The 567 will lock onto signals near $(2n + 1)f_o$ and will give an output for signals near $(4n + 1)f_o$ where $n = 0, 1, 2$, etc. Thus signals at $5f_o$ and $9f_o$ can cause an unwanted output. If such signals are anticipated, they should be attenuated before reaching the 567 input.

3. Maximum immunity from noise and outband signals is afforded in the low input level (below 200 mV RMS) and reduced bandwidth operating mode. However, decreased loop damping causes the worst-case lock-up time to increase.

4. Due to the high switching speeds (20 ns) associated with 567 operation, care should be taken in lead routeing. Lead length should be kept to a minimum. The power supply should be adequately bypassed close to the 567 with a 0.01 µF or greater capacitor; grounding paths should be carefully chosen to avoid ground loops and unwanted voltage variations. Another factor which must be considered is the effect of load energization on the power supply. For example, an incandescent lamp draws 10 times rated current at turn-on. This can cause supply voltage fluctuations which could, for example, shift the detection band of narrow-band systems sufficiently to cause momentary loss of lock. The result is a low-frequency oscillation into and out of lock. Such effects can be prevented by supplying heavy load currents from a separate supply or increasing the supply filter capacitor.

## Speed of operation

Minimum lock-up time is related to the natural frequency of the loop. The lower it is, the longer becomes the turn-on transient. Thus, maximum operating speed is obtained when $C_2$ is at a minimum. When the signal is first applied, the phase may be such as to initially drive the controlled oscillator away from the incoming frequency rather than towards it. Under this condition, which is of course unpredictable, the lock-up transient is at its worst and the theoretical minimum lock-up time is not achievable. We must simply wait for the transient to die out.

The following expressions give the values of $C_2$ and $C_3$ which allow highest operating speeds for various band centre frequencies. The minimum rate at which digital information may be detected without information loss due to the turn-on transient or output chatter is about 10 cycles per bit, corresponding to an information transfer rate of $f_o/10$ baud.

$C_2 = 130/f_o$ µF

$C_3 = 260/f_o$ µF

In cases where turn-off time can be sacrificed to achieve fast turn-on, the optional sensitivity adjustment circuit can be used to move the quiescent $C_3$ voltage lower (closer to the threshold voltage). However, sensitivity to beat frequencies, noise and extraneous signals will be increased.

## Optional controls

The 567 is designed such that for most applications no external adjustments are required. Certain applications, however, will be greatly facilitated if full advantage is taken of the added control possibilities available through the use of additional external components. In the figures given, typical values are suggested where applicable. For best results, the resistors used, except where noted, should have the same temperature coefficient. Ideally, silicon diodes would be low resistivity types, such as forward-biased transistor base-emitter junctions. However, ordinary low-voltage diodes should be adequate for most applications.

## Sensitivity adjustment

When operated as a very narrow-band detector (less than 8%), both $C_2$ and $C_3$ are made quite large in order to improve noise and out-band signal rejection. This will inevitably slow the response time. If, however, the output stage is biased closer to the threshold level, the turn-on time can be improved. This is accomplished by drawing additional current to terminal 1. Under this condition, the 567 will also give an output for lower-level signals (10 mV or lower).

By adding current to terminal 1, the output stage is biased further away from the threshold voltage. This is most useful when, to obtain maximum operating speed, $C_2$ and $C_3$ are made very small. Normally, frequencies just outside the detection band could cause false outputs under this condition. By desensitizing the output stage, the out-band beat notes do not feed through to the output stage. Since the input level must be somewhat greater when the output stage is made less sensitive, rejection of third harmonics or in-band harmonics (of lower frequency signals) is also improved.

## Chatter prevention*

Chatter occurs in the output stage when $C_3$ is relatively small, so that the lock transient and the AC components at the quadrature phase detector (lock detector) output cause the output stage to move through its threshold more than once. Many loads, for example lamps and relays, will not respond to the chatter.

---

* A method of obtaining clean digital output by eliminating chatter pulses from the output of a 567 tone decoder is explained in the design idea entitled 'One-shots tame tone decoder' (p.133).

However, logic may recognize the chatter as a series of outputs. By feeding the output stage back to its input (pin 1) the chatter can be eliminated. Three schemes for doing this are given in Figure 7.2.2. All operate by feeding the first output step (either on or off) back to the input, pushing the input past the threshold until the transient conditions are over. It is only necessary to ensure that the feedback time constant is not so large as to prevent operation at the highest anticipated speed. Although chatter can always be eliminated by making $C_3$ large, the feedback circuit will enable faster operation of the 567 by allowing $C_3$ to be kept small.

*Courtesy of Philips, The Netherlands*

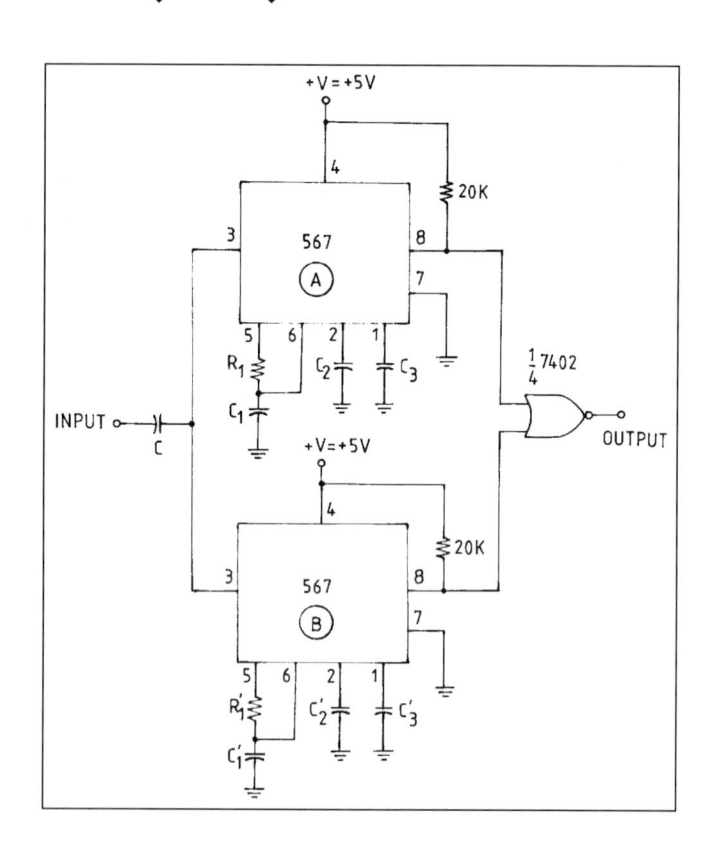

◆ ◆ ◆

## 7.3 Dual-tone decoder

**Figure 7.3.1** Dual-tone decoder

Two 567 tone decoders connected as shown in Figure 7.3.1 permit decoding of simultaneous or sequential tones. Both units must be turned ON before an output is given. $R_1 C_1$ and $R'_1 C'_1$ are chosen respectively for tones 1 and 2. If sequential tones (tone 1 followed by tone 2) are to be decoded, then $C_3$ is made very large to delay turn OFF of unit A until unit B has turned ON and the NOR gate is activated. Note that the wrong sequence (tone 2 followed by tone 1) will not provide an output since unit B will turn OFF before unit A comes ON. Figure 7.3.2

Phase locked loops 131

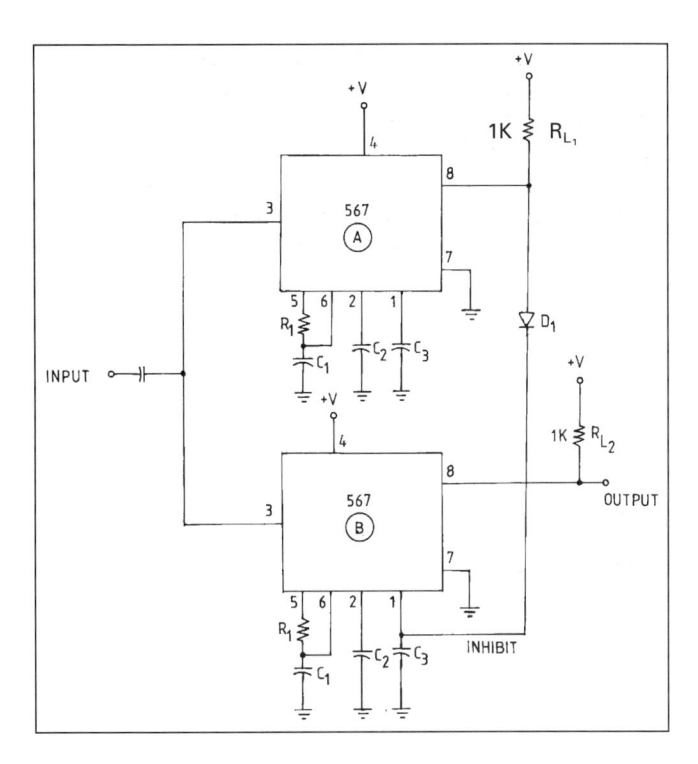

**Figure 7.3.2** Dual-tone decoder

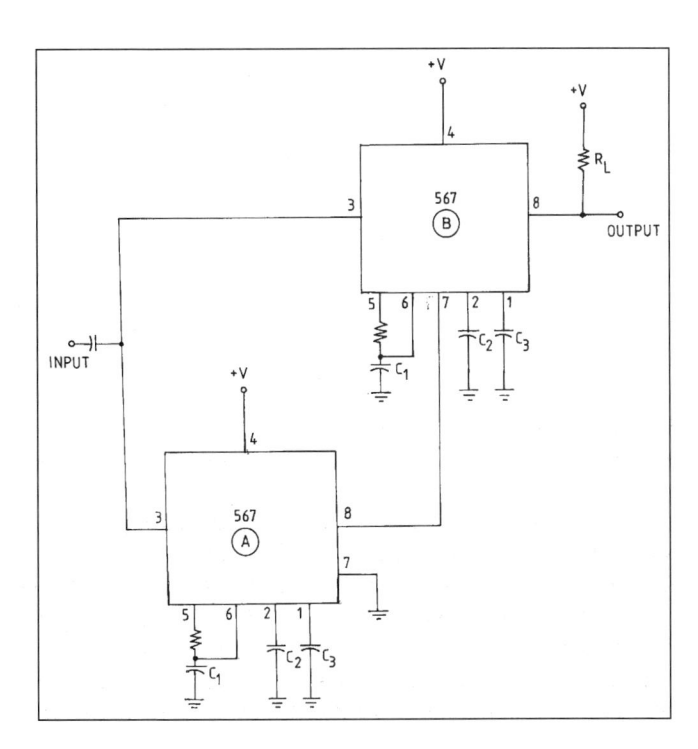

**Figure 7.3.3** Dual-tone decoder

shows a circuit variation which eliminates the NOR gate. The output is taken from unit B, but the unit B output stage is biased OFF by $R_{L_1}$ and $D_1$ until activated by tone 1. A further variation is given in Figure 7.3.3. Here, unit B is turned ON by the unit A output when tone 1 appears, reducing standby power to half. Thus, when unit B is ON, tone 1 is or was present. If tone 2 is now present, unit B comes ON also and an output is given. Since a transient output pulse may appear during unit A turn ON, even if tone 2 is not present, the load must be slow in response to avoid a false output due to tone 1 alone.

*Courtesy of Philips, The Netherlands*

## 7.4 Go/no-go frequency meter

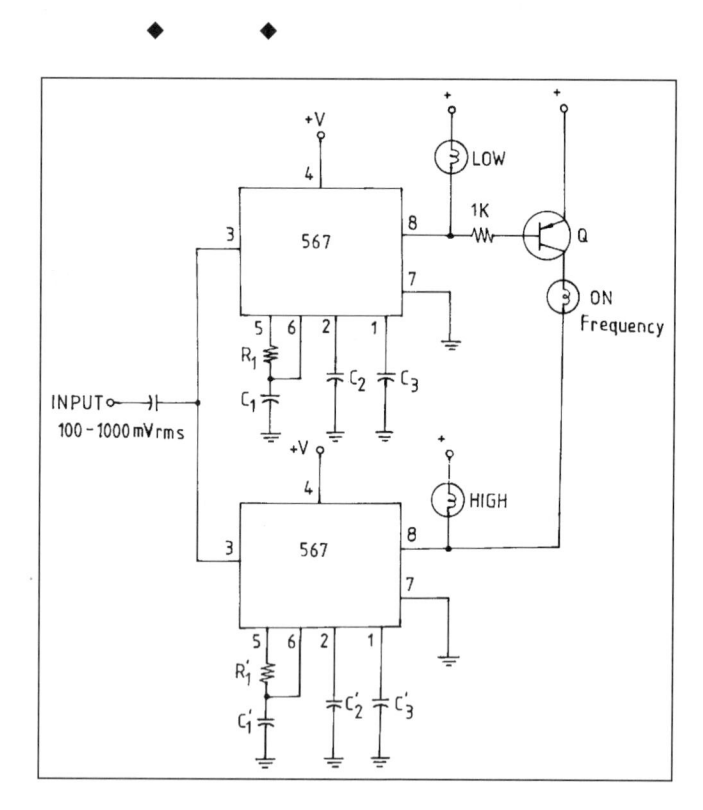

**Figure 7.4.1** Low-cost frequency indicator

A pair of 567 tone decoders (or a dual tone decoder) set up with overlapping detection bands can be used for a go/no-go frequency meter (Figure 7.4.1). One of the tone decoders is set 6% above the desired sensing frequency and the other is set 6% below the desired frequency. Now, if the incoming frequency is within 13% of the desired frequency, either one of the tone decoders will give an output. If both tone decoders are ON, it means that the incoming frequency is within 1% of the desired frequency. Three light bulbs and a transistor allow low cost read-out. The circuit can, therefore, be used as a low-cost frequency indicator.

*Courtesy of Philips, The Netherlands*

## 7.5 High-speed, narrow-band tone decoder

The circuit shown in Figure 7.3.1 of the design idea 'Dual-tone decoder' may be used to obtain a fast, narrow-band tone decoder. The detection bandwidth is achieved by overlapping the detection bands of the two tone decoders. Thus, only a tone within the overlap portion will result in an output. The input amplitude should be greater than 70 mV RMS at all times to prevent detection band shrinkage and $C_2$ should be between $130/f_o$ and $1300/f_o$ µF where $f_o$ is the nominal detection frequency. The small value of $C_2$ allows operation at the maximum speed so that worst-case output delay is only about 14 cycles.

*Courtesy of Philips, The Netherlands*

◆ ◆

## 7.6 One-shots tame tone decoder

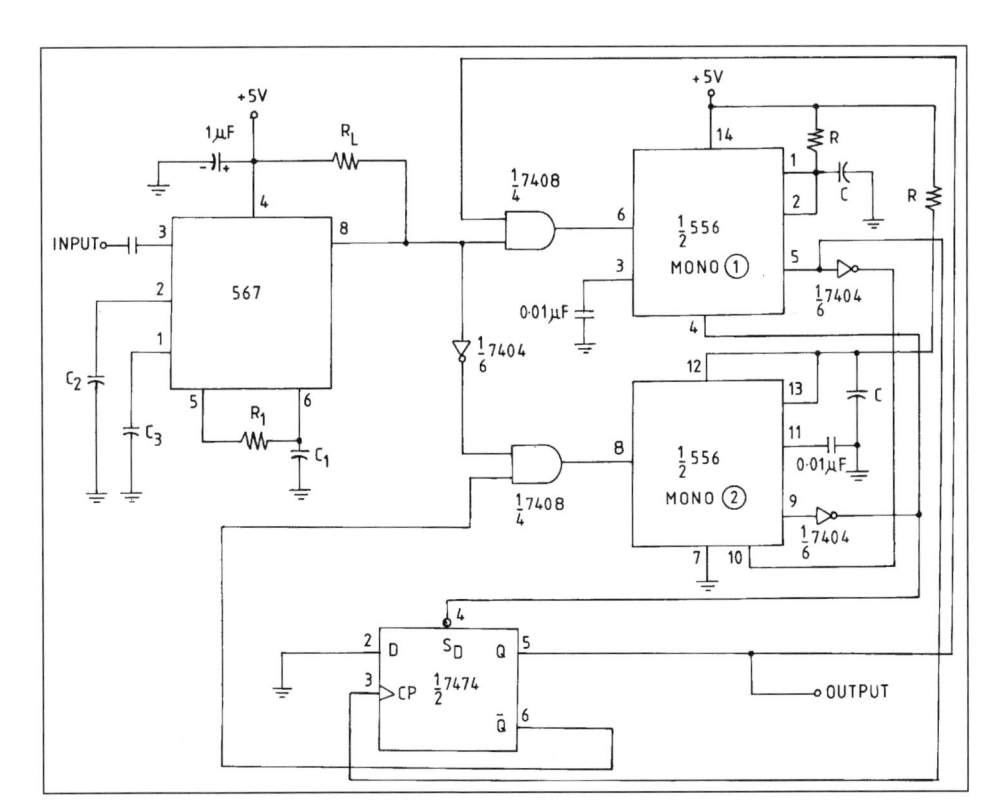

**Figure 7.6.1** One-shots tame tone decoder

Adding a pair of one-shots to the output of a 567 tone decoder renders it less sensitive to out-of-band signals and noise. Without the one-shots, the 567 is prone to spurious output chatter. Other protection schemes, such as feeding back outputs or using an input filter, do not work as well as the one-shots.

In Figure 7.6.1 circuit, the output of the 567 is HIGH in the absence of a tone and becomes LOW when it detects a tone. The tone decoder triggers a one-shot via an AND gate. The one-shot's period is set to slightly less than the duration of a tone burst. When the output of the tone decoder goes LOW, it triggers the second one-shot. The second one-shot's period is set to slightly less than the interval between tone bursts. The flip-flop enables and disables the inputs to one-shots such that spurious outputs from the tone decoder do not affect the output (Figure 7.6.2).

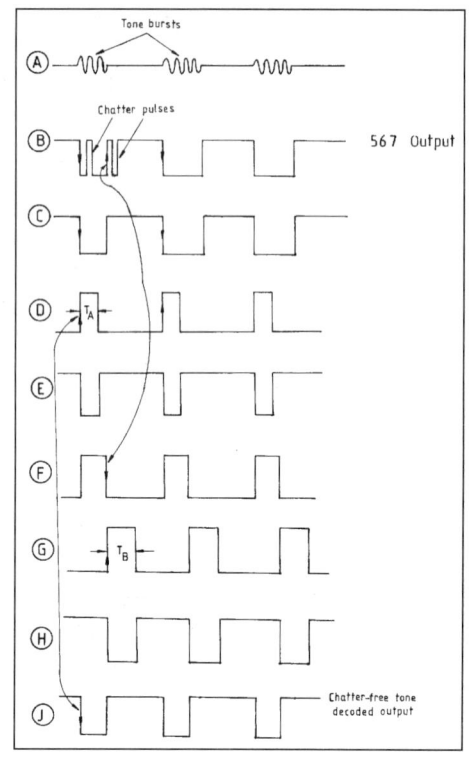

**Figure 7.6.2** Waveforms at different points in Figure 7.6.1

*Reprinted from EDN (September 29, 1988)*
*© 1992 CAHNERS PUBLISHING COMPANY*
*A Division of Reed Publishing USA*

## 7.7 PLL lock indicator

A D-type flip-flop and an op-amp can be used to detect the lock condition of a PLL, such as the XR-215 shown in Figure 7.7.1. This PLL consists of a balanced phase comparator, a highly stable VCO and a high-speed op-amp. It can operate with supplies from 5 to 26 V and frequencies from 0.5 Hz to 35 MHz. analog signals can be accommodated from 300 µV to 3 V and it can interface with DTL, ECL and TTL circuits. Tracking range is adjustable from +1 to 50% and the SNR is 65 dB.

Consider quadrature detection. In this condition, lock can be obtained within 90° phase difference between the input and output. A D-type flip-flop acts as a

Phase locked loops   135

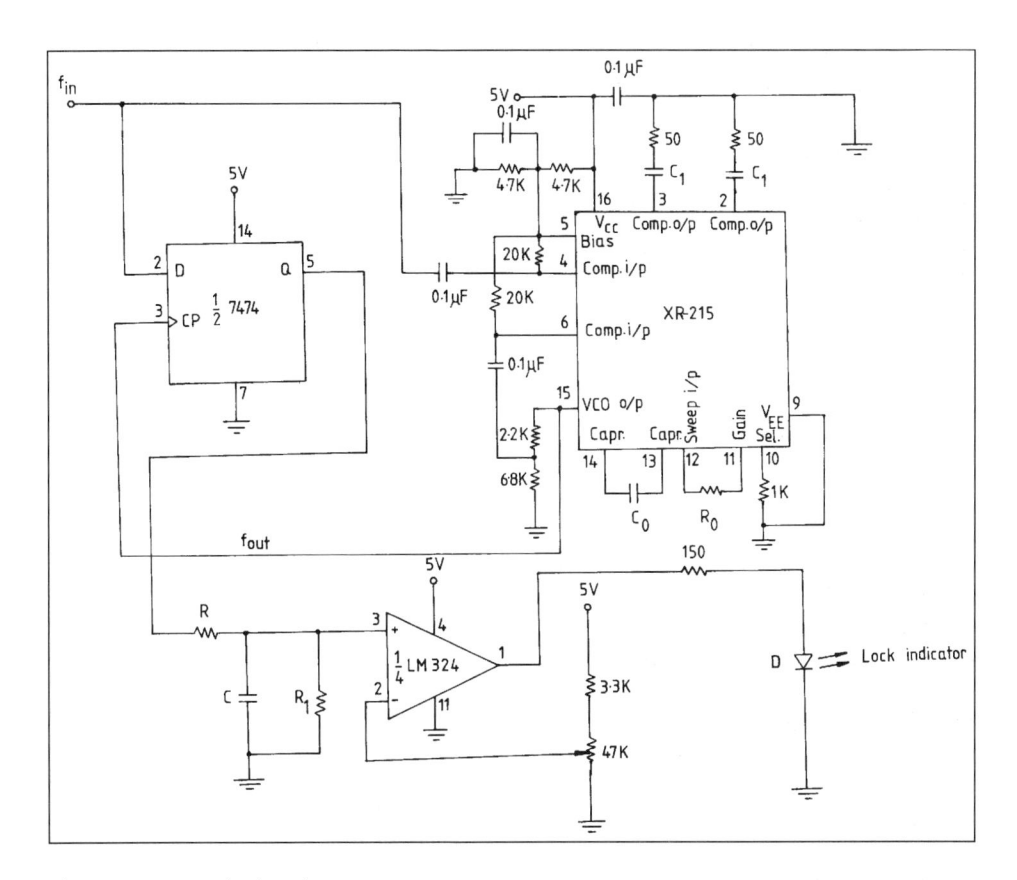

**Figure 7.7.1** PLL lock indicator

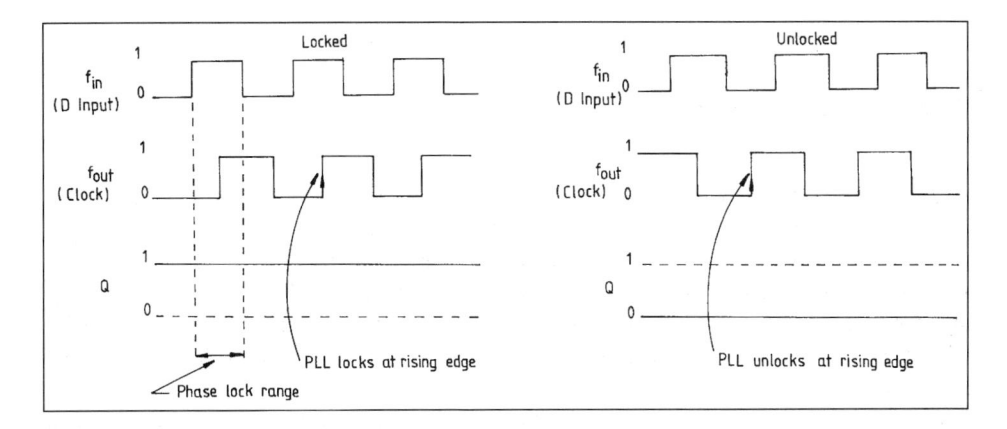

**Figure 7.7.2** Waveforms for circuit in Figure 7.7.1

phase comparator by clocking the state of the input frequency at the rising edge of the output waveform.

A steady state of 1 at the flip-flop indicates a lock condition, which will cause the integrator ($R$, $C$ and $R_1$) to make the op-amp output rise towards 5 V supply voltage, illuminating the LED.

If there is more than a quadrature phase difference between the input and output frequencies, the output of the flip-flop will be a train of pulses of different widths, indicating out-of-lock condition. Here, the output of the op-amp switches low and the LED is off (Figure 7.7.2).

*First published in Electronics and Wireless World, April 1990.*
*Reprinted with permission*

◆ ◆

## 7.8 Frequency multiplication using PLL 565

**Figure 7.8.1** Block diagram of PLL frequency multiplier

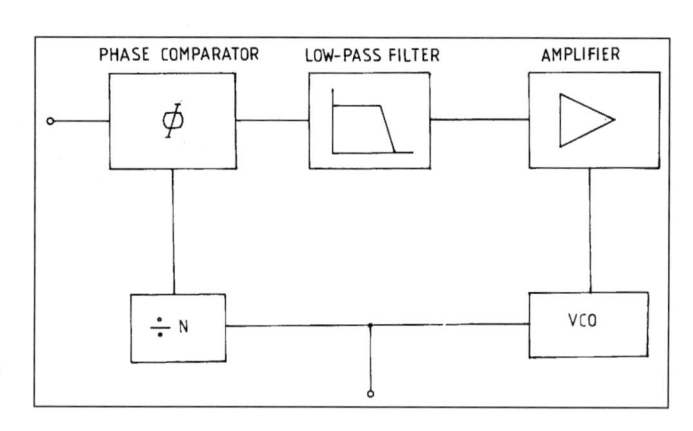

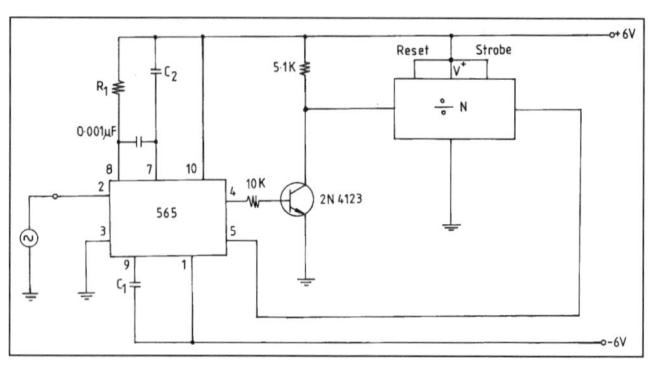

**Figure 7.8.2** PLL frequency multiplier

Multiplication of frequency of a signal can be achieved using PLL 565 by two methods:

1. Locking to a harmonic of the input signal.
2. Inclusion of a digital frequency divider or counter in the loop between the VCO and phase comparator.

Phase locked loops 137

The first method is the simplest and can be achieved by setting the free-running frequency of the VCO to a multiple of the input frequency. A limitation of this scheme is that the lock range decreases as successively higher and weaker harmonics are used for locking. If the input frequency is to be constant with little tracking required, the loop can generally be locked to any one of the first five harmonics. For higher orders of multiplication, or for cases where a large lock range is desired, the second scheme is more desirable. An example of this might be a case where the input signal varies over a wide frequency range and a large multiple of the input frequency is required.

A block diagram of the second scheme is shown in Figure 7.8.1. Here the loop is broken between the VCO and the phase comparator and a frequency divider is inserted. The fundamental of the divided VCO frequency is locked to the input frequency in this case, so that the VCO is actually at a multiple of the input frequency. The amount of multiplication is determined by the frequency of the divider. A typical connection scheme is shown in Figure 7.8.2. To set up the circuit, the frequency limits of the input signal must be determined. The free-running frequency of the VCO is then adjusted by means of $R_1$ and $C_1$ so that the output frequency of the divider is mid-way between the input frequency limits. The filter capacitor $C_2$ should be large enough to eliminate variations in the demodulated output voltage (at pin 7), in order to stabilize the VCO frequency. The output can now be taken as the VCO square wave output, and its fundamental frequency will be the desired multiple of the input frequency ($f_{IN}$) as long as the loop is in lock.

*Courtesy of Philips, The Netherlands*

## 7.9 Reduce distortion in mod–demod circuit

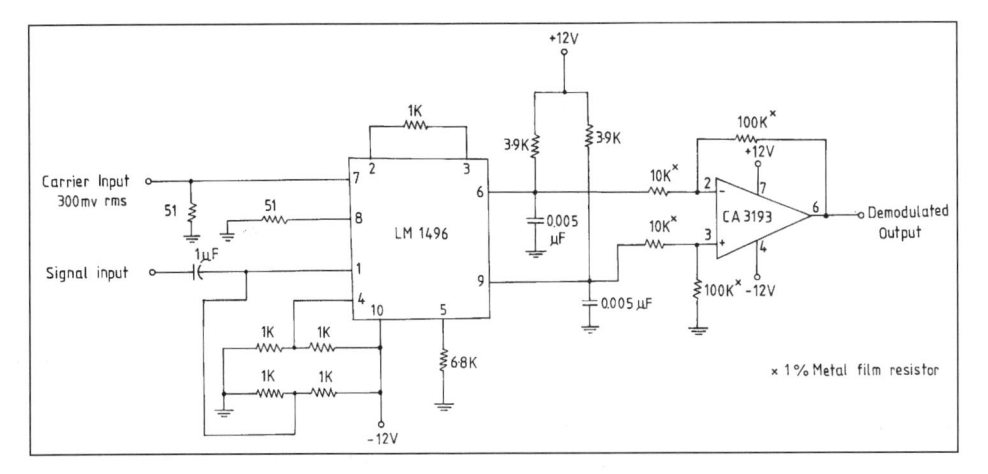

**Figure 7.9.1** Improved demodulator circuit reduces distortion

The LM 1496, a balanced modulator–demodulator, produces an output voltage proportional to the product of a signal, $f_s$, and a carrier, $f_c$, input. Unfortunately, the suppression of the carrier signal by the LM 1496 at its output is poor even within its specified band of operation, especially at high frequencies. Also, aliasing frequencies of even multiples of $f_c$ plus or minus even multiples of $f_s$ ($2f_c \pm 2f_s$) cause serious in-band spurious interference.

A CA 3193 BiCMOS op-amp, configured as a balanced differential amplifier at the output of the LM 1496, can eliminate these aliasing frequencies (see Figure 7.9.1). The circuit attenuates the aliasing products by more than 60 dB, thus preventing substantial distortion at the output. The balanced amplifier eliminates the need for a symmetrical carrier and highly symmetrical switching in the LM 1496 to suppress the aliasing frequencies. While the circuit could also use a balanced transformer instead, the balanced CA 3193 op-amp is the more cost-effective solution.

*Reprinted with permission from Electronic Design (Vol.37, No.17) August 10, 1989. Copyright 1989, Penton Publishing Inc.*

## 7.10 Phase modulator

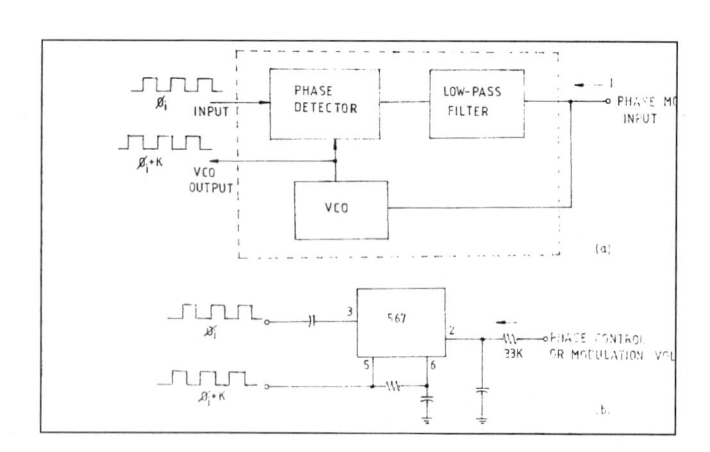

**Figure 7.10.1** Phase modulator using PLL 567

If a phase-locked loop is locked on to a signal at the free-running frequency, the phase of the VCO will be 90° with respect to the input signal. If a current is injected into the VCO terminal (the low-pass filter output), the phase will shift sufficiently to develop an opposing average current out of the phase comparator so that the VCO voltage is constant and lock is maintained. When the input signal amplitude is low enough so that the loop frequency is limited by the phase comparator output rather than the VCO swing, the phase can be modulated over the full range of 0 to 180°. If the input signal is a square wave, the phase will be a linear function of the injected current.

A block diagram of the phase modulator is given in Figure 7.10.1(a). The conversion factor K is a function of which loop is used, as well as the input square wave amplitude. Figure 7.10.1(b) shows an implementation of this circuit using the 567 tone decoder.

*Courtesy of Philips, The Netherlands*

## 7.11 Phase detector

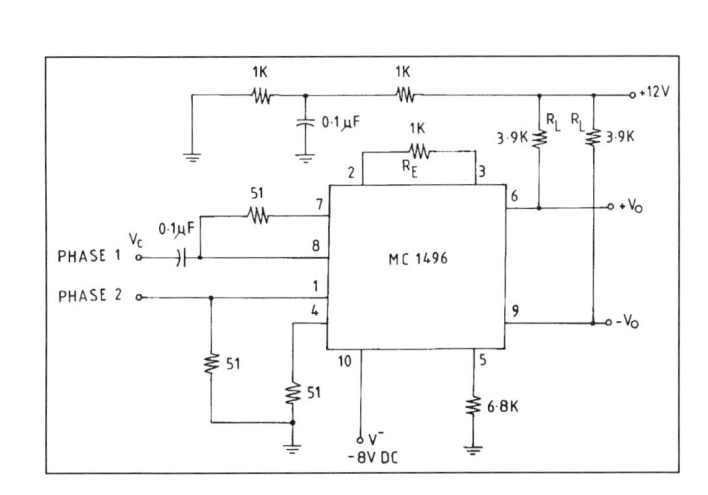

**Figure 7.11.1** Phase comparator

The MC 1496 is a monolithic transistor array arranged as a balanced modulator–demodulator. The device takes advantage of the excellent matching qualities of monolithic devices to provide superior carrier and signal rejection. Carrier suppressions of 50 dB at 10 MHz are typical with no external balancing networks required.

The versatility of the balanced modulator or multiplier also allows the device to be used as a phase detector. The output of the detector contains a term related to the cosine of the phase angle. Two signals of equal frequency are applied to the inputs as shown in Figure 7.11.1. The frequencies are multiplied together producing sum and difference frequencies. Equal frequencies cause the difference component to become DC while the undesired sum component is filtered out. The DC component is related to the phase angle. At 90° the cosine becomes zero, while being at maximum positive or maximum negative at 0° and 180° respectively.

The advantage of using the balanced modulator over other types of phase comparators is the excellent linearity of conversion. This configuration also provides a conversion gain rather than a loss for greater resolution. Used in conjunction with a phase locked loop, for instance, the balanced modulator provides a very low distortion FM demodulator.

Figure 7.11.1 shows the connections for a phase comparator using MC 1496.

*Courtesy of Philips, The Netherlands*

140  Phase locked loops

## 7.12 Frequency doubler

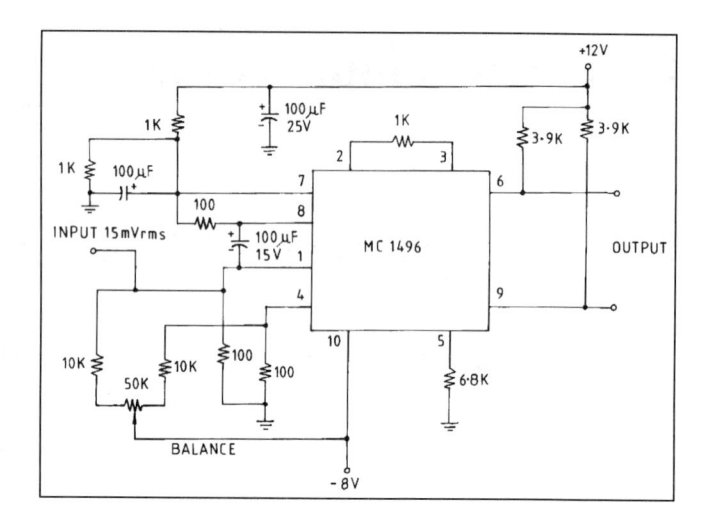

**Figure 7.12.1** Low frequency doubler

Very similar to the phase detector of Figure 7.11.1, a frequency doubler schematic is shown in Figure 7.12.1. Departure from Figure 7.11.1 is primarily the removal of the low-pass filter. The output then contains the sum component which is twice the frequency of the input, since both input signals are of the same frequency.

*Courtesy of Philips, The Netherlands*

# 8 Power supply circuits

## 8.1 Transformer-isolated 5 V power supply

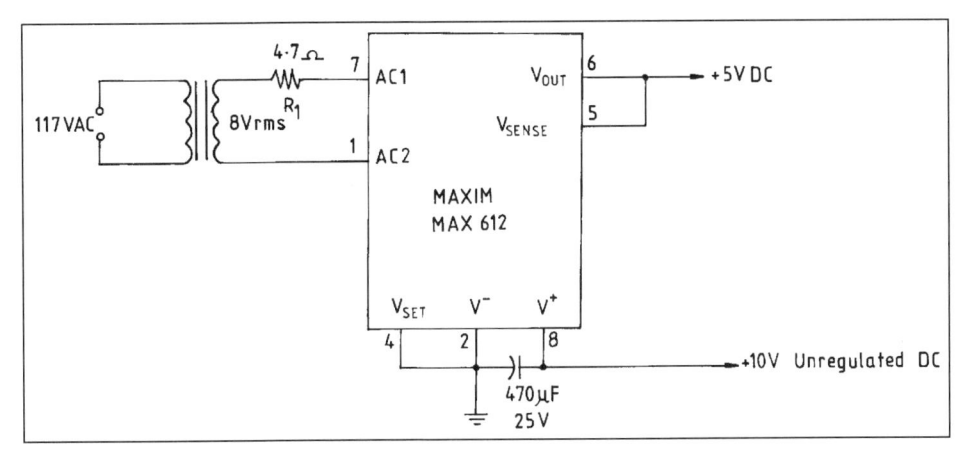

**Figure 8.1.1** Transformer-isolated 5 V power supply

The MAX 612 from MAXIM Integrated Products is a linear voltage regulator which can accept an input voltage in the range 6 V to 9 V AC and can give an output of 5 V DC (fixed) or 1.3 V to 8 V DC. It can be used as shown in the circuit of Figure 8.1.1 to derive 5 V DC isolated from the AC mains supply. The AC input voltage can range from 80 V RMS to 160 V RMS with the 8 V RMS nominal transformer voltage shown. The power dissipation of the MAX 612 is approximately $(V_{IN(peak)} - V_{OUT}) \times I_{LOAD}$. With the 8 V RMS transformer shown, the power dissipated in the MAX 612 limits the maximum output current to 100 mA at 25°C ambient and 30 mA at 70°C. If the 8 V transformer is replaced by a 6.3 V transformer, the maximum output current increases to 150 mA at 25°C but the minimum input voltage to maintain output voltage regulation is increased to 100 V RMS. When using a 6.3 V RMS transformer, the filter capacitor connected to $V^+$ must be increased to 2200 μF to ensure that the minimum voltage at $V^+$ is greater than 6 V throughout each line cycle. Resistor $R_1$ limits the peak input current, but is not needed if the transformer impedance limits the peak current to a suitable value. As a rule of thumb, $R_1$ can be omitted if the short-circuit output current of the transformer is less than 2 A. The typical quiescent current of MAX 612 is 70 μA and the maximum quiescent current is 150 μA.

*Courtesy of Maxim Integrated Products, Sunnyvale, CA*

◆ ◆

## 8.2 Uninterruptible +5 V supply

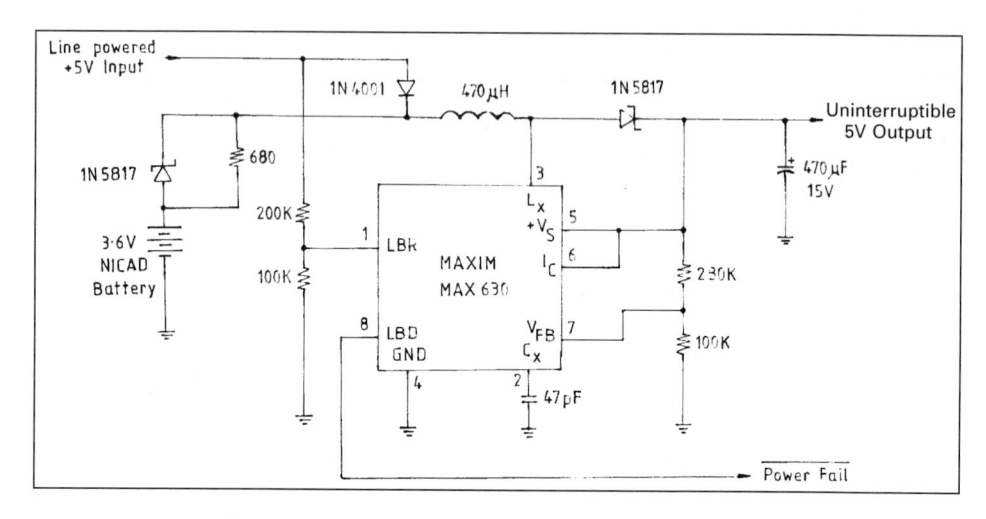

**Figure 8.2.1** Uninterruptible +5 V supply

The circuit shown in Figure 8.2.1, using the MAX 630 micropower step-up switching regulator provides a continuous supply of +5 V, with automatic switch-over between line power and battery backup. When the line powered input voltage is at +5 V, it provides 4.4 V to the MAX 630 and trickle charges the battery. If the line powered input falls below the battery voltage, the 3.6 V battery supplies power to the MAX 630, which boosts the battery voltage up to +5 V, thus maintaining a continuous supply to the uninterruptible +5 V bus. Since the +5 V output is always supplied through the MAX 630, there are no power spikes or glitches during power transfer.

The MAX 630's low battery detector monitors the line powered +5 V, and the LBD output can be used to shut down unnecessary sections of the system during power failures. Alternatively, the low battery detector could monitor the Nicad battery voltage and provide warning of power loss when the battery is nearly discharged.

Unlike battery backup systems that use 9 V batteries, this circuit does not need +12 V or +15 V to recharge the battery. Consequently, it can be used to provide +5 V backup on modules or circuit cards which only have 5 V available.

*Courtesy of Maxim Integrated Products, Inc., Sunnyvale, CA*

## 8.3 +5 V to ±10 V voltage converter

**Figure 8.3.1** (a) Positive charge pump; (b) negative charge pump

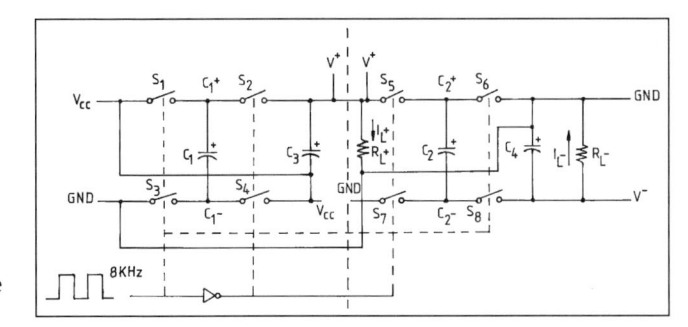

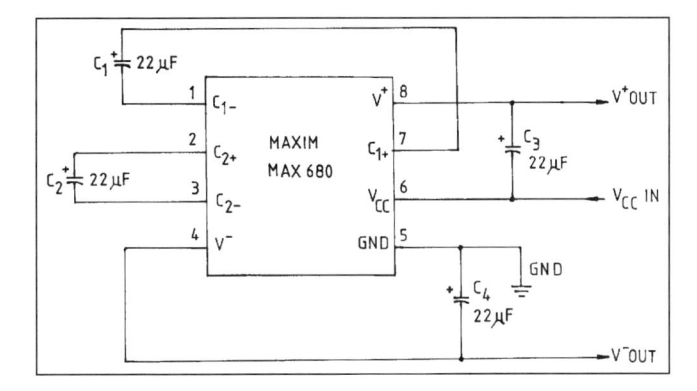

**Figure 8.3.2** Positive and negative converter

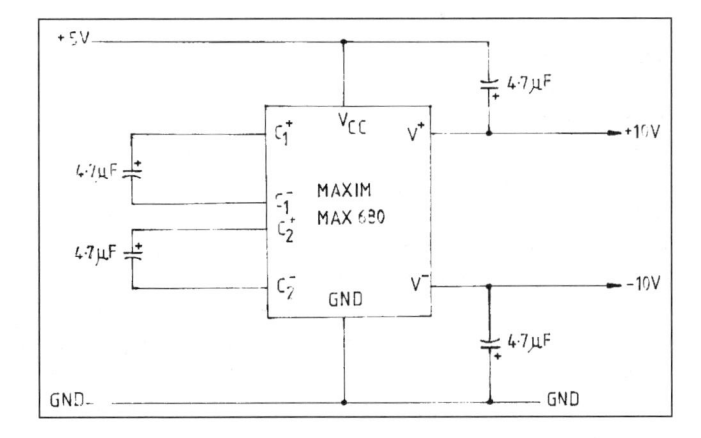

**Figure 8.3.3** +5 V to ±10 V converter

The MAX 680 is a monolithic CMOS dual charge pump voltage converter that provides ±10 V outputs from an input voltage of +5 V, using four external capacitors, which could be inexpensive electrolytic capacitors with values in the range of 0.1 µF to 100 µF. The MAX 680 provides both a positive step-up charge pump to develop +10 V from the +5 V input and an inverting charge pump to generate the −10 V output. The MAX 680 includes an on-chip 8 kHz oscillator and all the necessary circuitry (except the four capacitors) to produce both positive and negative voltages from a single positive supply.

The output source impedances are typically 150 ohms, providing useful output currents up to 10 mA. The low quiescent current and high efficiency make this device suitable for a variety of applications that need both positive and negative voltages generated from a single supply.

## Description of device operation

Figure 8.3.1(a) illustrates the idealized operation of the positive voltage converter. The on-chip oscillator generates a 50% duty cycle clock signal. During the first half of the cycle, switches $S_2$ and $S_4$ are open, switches $S_1$ and $S_3$ are closed, and the capacitor is charged to the input voltage $V_{CC}$. During the second half cycle, switches $S_1$ and $S_3$ are open, $S_2$ and $S_4$ are closed, and the capacitor $C_1$ is translated upwards by $V_{CC}$ volts. Assuming ideal switches and no load on $C_3$, charge is transferred onto $C_3$ from $C_1$ such that the voltage on $C_3$ will be $2V_{CC}$, generating the positive supply.

Figure 8.3.1(b) illustrates the negative converter. The switches of the negative converter are out of phase from the positive converter. During the second half of the clock cycle, $S_6$ and $S_8$ are open, $S_5$ and $S_7$ are closed, thus charging $C_2$ from $V^+$ (pumped up to $2V_{CC}$ by the positive charge pump) to GND. In the first half of the clock cycle, $S_5$ and $S_7$ are open, $S_6$ and $S_8$ are closed, and the charge on $C_2$ is transferred to $C_4$, generating the negative supply. The eight switches are CMOS power MOSFETs. Swtiches $S_1$, $S_2$, $S_4$ and $S_5$ are P-channel devices while switches $S_3$, $S_6$, $S_7$ and $S_8$ are N-channel devices.

Larger values of reservoir capacitors $C_3$ and $C_4$ will reduce output ripple. There will be a substantial voltage difference between $(V^+ - V_{CC})$ and $V_{CC}$ for the positive pump and between $V^+$ and $V^-$ if the impedances of the pump capacitors $C_1$ and $C_2$ are high with respect to their respective output loads. The MAX 680 has on-chip Zener diodes that clamp $V_{CC}$ to approximately 6.2 V, $V^+$ to 12.4 V and $V^-$ to −12.4 V.

A simple dual charge pump voltage converter which provides positive and negative outputs of two times a positive input voltage is shown in Figure 8.3.2. Capacitors $C_1$ and $C_3$ are for the positive pump and $C_2$ and $C_4$ are for the negative pump. In most applications all four capacitors are low-cost 10 µF or 22 µF polarized electrolytics. For applications where PC board space is at a premium and low currents are being drawn from the MAX 680, 1 µF reservoir capacitors may be used for the pump capacitors $C_1$ and $C_2$, with 4.7 µF reservoir capacitors $C_3$ and $C_4$. Capacitors $C_1$ and $C_3$ must be rated at 6 V or higher, and capacitors $C_2$ and $C_4$ must be rated at 12 V or higher.

The MAX 680 is not a voltage regulator; the output source resistance of either charge pump is approximately 150 ohms at room temperature with $V_{CC}$ at 5 V. This means that with an input $V_{CC}$ of 5 V, under light load $V^+$ will approach + 10 V and $V^-$ will be at −10 V, but both $V^+$ and $V^-$ will droop towards GND as the current drawn from either $V^+$ or $V^-$ increases, since the negative converter draws its power from the output of the positive converter. To predict the output voltages, treat the chip as two separate converters and analyse them separately.

First, the droop of the negative supply ($V_{DROP^-}$) equals the current drawn from $V^-$ – ($I_{L^-}$) times the source resistance of the negative converter ($R_{S^-}$):

$$V_{DROP^-} = I_L \times R_{S^-}$$

Likewise, the droop of the positive supply ($V_{DROP^+}$) equals the current drawn from the positive supply ($I_{L^+}$) times the source resistance of the positive converter ($R_{S^+}$), except that the current drawn from the positive supply is the sum of the current drawn by the load on the positive supply ($I_{L^+}$) plus the current drawn by the negative converter ($I_{L^-}$):

$$V_{DROP^+} = I_{L^+} \times R_{S^+} = (I_{L^+} + I_{L^-}) \times R_{S^+}$$

The positive output voltage will be

$$V^+ = 2V_{CC} - V_{DROP^+}$$

The negative output voltage will be

$$V^- = -(V^+ - V_{DROP^-})$$
$$= -(2V_{CC} - V_{DROP^+} - V_{DROP^-})$$

The positive and negative charge pumps are tested and specified separately to provide the separate values of output source resistance for use in the above formulas. When the positive charge pump is tested, the negative charge pump is unloaded, and when the negative charge pump is tested, the positive supply $V^+$ is from an external source, isolating the negative charge pump.

The ripple voltage on either output can be calculated by noting that the current drawn from the output is supplied by the reservoir capacitor alone during one half-cycle of the clock. This results in a ripple given by

$$V_{RIPPLE} = (1/2)(I_{OUT})(1/f_{PUMP})(1/CR)$$

For the nominal value $f_{PUMP}$ of 8 kHz with 10 µF reservoir capacitors, the ripple will be 30 mV with $I_{OUT}$ at 5 mA. Remember that in most applications, the $I_{OUT}$ of the positive charge pump is the load current PLUS the current taken by the negative charge pump.

Figure 8.3.3 shows a +5 V to ±10 V converter.

*Courtesy of Maxim Integrated Products, Sunnyvale, CA*

◆ ◆

## 8.4 ± 5 V supply from a single 9 V battery

The circuit in Figure 8.4.1 shows a complete ±5 V power supply using a 9 V battery. The ICL 7660 inverts the +9 V input voltage into −9 V which is then regulated by the ICL 7664 negative regulator to a constant −5 V output. The ICL

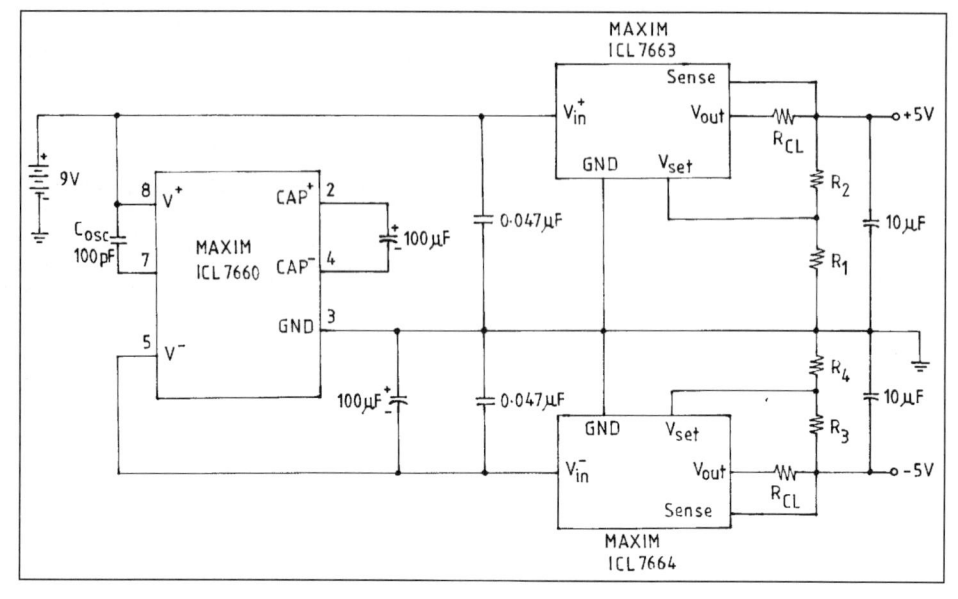

**Figure 8.4.1** Regulated ±5 V using a 9 V battery

7663 positive voltage regulator uses the +9 V input directly to generate a regulated +5 V output. The combined quiescent current of the Maxim ICL 7660 and the two regulators is less than 100 µA, while the output current capability is 40 mA.

*Courtesy of Maxim Integrated Products, Inc. Sunnyvale, CA)*

◆  ◆  ◆

## 8.5 +3 V battery to +5 V DC–DC converter

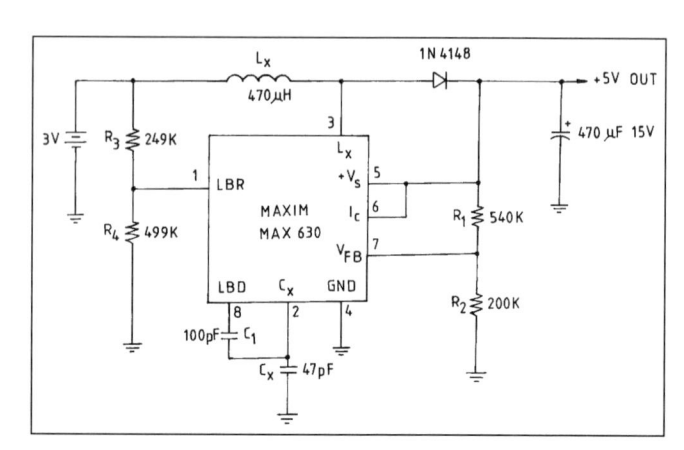

**Figure 8.5.1** +3 V battery to +5 V DC–DC converter

A common power supply requirement involves conversion of a 2.4 or 3 V battery voltage to a 5 V logic supply. The circuit shown in Figure 8.5.1 converts 3 V to 5 V at 40 mA with 85% efficiency. The circuit uses the micropower step-up switching regulator MAX 630 which requires an operating current of only 70 µA nearly independent of output switch current or duty cycle. In Figure 8.5.1, when $I_C$ (pin 6) is driven low, the output voltage will be the battery voltage minus the drop across diode $D_1$.

The optional circuitry using $C_1$, $R_3$ and $R_4$ lowers the oscillator frequency when the battery voltage falls to 2 V. This lower frequency maintains the output power capability of the circuit by increasing the peak inductor current, compensating for the reduced battery voltage.

*Courtesy of Maxim Integrated Products, Inc., Sunnyvale, CA*

## 8.6 Low-cost DC voltage booster

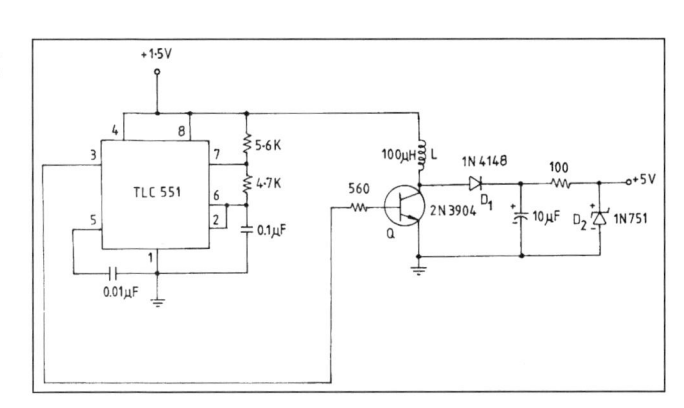

**Figure 8.6.1** DC voltage booster

You can derive 5 V DC from a single 1.5 V cell using the circuit shown in Figure 8.6.1. This will be useful in many applications where 5 V is required for a few components (such as logic ICs) and a single cell operation is sufficient for other parts of the circuit.

The circuit uses the LinCMOS timer TLC 551 (pin compatible with 555 timer) which can operate even at 1 V, consuming less than 1 mW. The TLC 551 is used as an astable operating at a frequency of about 100 kHz. The output of the 551 timer drives transistor Q ON and OFF at this high frequency. A 100 µH coil (L) is connected to the collector of the transistor. The switching ON and OFF of the transistor causes voltage spikes of about four times the supply voltage (4 × 1.5 V) to appear at the collector of Q because of the inductive load L. These voltage spikes are rectified by diode $D_1$ and filtered by the 10 µF capacitor. A Zener diode ($D_2$) is used to clamp the output voltage to 5 V. Thus the circuit affords a cost-effective method of deriving 5 V from 1.5 V. This circuit technique can also be used for up-converting any other DC voltage to a higher value by a suitable choice of components.

## 8.7 DC voltage splitter

**Figure 8.7.1** Supply voltage splitter

$V_{OUT} = \dfrac{V^+ - V^-}{2}$

The monolithic voltage converter ICL 7660 can be used to split a DC voltage $V$ into two halves – a positive voltage of $V/2$ and a negative voltage of $V/2$ as shown in Figure 8.7.1. The combined load will be evenly shared between the positive and negative sides. Because the switches share the load in parallel, the output impedance is much lower than in the standard circuits, and higher currents can be drawn from the device. By using this circuit, for example, a 15 V single supply can be split into + 7.5 V and −7.5 V supplies.

*Courtesy of Harris Semiconductor, Melbourne, FL*

## 8.8 Supply voltage splitter

Dual supply op-amps and comparators can be operated from a single supply by creating an artificial ground at half the supply voltage. The supply splitter shown in Figure 8.8.1 can source or sink 150 mA. The output capacitor $C_2$ can be made as large as necessary to absorb current transients. An input capacitor is also used on the buffer to avoid high frequency instability that can be caused by high source impedance.

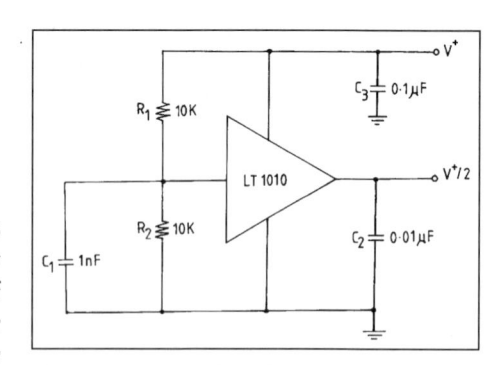

**Figure 8.8.1** Supply voltage splitter

*Courtesy of Linear Technology Corporation, Milpitas, CA*

## 8.9 Step-up negative converter

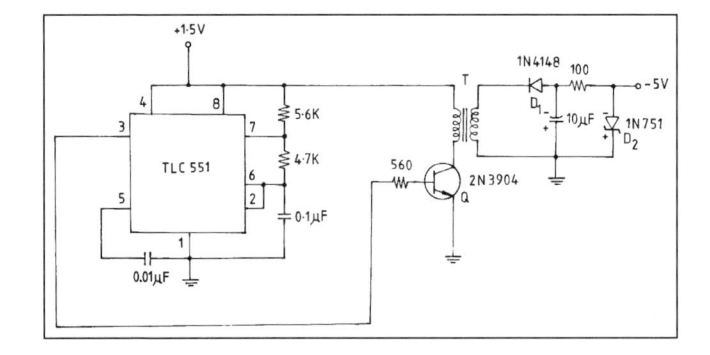

**Figure 8.9.1** Step-up negative converter

You can derive −5 V DC from a single 1.5 V cell using the circuit shown in Figure 8.9.1. This technique will be useful in applications where a dual supply is required for operating devices, such as op-amps, and only a single supply voltage is available.

The circuit uses the LinCMOS timer TLC 551 (pin compatible with 555 timer) which can operate even at 1 V, consuming less than 1 mW. The TLC 551 is used as an astable operating at a frequency of about 100 kHz. The output of the 551 timer drives transistor Q ON and OFF at this high frequency. The primary of a 100 µH pulse transformer (T) is connected to the collector of the transistor. The switching ON and OFF of the transistor causes voltage spikes of about four times the supply voltage (4 × 1.5 V) to appear at the collector of Q because of the inductive load, which is reflected at the secondary of the transformer. The voltage spikes at the secondary of the transformer are rectified by diode $D_1$ and filtered by the 10 µF capacitor. A zener diode ($D_2$) is used to clamp the output voltage to −5 V. The circuit affords a cost-effective method of deriving −5 V from 1.5 V. This circuit technique can also be used for deriving a higher negative voltage by up-converting any other DC voltage by a suitable choice of components.

## 8.10 Positive voltage doubler

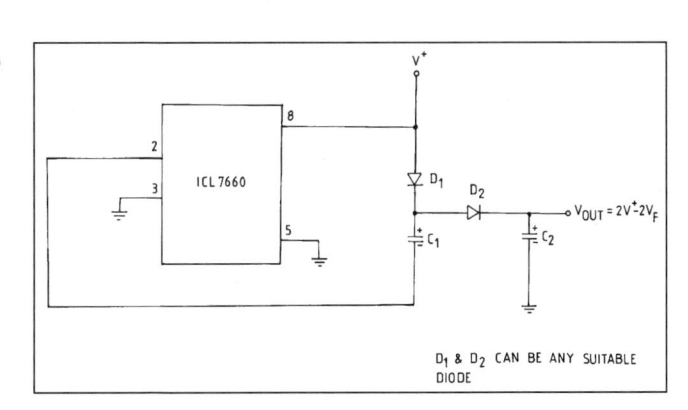

**Figure 8.10.1** Positive voltage doubler

The monolithic voltage converter ICL 7660 may be employed to achieve positive voltage doubling using the circuit shown in Figure 8.10.1. In this application, the pump inverter switches of the ICL 7660 are used to charge capacitor $C_1$ to a voltage level of $(V^+ - V_F)$, where $V^+$ is the supply voltage and $V_F$ is the forward voltage drop of diode $D_1$. On the transfer cycle, the voltage on $C_1$ plus the supply voltage $(V^+)$ is applied through diode $D_2$ to capacitor $C_2$. The voltage thus created on $C_2$ becomes $(2V^+) - (2V_F)$ or twice the supply voltage minus the combined forward voltage drops of diodes $D_1$ and $D_2$.

The source impedance of the output $(V_{OUT})$ will depend on the output current, but for $V^+ = 5\,V$, and an output current of $10\,mA$ it will be approximately 60 ohms.

*Courtesy of Harris Semiconductor, Melbourne, FL*

## 8.11 Circuit gives constant DC output with selectable AC input

**Figure 8.11.1**
220 V/110 V supply voltage selector

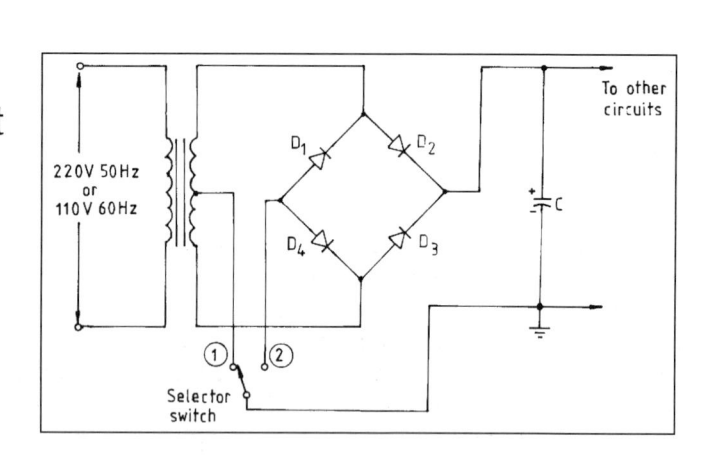

The circuit shown in Figure 8.11.1 can be used to derive a DC output voltage which is independent of the input AC supply which could be either 220 V or 110 V. The circuit consists of a centre-tapped transformer connected to a bridge rectifier and a capacitor filter. The ground terminal of the capacitor is connected to the selector contact of a SPDT switch. The centre-tap of the transformer and the common anode end of the diodes in the bridge are connected to the two positions of the switch. If the selector is in position 1, the circuit functions as a full-wave rectifier with a centre-tapped transformer and gives half the output of a full-wave bridge rectifier. If the selector is in position 2, the circuit functions as a normal full-wave bridge rectifier. Therefore, with the switch in position 1 the DC output voltage is half that with the switch in position 2. The filter capacitor should be chosen to withstand the higher of the two DC voltages. This technique is useful when dual voltages ($V$ and $V/2$) are required from a given AC supply voltage and for power supply design of equipment which should operate on 220 V as well as 110 V AC supplies.

## 8.12 Power-fail alarm

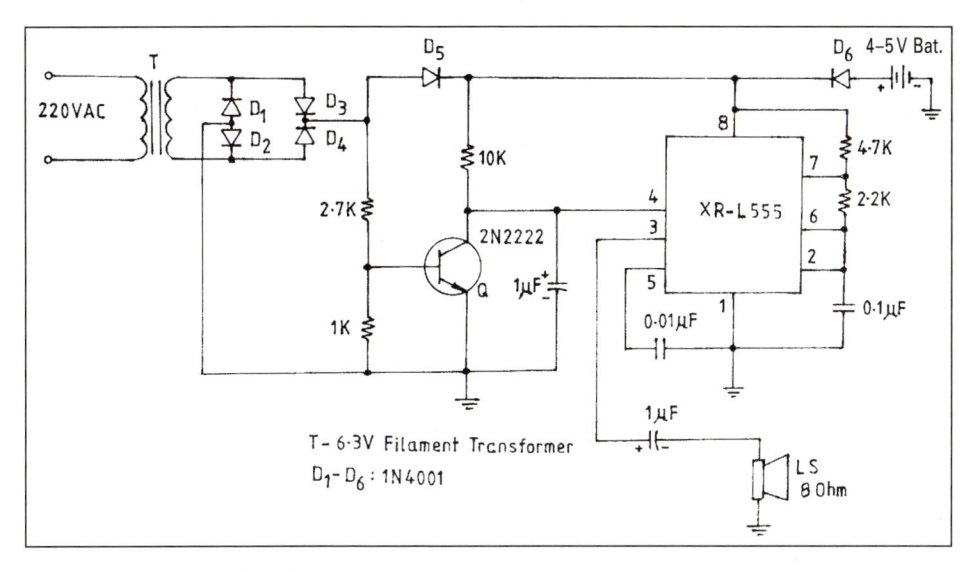

**Figure 8.12.1** Power-fail alarm

The circuit shown in Figure 8.12.1 can be used as a power-fail alarm giving an audible warning when the AC mains power fails. The circuit consists of a low-cost 6.3 V filament transformer connected to the AC mains supply, a transistor switch and a low-power timer – the XR-L555 whose output drives an 8 ohm loudspeaker. The secondary of the filament transformer feeds a full-wave rectifier. The output of the rectifier keeps the NPN transistor Q ON as long as the mains AC power is available. So long as Q is ON, the voltage across the 1 µF capacitor connected across the collector and emitter, i.e. $V_{CE}$, is equal to about 0.3 V and the Reset terminal (pin 4) of the timer is pulled LOW. This inhibits the 555 astable from oscillating. Since the full-wave rectifier output is greater than the battery voltage (4.5 V), the diode $D_6$ does not conduct and the timer is powered from the DC supply derived from the AC mains. The 555 timer is configured as an astable with a frequency given by

$$f = \{1.44/\{[4.7k + 2(2.2k)](0.1 \times 10^{-6})\}\} \text{ Hz}$$

which works out to about 1.6 kHz.

When the AC mains power fails, the diode $D_6$ becomes forward-biased and supplies power to the timer. The Reset terminal (pin 4) of the timer is now pulled up to the battery supply voltage through the 10k resistor since the transistor Q is now OFF (there being no bias current flowing into its base); the timer now oscillates and produces a 1.6 kHz output which activates the loudspeaker. Diode $D_5$ prevents the transistor Q from deriving base current drive from the battery through the 2.7k resistor when the mains supply fails. The advantage of the XR-

L555 timer is that it can operate at 1/15 the power of a normal 555 timer at a voltage as low as 2.7 V and it has a typical power dissipation of 1 mW at 5 V; it can operate for 1500 hours with two 300 mAh batteries and its output can source up to 100 mA current which is sufficient to drive a loudspeaker.

This circuit affords a low-cost method for obtaining audible power fail indication.

## 8.13 AC power fail and brownout detector

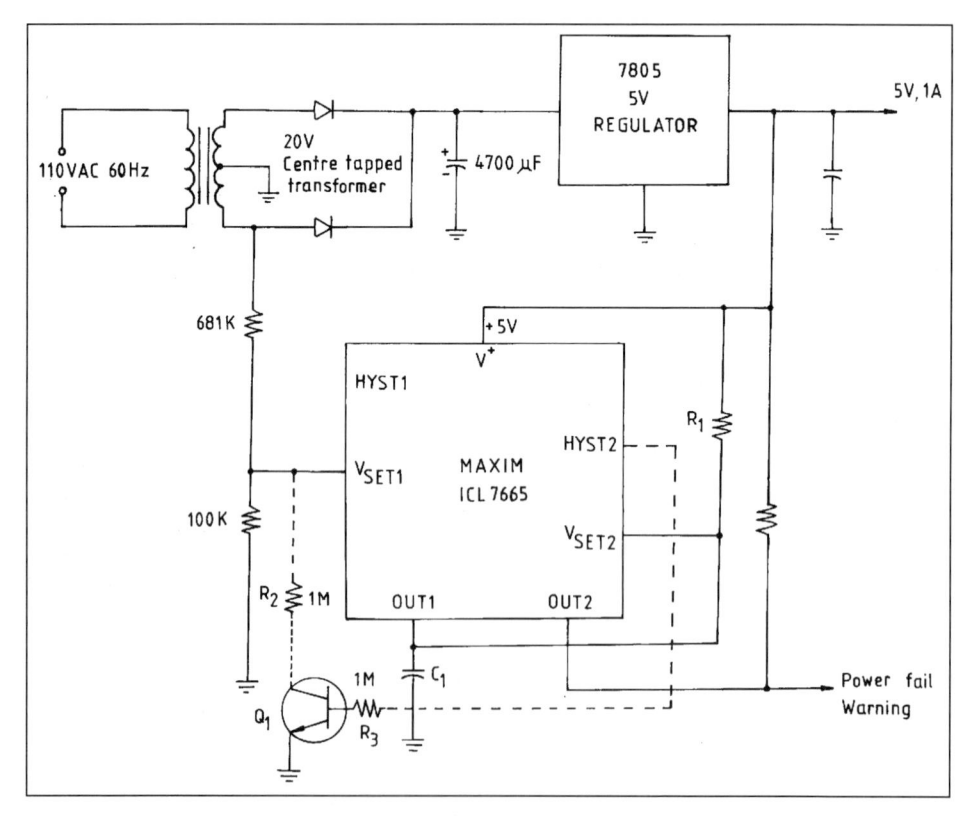

**Figure 8.13.1** AC power fail and brownout detector

By monitoring the secondary of the transformer, the circuit shown in Figure 8.13.1 performs a power failure warning function. It uses the dual over/under voltage detector ICL 7665. The ICL 7665 combines a 1.3 V reference with two comparators, two open drain n-channel outputs, and two open drain p-channel hysteresis outputs. The reference and comparator are very low power linear CMOS circuits, with a total operating current of 10 µA maximum, 3 µA typical. The OUT1 is an inverting output, all other outputs are non-inverting. HYST1 and HYST2 are p-channel current sources whose sources are connected to $V^+$. OUT1 and OUT2 are n-channel current sinks with their sources connected to ground.

With a normal 110 V AC input to the transformer, OUT1 will discharge $C_1$ every 16.7 ms when the peak transformer secondary voltage exceeds 10.2 V. When the 110 V AC power line voltage is either interrupted or reduced so that the peak voltage is less than 10.2 V, $C_1$ will be charged through $R_1$. OUT2, the power-fail warning output goes high when the voltage on $C_1$ reaches 1.3 V. The time constant $R_1 \times C_1$ determines the delay time before the power-fail warning signal is activated, in this case 42 mS or (2½) line cycles. Optional components $R_2$, $R_3$ and $Q_1$ add hysteresis by increasing the peak secondary voltage required to discharge $C_1$ once the power-fail warning is active.

*Courtesy of Maxim Integrated Products, Sunnyvale, CA*

◆ ◆

## 8.14 Power-fail warning and power-up/power-down reset

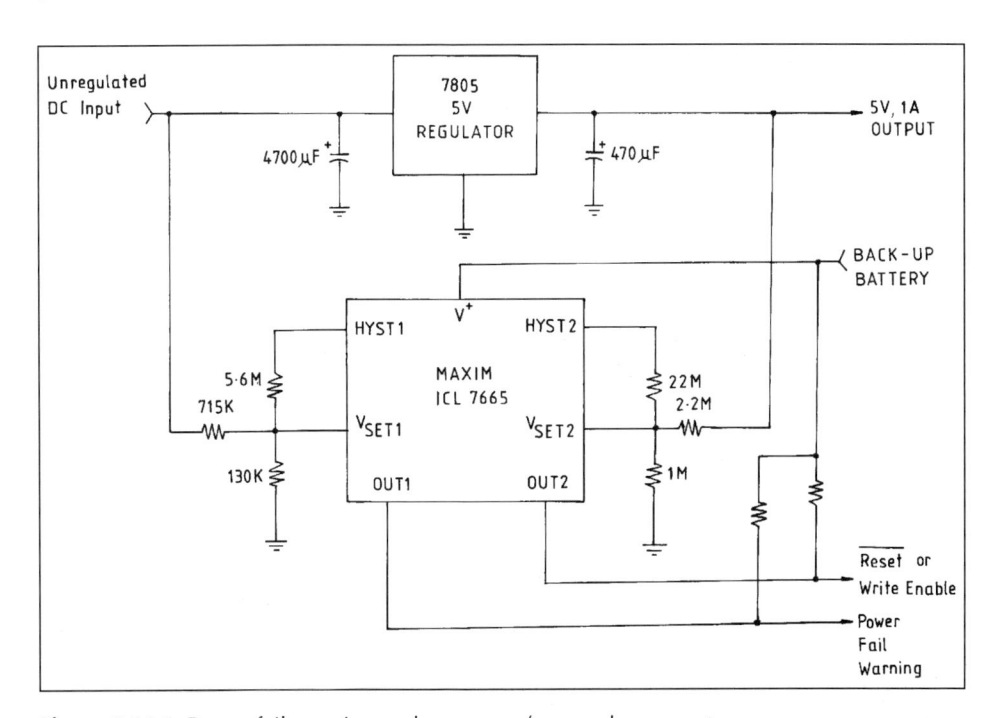

**Figure 8.14.1** Power-fail warning and power up/power down reset

Figure 8.14.1 illustrates a power fail warning circuit which monitors raw DC input voltage to the 7805 three-terminal 5 V regulator. The power-fail warning signal goes high when the unregulated DC input falls below 8.0 V. The circuit uses the dual over/under voltage detector ICL 7665. When the raw DC power source is disconnected or the AC power fails, the voltage on the input of the 7805 decays at a rate of $I_{OUT}/C$ (in this case 200 mV/ms). Since the 7805 will continue to provide 5 V out at 1 A until $V_{IN}$ is less than 7.3 V, this circuit will give at least 3.5 ms of warning before the 5 V output begins to drop. If additional warning time is needed, either the trip voltage or filter capacitance should be increased or the output current should be decreased.

The ICL 7665 OUT2 is set to trip when the 5 V output has decreased to 3.9 V. This output can be used to prevent the microprocessor from writing spurious data to a CMOS battery backup memory, or can be used to activate a battery backup system.

*Courtesy of Maxim Integrated Products, Sunnyvale, CA*

## 8.15 Blown-fuse indicator

The simple circuit shown in Figure 8.15.1 can give indication of a blown fuse in electronic equipment. As long as the fuse is intact, the transistor Q is ON due to the base-current drive provided through the resistor $R_2$. This keeps the green LED $D_2$ ON. The forward voltage drop across LED $D_2$, which is about 1.2 V, plus the $V_{CE(ON)}$ of transistor Q, which is about 0.3 V, summing up to about 1.5 V is insufficient to keep the red LED $D_1$ ON after providing for the forward voltage drop (0.6 V) for silicon diode $D_3$ and, therefore, the red LED is OFF. When the fuse blows, the base-current drive to transistor Q is cut off, turning it OFF; under this condition, the red LED $D_1$ and the diode $D_3$ are both forward-biased making the red LED ON. The values of resistors $R_1$ and $R_2$ are chosen such that the transistor Q goes into saturation when the fuse is intact and when the fuse blows, the current through the red LED $D_1$ is sufficient to give optimum LED illumination. The values of $R_1$ and $R_2$ are given by

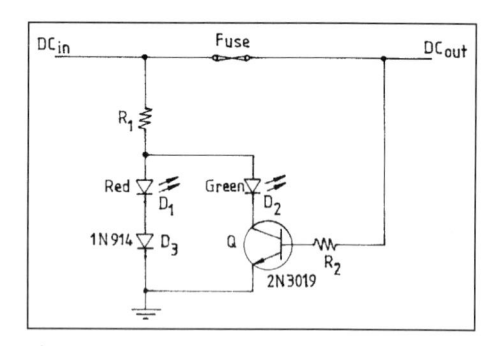

$$R_1 = (V_{DCin} - V_{D_3} - V_{D_1})/I_{R_1}$$
$$R_2 = (V_{DCout} - V_{BE})/I_B$$

**Figure 8.15.1** Fuse-blown indicator

# 9 Voltage regulator circuits

## 9.1 Positive adjustable voltage regulator

**Figure 9.1.1** Adjustable voltage regulator (1.2 V–25 V)

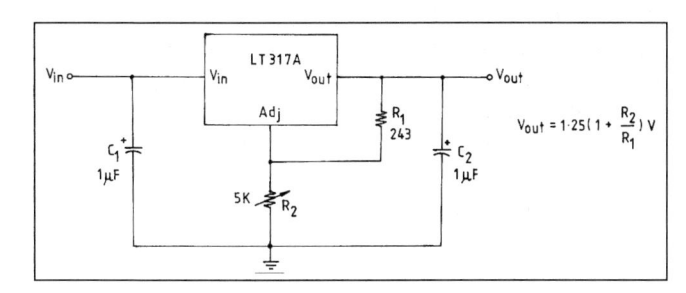

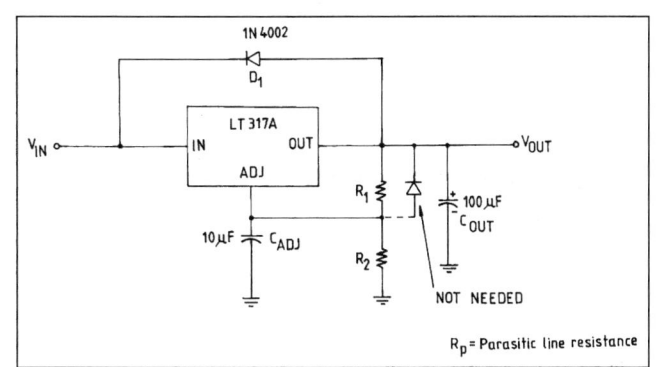

$$V_{out} = 1\cdot 25(1 + \frac{R_2}{R_1}) \text{ V}$$

**Figure 9.1.2** Protection diodes

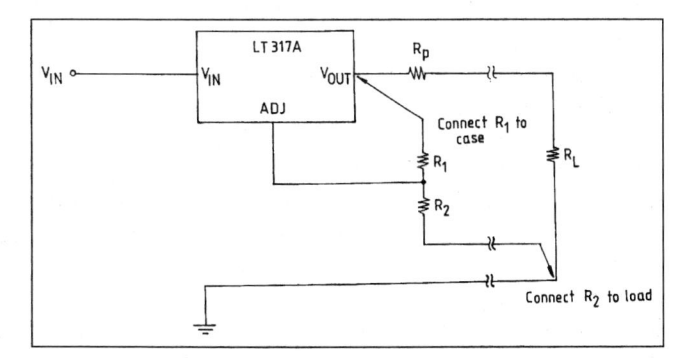

$R_p$ = Parasitic line resistance

**Figure 9.1.3** Connections for best load regulation

The LT 317A from Linear Technology Corporation is a three-terminal positive adjustable voltage regulator which offers improved performance over earlier devices. A major feature of the LT 317A is that the output voltage tolerance is guaranteed at a maximum of ± 1%, allowing an overall power supply tolerance to be better than 3% using inexpensive 1% resistors. Line and load regulation performance is better than in earlier devices. Additionally, the LT 317A reference voltage is guaranteed not to exceed 2% when operating over the full load, line and power dissipation conditions. The LT 317A adjustable regulator offers an improved solution for all positive voltage regulator requirements with load currents up to 1.5 A.

### Device operation and design equation

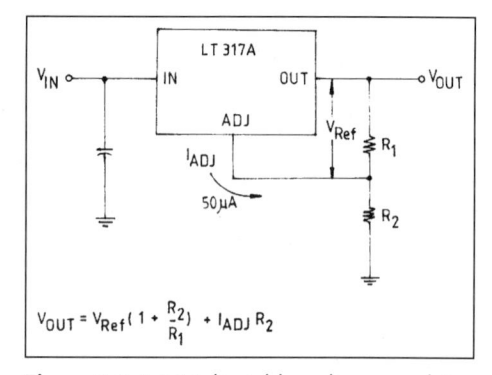

**Figure 9.1.4** (a) Adjustable voltage regulator

The LT 317A develops a 1.25 V reference voltage between the output and the adjustable terminal (see Figure 9.1.1). By placing a resistor, $R_1$, between these two terminals, a constant current is caused to flow through $R_1$ and down through $R_2$ to set the overall output voltage. Normally this current is the specified minimum load current of 5 mA or 10 mA.

Because $I_{ADJ}$ is very small and constant when compared with the current through $R_1$, it represents a small error and can usually be ignored.

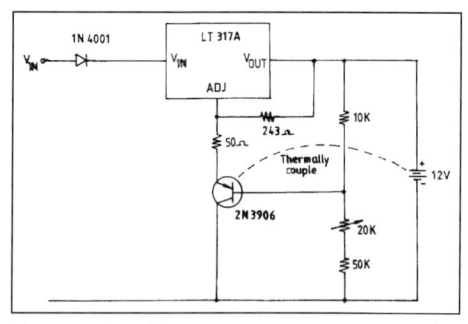

**Figure 9.1.4** (b) temperature-compensated lead acid battery charger

The output voltage is given by

$$V_{OUT} = V_{REF}[1 + (R_2/R_1)] + I_{ADJ}R_2$$

It is easily seen from the above equation that even if the resistors were of exact value, the accuracy of the output is limited by the accuracy of $V_{REF}$. Earlier adjustable regulators had a reference tolerance of ±4%. This tolerance is dangerously close to the ±5% supply tolerance required in many logic and analog systems. Further, many 1% resistors can drift 0.01%/°C, adding another 1% to the output voltage tolerance. For example, using 2% resistors and ±4% tolerance for $V_{REF}$, calculations will show that the expected range of a 5 V regulator design would be 4.66 V ≤ $V_{OUT}$ ≤ 5.36 V or approximately ±7%. If the same example were used for a 15 V regulator, the expected tolerance would be ±8%. With these results, most applications require some method of trimming, usually a trim pot. This solution is both expensive and not conducive to volume production. One of the enhancements of Linear Technology's adjustable regulators over existing devices is tightened initial tolerance. This allows relatively inexpensive 1% or 2% film resistors to be used for $R_1$ and $R_2$ while setting output voltage within an acceptable tolerance range.

## Bypass capacitors

Input bypassing using a 1 µF tantalum or 25 µF electrolytic capacitor is recommended when the input filter capacitors are more than 5 inches from the device. Improved ripple rejection (80 dB) can be accomplished by adding a 10 µF capacitor from the adjust pin to ground. Increasing the size of the capacitor to 20 µF will help ripple rejection at low output voltage since the reactance of this capacitor should be small compared to the voltage setting resistor, $R_2$. For improved AC transient response and to prevent the possibility of oscillation due to unknown reactive load, a 1 µF capacitor is also recommended at the output. Because of their low impedance at high frequencies, the best type of capacitor to use is solid tantalum.

## Protection diodes

The LT 317A does not require a protection diode from the adjustment terminal to the output (see Figure 9.1.2). Improved internal circuitry eliminates the need for this diode when the adjustment pin is bypassed with a capacitor to improve ripple rejection.

If a very large output capacitor is used, such as a 100 µF shown in Figure 9.1.2, the regulator could be damaged or destroyed if the input is accidentally shorted to ground or crowbarred. This is due to the output capacitor discharging into the output terminal of the regulator. To prevent damage a diode $D_1$ is recommended to safely discharge the capacitor.

## Load regulation

Because the LT 317A is a three-terminal device, it is not possible to provide true remote load sensing. Load regulation will be limited by the resistance of the wire connecting the regulator to the load. Best load regulation is obtained when the top of the divider is connected directly to the case, not to the load. This is illustrated in Figure 9.1.3. If $R_1$ were connected to the load, the effective resistance between the regulator and the load would be

$$R_p \times [(R_2 + R_1)/R_1] \quad (R_p = \text{parasitic line resistance})$$

Connected as shown, $R_p$ is not multiplied by the divider ratio. $R_p$ is about 0.004 ohm per foot using 16 gauge wire. This translates to 4 mV/ft at 1 A load current, so it is important to keep the positive lead between regulator and load as short as possible.

Figure 9.1.4 lists some applications of LT 317A.

*Courtesy of Linear Technology Corporation, Milpitas, CA.*

158  Voltage regulator circuits

## 9.2  5 A positive adjustable voltage regulator

**Figure 9.2.1** Basic adjustable regulator

The LT 338A from Linear Technology Corporation is a three-terminal positive adjustable voltage regulator which offers improved performance over earlier devices, with improved circuit design and produced with advanced process techniques to provide superior performance and reliability. The internal voltage reference is trimmed to less than 1%, enabling a very tight output voltage. In addition to excellent line and load regulation, with full overload protection, the LT 338A incorporates new current limiting circuitry allowing large transient load currents to be handled for short periods. Transient load currents of up to 2 A can be supplied without limiting, eliminating the need for a large output capacitor.

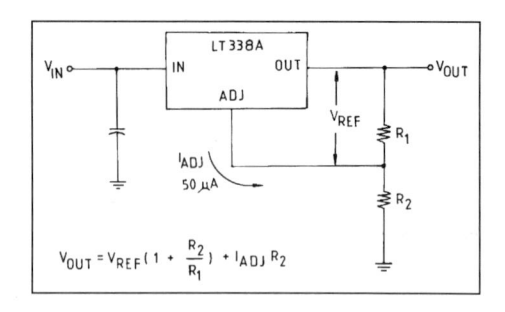

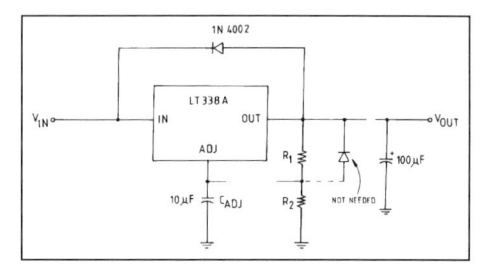

**Figure 9.2.2** Protection diodes

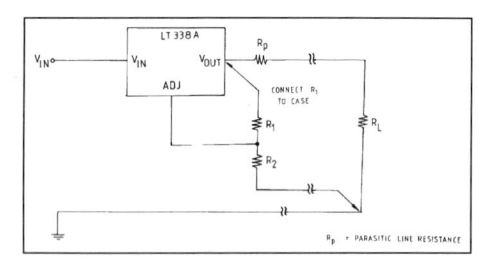

**Figure 9.2.3** Connections for best load regulation

### Device operation and design equation

The LT 338A develops a 1.25 V reference voltage between the output and the adjustable terminal (see Figure 9.2.1). By placing a resistor, $R_1$, between these two terminals, a constant current is caused to flow through $R_1$ and down through $R_2$ to set the overall output voltage. Normally this current is the specified minimum load current of 5 mA or 10 mA. Because $I_{ADJ}$ is very small and constant when compared with the current through $R_1$, it represents a small error and can usually be ignored.

The output voltage is given by

$$V_{OUT} = V_{REF}[1 + (R_2/R_1)] + I_{ADJ}R_2$$

It is easily seen from the above equation, that even if the resistors were of exact value, the accuracy of the output is limited by the accuracy of $V_{REF}$. Earlier

# Voltage regulator circuits 159

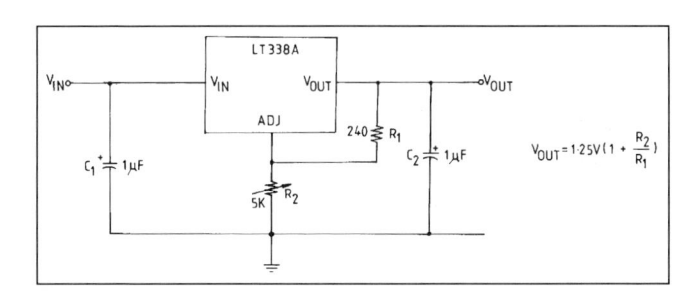

**Figure 9.2.4** (a) 1.2 V–25 V adjustable regulator

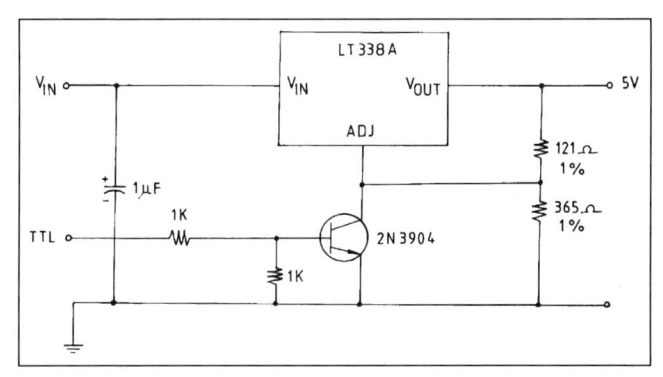

**Figure 9.2.4** (b) 5 V regulator with shutdown

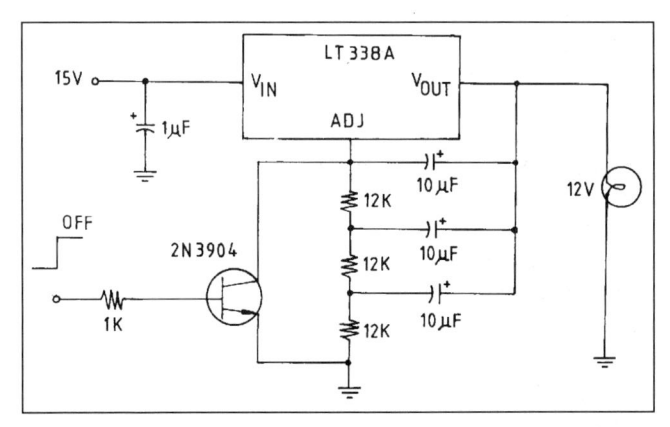

**Figure 9.2.4** (c) lamp flasher

adjustable regulators had a reference tolerance of ±4%. This tolerance is dangerously close to the ±5% supply tolerance required in many logic and analog systems. Further, many 1% resistors can drift 0.01%/°C, adding another 1% to the output voltage tolerance. For example, using 2% resistors and ±4% tolerance for $V_{REF}$, calculations will show that the expected range of a 5 V regulator design would be 4.66 V ≤ $V_{OUT}$ ≤ 5.36 V or approximately ±7%. If the same example were used for a 15 V regulator, the expected tolerance would be ±8%. With these results, most applications require some method of trimming, usually a trim pot. This solution is both expensive and not conducive to volume production. One of the enhancements of Linear Technology's adjustable regulators over existing devices is tightened initial tolerance. This allows relatively inexpensive 1% or 2%

film resistors to be used for $R_1$ and $R_2$ while setting output voltage within an acceptable tolerance range.

## Bypass capacitors

Input bypassing using a 1 µF tantalum or 25 µF electrolytic capacitor is recommended when the input filter capacitors are more than 5 inches from the device. Improved ripple rejection (80 dB) can be accomplished by adding a 10 µF capacitor from the adjust pin to ground. Increasing the size of the capacitor to 20 µF will help ripple rejection at low output voltage since the reactance of this capacitor should be small compared to the voltage setting resistor, $R_2$. For improved AC transient response and to prevent the possibility of oscillation due to unknown reactive load, a 1 µF capacitor is also recommended at the output. Because of their low impedance at high frequencies, the best type of capacitor to use is solid tantalum.

## Protection diodes

The LT 338A does not require a protection diode from the adjustment terminal to the output (see Figure 9.2.2). Improved internal circuitry eliminates the need for this diode when the adjustment pin is bypassed with a capacitor to improve ripple rejection.

If a very large output capacitor is used, such as a 100 µF shown in Figure 9.2.2, the regulator could be damaged or destroyed if the input is accidentally shorted to ground or crowbarred. This is due to the output capacitor discharging into the output terminal of the regulator. To prevent damage a diode $D_1$ is recommended to safely discharge the capacitor.

## Load regulation

Because the LT 338A is a three-terminal device, it is not possible to provide true remote load sensing. Load regulation will be limited by the resistance of the wire connecting the regulator to the load. Best load regulation is obtained when the top of the divider is connected directly to the case, not to the load. This is illustrated in Figure 9.2.3. If $R_1$ were connected to the load, the effective resistance between the regulator and the load would be

$$R_p \times [(R_2 + R_1)/R_1] \quad R_p = \text{parasitic line resistance}$$

Connected as shown, $R_p$ is not multiplied by the divider ratio. $R_p$ is about 0.004 ohm per foot using 16 gauge wire. This translates to 4 mV/ft at 1 A load current, so it is important to keep the positive lead between regulator and load as short as possible.

Figure 9.2.4 shows some applications of LT 338A.

*Courtesy of Linear Technology Corporation, Milpitas, CA*

# 9.3 Negative adjustable voltage regulator

**Figure 9.3.1** Negative adjustable regulator

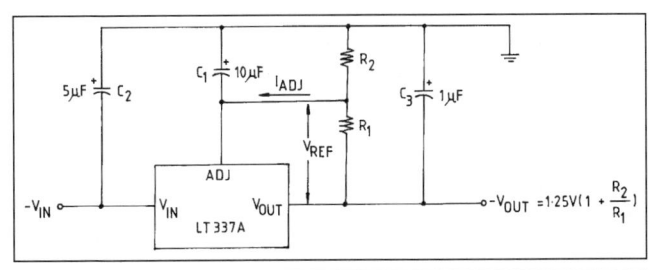

$$-V_{OUT} = 1.25V \left(1 + \frac{R_2}{R_1}\right)$$

**Figure 9.3.2** Connection of protection diodes

**Figure 9.3.3** Proper connection of divider resistors

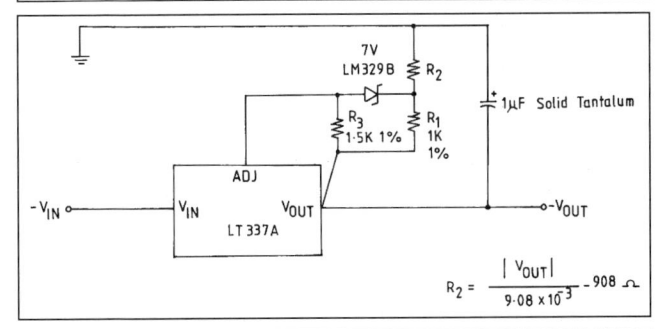

**Figure 9.3.4** (a) High stability regulator

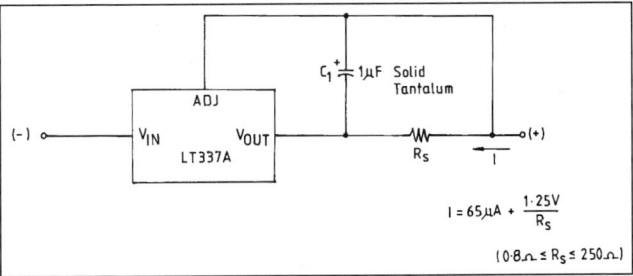

$$R_2 = \frac{|V_{OUT}|}{9.08 \times 10^{-3}} - 908 \, \Omega$$

**Figure 9.3.4** (b) current regulator

$$I = 65\mu A + \frac{1.25V}{R_S}$$

$$(0.8\,\Omega \leq R_S \leq 250\,\Omega)$$

162  Voltage regulator circuits

The LT 337A from Linear Technology Corporation is a three-terminal negative adjustable voltage regulator which offers improved performance over earlier devices. The LT 337A can deliver up to 1.5 A output current over an output voltage range of −1.2 V to −37 V. The device has better line and load regulation than previous devices and a maximum output voltage error of 1%. The device has internal current and power limiting coupled with true thermal limiting which prevents device damage due to overloads or shorts, even if the regulator is not fastened to a heat sink. Maximum reliability is attained with Linear Technology's advanced processing techniques combined with a 100% burn-in in the thermal limit mode. This ensures that all device protection circuits are working and eliminates field failures experienced with other regulators that receive only standard electrical testing.

## Device operation and design equation

The output voltage is determined by two external resistors, $R_1$ and $R_2$ (see Figure 9.3.1). The exact formula for the output voltage is

$$V_{OUT} = V_{REF}[1 + (R_2/R_1)] + I_{ADJ}R_2$$

where $V_{REF}$ = reference voltage, $I_{ADJ}$ = adjustment pin current. In most applications, the second term is small enough to be ignored, typically about 0.5% of $V_{OUT}$. In more critical applications, the exact formula should be used, with $I_{ADJ}$ equal to 65 µA. Solving for $R_2$ yields:

$$R_2 = (V_{OUT} - V_{REF})/[(V_{REF}/R_1) + I_{ADJ}]$$

Smaller values of $R_1$ and $R_2$ will reduce the influence of $I_{ADJ}$ on the output voltage, but the no-load current drain on the regulator will be increased. Typical values for $R_1$ are between 100 ohms and 300 ohms, giving 12.5 mA and 4.2 mA no-load current, respectively. There is an additional consideration in selecting $R_1$, the minimum load current specification of the regulator. The operating current of the LT 337A flows from input to output. If this current is not absorbed by the load, the output of the regulator will rise above the regulated value. The current drawn by $R_1$ and $R_2$ is normally high enough to absorb the current, but care must be taken in no-load situations where $R_1$ and $R_2$ have high values. The maximum value for the operating current, which must be absorbed, is 5 mA for the LT 337A. If input–output voltage differential is less than 10 V, the operating current that must be absorbed drops to 3 mA.

### Examples
1. A precision 10 V regulator to supply up to 1 A load current.
      a. Select $R_1$ = 100 ohms to minimize effect of $I_{ADJ}$.
      b. Calculate $R_2$ = $(V_{OUT} - V_{REF})/[(V_{REF}/R_1) - I_{ADJ}]$
                    = (10 V − 1.25 V)/[(1.25 V/100 ohms) − 65 µA]
                    = 704 ohms

2. A 15 V regulator to run off batteries and supply 50 mA, $V_{INMAX} = 25\,V$.
   a. To minimize battery drain, select $R_1$ as high as possible $R_1 = (1.25\,V/3\,mA)$ = 417 ohms, use 404 ohms, 1%.
   b. The high value for $R_1$ will exaggerate the error due to $I_{ADJ}$, so the exact formula to calculate $R_2$ should be used.

$$R_2 = (V_{OUT} - V_{REF})/[(V_{REF}/R_1) - I_{ADJ}]$$
$$= (15\,V - 1.25\,V)/[(1.25\,V/404\text{ ohms}) - 65 \times 10^{-6}]$$
$$= 4539\text{ ohms}$$
$$\text{use } R_2 = 4530\text{ ohms}$$

## Capacitors and protection diodes

An output capacitor, $C_3$ is required to provide proper frequency compensation of the regulator feedback loop. A 1 µF or larger solid tantalum capacitor is generally sufficient for this purpose if the 1 MHz impedance of the capacitor is 2 ohms or less. High Q capacitors, such as Mylar, are not recommended because they tend to reduce the phase margin at light load currents. Aluminium electrolytic capacitors may also be used, but the minimum value should be 10 µF to ensure a low impedance at 1 MHz. The output capacitor should be located within a few inches of the regulator to keep lead impedance to a minimum. The following caution should be noted: if the output voltage is greater than 6 V and an output capacitor greater than 20 µF has been used, it is possible to damage the regulator if the input voltage becomes shorted, due to the output capacitor discharging into the regulator. This can be prevented by using diode $D_1$ (see Figure 9.3.2) between the input and the output.

The input capacitor $C_2$, is only required if the regulator is more than 4 inches from the raw supply filter capacitor.

## Bypassing the adjustment pin

The adjustment pin of the LT 337A may be bypassed with a capacitor to ground, $C_1$, to reduce output ripple, noise, and impedance. These parameters scale directly with output voltage if the adjustment pin is not bypassed. A bypass capacitor reduces ripple, noise, and impedance to that of a 1.25 V regulator. In a 15 V regulator for example, these parameters are improved by $15\,V/1.25\,V = 12$ to 1. This improvement holds only for those frequencies where the impedance of the bypass capacitor is less than $R_1$. 10 µF is generally considered sufficient for 60 Hz power line applications where the ripple frequency is 120 Hz, since $X_C$ = 130 ohms. The capacitor should have a voltage rating at least as high as the output voltage of the regulator. Values larger than 10 µF may be used, but if the output is larger than 25 V, a diode, $D_2$ should be added between the output and adjustment pins (see Figure 9.3.2).

## Proper connection of divider resistors

The LT 337A has an excellent load regulation specification of 0.5% and is measured at a point 1/8" from the bottom of the package. To prevent degradation of load regulation, the resistors which set output voltage, $R_1$ and $R_2$ must be connected as shown in Figure 9.3.3. Note that the positive side of the load has a true force and sense (Kelvin) connection but the negative side of the load does not.

$R_1$ should be connected directly to the output lead of the regulator, as close as possible to the specified point 1/8" away from the case. $R_2$ should be connected to the positive side of the load separately from the positive (ground) connection to the raw supply. With this arrangement, load regulation is degraded only by the resistance between the regulator output pin and the load. If $R_1$ is connected to the load, regulation will be degraded.

Figure 9.3.4 shows some application circuits using LT 337A.

*Courtesy of Linear Technology Corporation, Milpitas, CA*

◆ ◆

## 9.4 Automobile voltage regulator

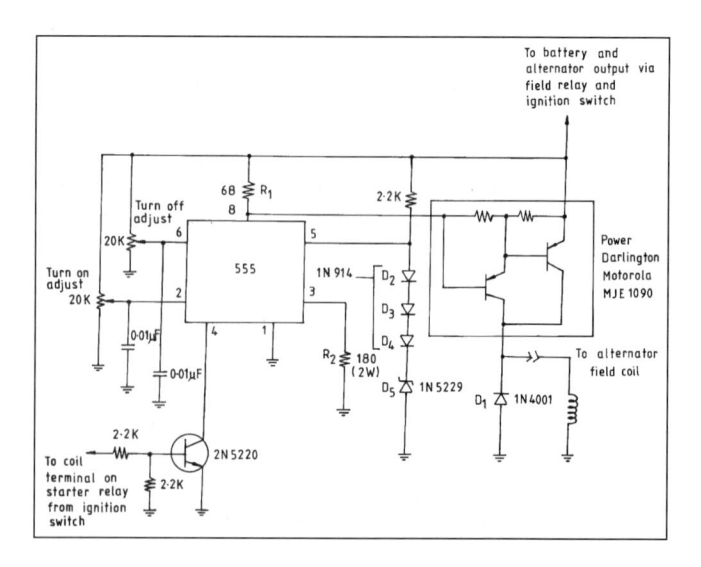

**Figure 9.4.1** Automobile voltage regulator

The 555 timer is the heart of the simple automobile voltage regulator shown in Figure 9.4.1. When the timer is OFF so that its output (pin 3) is low, the power Darlington transistor pair is OFF. If battery voltage becomes too low (less than 14.4 V in this case), the timer turns ON and the Darlington pair conducts.

*Courtesy of Philips, The Netherlands*

## 9.5 Voltage/current regulator

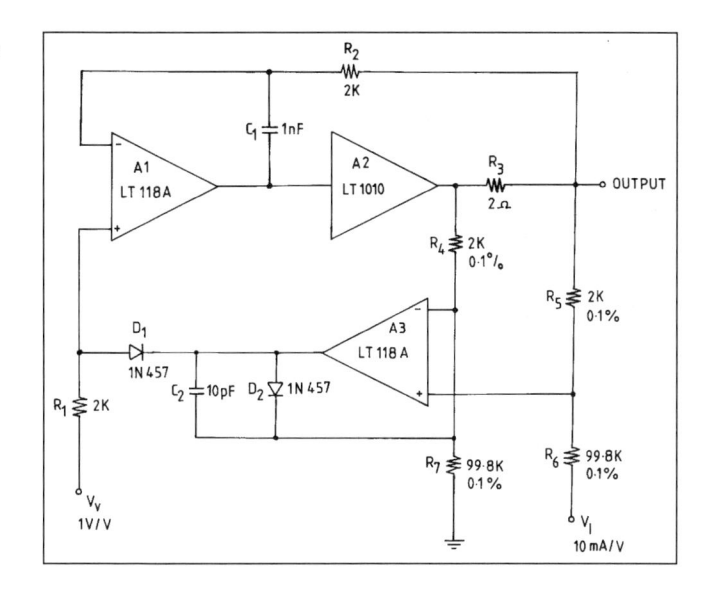

**Figure 9.5.1** Voltage/current regulator

The circuit shown in Figure 9.5.1 regulates the output voltage at $V_V$ until the load current reaches a value programmed by $V_I$. For heavier loads, it is a precision current regulator.

With output currents below the current limit, the current regulator is disconnected from the loop by $D_1$, with $D_2$ keeping its output out of saturation. This output clamp enables the current regulator to get control of the output current from the buffer current limit within a microsecond for an instantaneous short.

In the voltage regulation mode, $A_1$ and $A_2$ act as a fast voltage follower using a capacitive load isolation technique. Load transient recovery as well as capacitive load stability are determined by $C_1$. Recovery from short circuit is clean.

Bidirectional current limit can be obtained by adding another op-amp connected as a complement to $A_3$.

*Courtesy of Linear Technology Corporation, Milpitas, CA*

◆ ◆

## 9.6 Voltage inverting switching regulator

The MAX 634 from Maxim Integrated Products is a CMOS DC–DC regulator designed for simple, efficient, inverting DC–DC converter circuits. The MAX 634 provides all control and power handling functions in a compact 8 pin package: a 1.25 V bandgap reference, an oscillator, a comparator for output voltage regulation and a 525 mA P-channel output MOSFET. A second comparator is also provided for convenient low battery detection.

166  Voltage regulator circuits

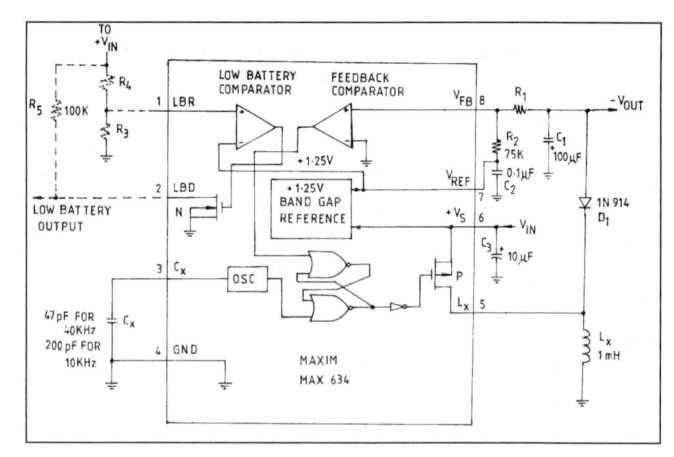

**Figure 9.6.1** Standard application circuit

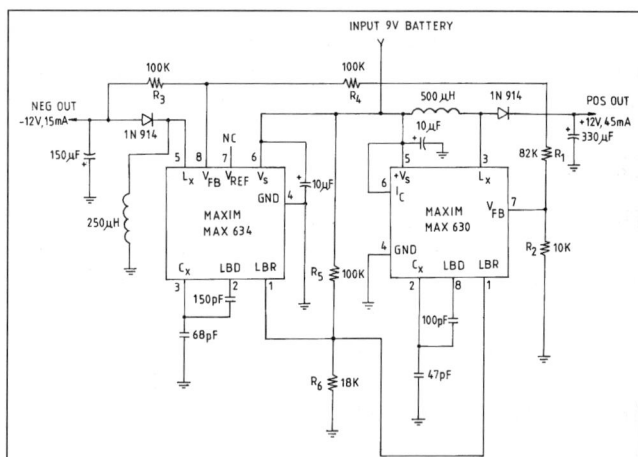

**Figure 9.6.2** (a) ± 12 V dual tracking regulator

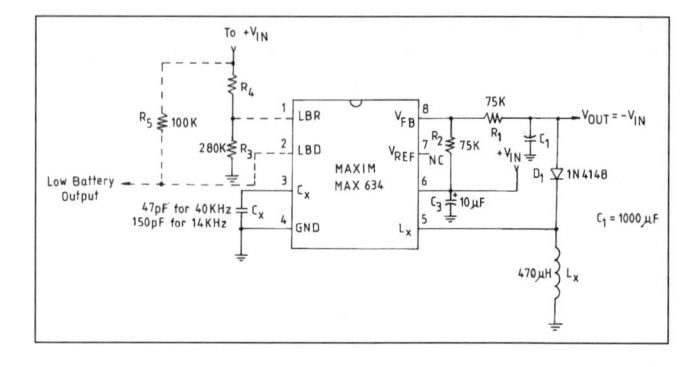

**Figure 9.6.2** (b) regulated voltage inverter

The operating current of the device is typically 100 µA and is nearly independent of output switch current and duty cycle, thus ensuring high efficiency even in low power battery operated systems. Operating in the inverting configuration, the MAX 634 can convert a positive input voltage in the range of +3 V to 16.5 V to any negative output voltage up to −20 V.

# Voltage regulator circuits

Figure 9.6.1 shows the standard circuit for converting a positive input voltage into a negative voltage. When the feedback voltage at pin 8 is above ground, the P-channel MOSFET at pin 5 turns ON during the next LOW-going period of the oscillator. The P-channel MOSFET delivers current to the external inductor, storing energy in its magnetic field. When the oscillator output goes HIGH, the P-channel MOSFET turns OFF, and the 'kickback' of the inductor pulls current through diode $D_1$, negatively charging the output filter capacitor, $C_1$. This cycle repeats until the output voltage pulls the feedback input, pin 8, below ground. The NOR gate latch prevents high frequency oscillations by not allowing $L_x$ to switch repeatedly during an oscillator cycle.

The output voltage is determined by the internal 1.25 V reference and the ratio of the resistors $R_1$ and $R_2$.

$$V_{OUT} = 1.25\,\text{V} \times (R_1/R_2)$$

Capacitor $C_1$ is the output filter capacitor. The capacitance and ESR (equivalent series resistance) of $C_1$ determine the output ripple. $C_2$ and $C_3$ are bypass capacitors, while $C_x$ sets the oscillator frequency.

## Oscillator

The MAX 634 oscillator uses only one external component, a capacitor $C_x$ connected between pin 3 and ground. A value of 47 pF sets the oscillator frequency to approximately 40 kHz. The oscillator can also be externally driven with a CMOS gate which swings from ground to $+V_s$. The $L_x$ output is always OFF when the $C_x$ pin is externally driven HIGH.

## Low-battery detector

The low-battery detector (LBD) output (pin 2, Figure 9.6.1) sinks current whenever the input voltage at low-battery resistor (pin 1) is less than + 1.25 V. The LBR input is a high impedance CMOS input with less than 10 nA leakage current. The LBD output is an open drain N-channel MOSFET with about 500 ohms of output resistance. The operating voltage of the low battery detector can be adjusted using an external voltage divider as shown in Figure 9.6.1. If hysteresis is desired, add a resistor between pins 1 and 2.

$$V_{LOBATT} = 1.25\,\text{V} \times [1 + (R_4/R_3)]$$

i.e.

$$R_4 = R_3 \times [(V_{LOBATT}/1.25\,\text{V}) - 1]$$

where $V_{LOBATT}$ is the operating voltage of the low battery detector, and $R_3$ is usually between 10k and 10M with a typical value being 470k.

## Selection of inductor value

The available output current from an inverting DC–DC voltage converter is determined by the value of the external inductor, the output voltage, the input voltage and the operating frequency. The inductor must have the following characteristics:

1. it should have the correct inductance,
2. it should be able to handle the peak currents, and
3. it should have acceptable series resistance and core losses.

The maximum and minimum values of the inductor $L$ are given by

$$L_{MAX} = (V_{IN} T_{ON})^2 f / (2 P_{OUT})$$

$$L_{MIN} = V_{IN} T_{ON} / I_{MAX}$$

where $I_{MAX}$ is the maximum allowable peak $L_x$ current (525 mA).

Contrary to what one would expect at first glance, reducing the inductor value increases the available output current; lower $L$ increases the peak current, thereby increasing the available power. If the inductance is too high, the MAX 634 will not be able to deliver the desired output power; even with the $L_x$ output turned ON with each oscillator cycle. The available output power can be increased by either decreasing the inductance or by decreasing the frequency. Decreasing the frequency increases the ON period of the $L_x$ output, thereby increasing the peak inductor current, which in turn increases the available output power since the output power is proportional to the square of the peak inductor current. The types of inductors which can be used with the MAX 634 include moulded inductors, potted toroidal inductors, ferrite cores (pot cores) and toroidal cores.

## *External diode*

In most circuits using the MAX 634, the inductor current returns to zero before $L_x$ turns ON for the next output pulse. This allows the use of slow turn-off diodes. On the other hand, the diode current abruptly goes from zero to full peak current each time $L_x$ switches OFF (Figure 9.6.1, $D_1$). To avoid excessive losses during turn-on the diode must have a fast turn-on time.

The 1N 914 or 1N 4148 is suitable for low power applications. The 1N 5817 series of Schottky diodes or their equivalent are suitable for higher power applications. Rectifier diodes such as the 1N 4001 series are unacceptable since their slow turn-on results in excessive losses.

## *Filter capacitor*

The output filter capacitor ($C_1$ in Figure 9.6.1) stores the energy delivered by the inductor and delivers current to the load. The output voltage ripple is directly affected by the capacitance and the equivalent series resistance (ESR) of the output filter capacitor.

## *Oscillator capacitor ($C_x$)*

The oscillator capacitor can be a low cost ceramic capacitor. If the circuit will be operated over a wide temperature range, a capacitor with a low temperature coefficient of capacitance should be used.

The value of $C_x$ can be calculated using the formula:

$$C_x = (2.14 \times 10^{-6}/f) - C_{INT}$$

where $f$ is the desired operating frequency in Hz and $C_{INT}$ is the sum of the stray capacitance on the $C_x$ pin and the internal capacitance of the package. The internal capacitance is about 1 pF for the plastic package and 3 pF for the CERDIP package. Typical stray capacitance is about 3 pF for normal printed circuit board layouts, but will be significantly higher if a socket is used.

## Application hints

When using off-the-shelf inductors, make sure that the peak current rating is observed. When designing your own inductors, observe the core manufacturer's ampere-turns or NI ratings. Failure to observe the peak current or NI ratings may lead to saturation of the inductor, especially in circuits with external current boosting transistors. Inductor saturation leads to very high current levels through the external boost transistors, causing excessive power dissipation, poor efficiency, and possible damage to the inductor and the external transistor. Test for saturation by applying the maximum load, the maximum input voltage and (for a safety margin) lowering the clock frequency by 25%. Monitor the inductor current using a current probe. The normal inductor current waveform is a sawtooth with a linear current ramp. Saturation creates a non-linear current waveform with a very rapid increase in current once the inductor saturates. It is this rapid current increase and the resultant high peak currents that can damage the inductor and the external boost transistor.

## Bypassing and compensation

The high operating current pulses in the $L_x$ output and the external inductor can cause erratic operation unless the MAX 634 is properly bypassed. Connect a 10 µF bypass capacitor directly across the MAX 634 between pin 6 ($+V_s$) and pin 4 (ground) to minimize the inductance and high frequency impedance of the power source. Make sure that the high current ground path of the inductor does not cause a voltage drop in the MAX 634 ground line. The reference voltage output pin 7 should also be bypassed to ground to avoid coupling to the high current path that includes $L_x$ output, the inductor, and its ground return.

With light loads, coupling from the high power circuit into the control circuitry may cause the output pulses to occur in bursts, thereby increasing low frequency ripple and degrading the line and load regulation. Normal operation with evenly distributed output pulses can be restored by adding a 100 pF to 10 nF compensation capacitor across the feedback resistor, $R_1$. Minimizing the stray capacitance on the $V_{FB}$ terminal will often eliminate the need for this compensation capacitor.

Figure 9.6.2 shows some application circuits using the MAX 634 inverting switching regulator.

*Courtesy of Maxim Integrated Products, Sunnyvale, CA*

## 9.7 Voltage converter

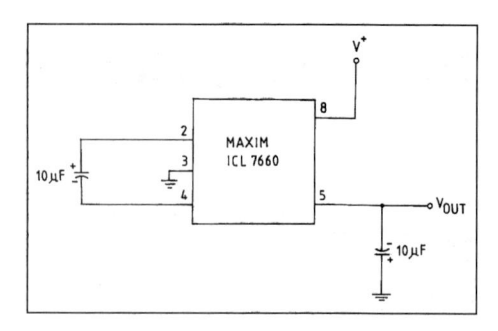

**Figure 9.7.1** Negative voltage converter

The Maxim ICL 7660 is a monolithic charge pump converter that will convert a positive voltage in the range of +1.5 V to +10 V to the corresponding negative voltage in the range of −1.5 V to −10 V. The ICL 7660 provides performance far superior to other charge pump voltage inverters by combining low quiescent current with high efficiency and by eliminating diode drop voltage losses. A common application of the ICL 7660 is to generate a −5 V supply for use with analogue circuitry. Another popular usage is to convert a +9 V battery voltage to −9 V, which can then be regulated to −5 V by the Maxim ICL 7664 micropower voltage regulator. It can also be used to double the output voltage of a battery, generating a 3 V total supply voltage from a single 1.5 V flashlight battery or generating a 6 V total supply voltage from a single lithium cell.

The circuit shown in Figure 9.7.1 shows a simple negative voltage converter using the ICL 7660. As seen from the figure, only two external components $C_1$ and $C_2$ are needed. In most applications $C_1$ and $C_2$ are low-cost 10 µF electrolytic capacitors. The ICL 7660 is not a voltage regulator and the output source resistance is approximately 70 ohms with +5 V input. This means that with an input voltage of +5 V, the output voltage will be −5 V under light load, but will decrease to about −4.3 V with a 10 mA load current. The output impedance of the complete circuit is the sum of the ICL 7660 output resistance and the impedance of the pump capacitors at the pump frequency.

The ripple voltage on the output can be calculated by noting that the output current is supplied solely from capacitor $C_2$ during one-half of the charge pump cycle. This introduces a ripple

$$V_{ripple} = I_{OUT}/2(F_{pump}C_2)$$

For the nominal $F_{pump}$ of 5 kHz (one-half of the nominal 10 kHz oscillator frequency) and a 10 µF $C_2$, the ripple will be approximately 100 mV with an output current of 10 mA.

*Courtesy of Maxim Integrated Products, Inc., Sunnyvale, CA*

# 10 Battery circuits

## 10.1 Simple battery charger

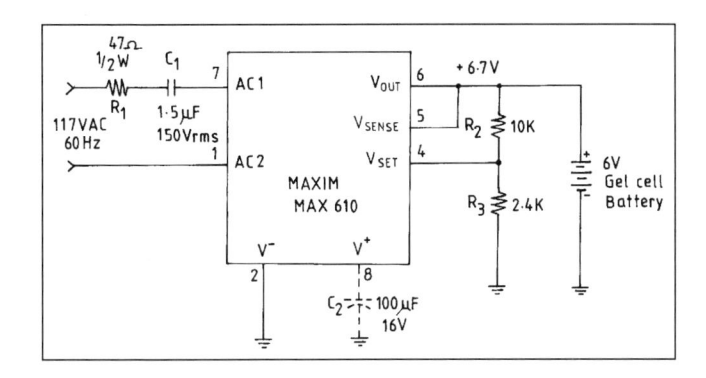

**Figure 10.1.1** Simple battery charger

A simple battery charger can be built using the MAX 610 AC–DC regulator. In Figure 10.1.1, the open circuit or float voltage of +6.7 V is set by $R_2$ and $R_3$:

$$V_{out} = 1.3[1 + (R_2/R_3)] \text{ V}$$

The maximum charging current of 60 mA is set by the value of $C_1$. Since, unlike transformer-driven battery chargers, $C_1$ conducts current throughout most of each line cycle, the ratio of the RMS charging current to the average charging current is only about 1.2:1 and capacitor $C_2$ is optional.

$$I_{avg(max)} = V_{in} \times 5.56 \, F_{in} \times C \text{ (maximum charging current)}$$

where $F_{in}$ = input frequency.

$$I_{rms} = 1.2 \, I_{avg} \text{ (without } C_2\text{)}$$
$$I_{rms} = I_{avg} \text{ (with } C_2\text{)}$$

*Courtesy of Maxim Integrated Products, Inc., Sunnyvale, CA*

◆ ◆

## 10.2 Wind-powered battery charger

A simple wind-powered battery charger can be constructed using the LTC 1042 (which is a window comparator from Linear Technology Corporation), a 12 V DC permanent magnet motor, and a low-cost power FET as shown in Figure 10.2.1.

172  Battery circuits

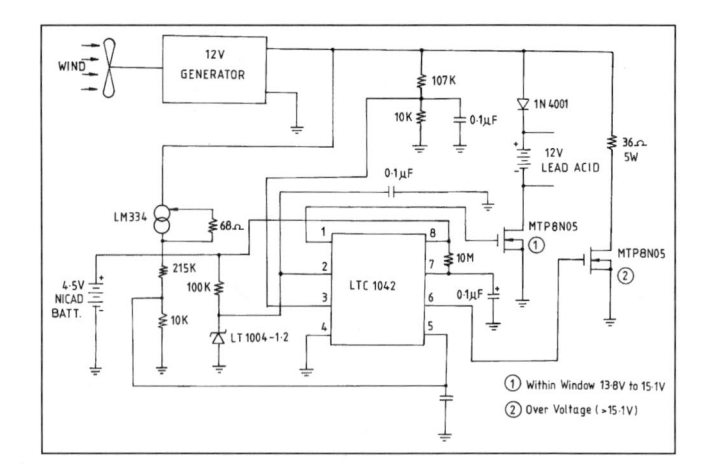

**Figure 10.2.1** Wind-powered battery charger

The DC motor is used as a generator with the voltage output being proportional to its RPM. The LTC 1042 monitors the voltage output and provides the following control functions:

(1) If the generator voltage output is below 13.8 V, the control circuit is active and the NiCad battery charges through the LM 334 current source. The lead acid battery is not being charged at this time.
(2) If the generator voltage output is between 13.8 V and 15.1 V, the 12 V lead acid battery is charged at a rate of about 1 A/hour (limited by the power FET).
(3) If the generator voltage exceeds 15.1 V (a condition caused by excessive wind speed or the 12 V battery being fully charged), then a fixed load is connected thus limiting the generator RPM to prevent damage.

This charger can be used as a remote source of power where wind energy is plentiful such as on sailboats or remote radio repeater sites. Unlike solar powered panels, this system will function in bad weather and at night.

*Courtesy of Linear Technology Corporation, Milpitas, CA*

◆   ◆

## 10.3  Battery-status indicator

The circuit shown in Figure 10.3.1 indicates the low-voltage condition of a battery by flickering the led D.

Normally, the supply voltage is high enough to maintain the transistor in its ON state and the micropower timer does not receive enough voltage to operate, the led being OFF. As the supply voltage falls to about 3 V, the transistor base current becomes too small for conduction, its collector voltage now being high enough to operate the timer, which will work with a 2.7 V supply and which is

Battery circuits 173

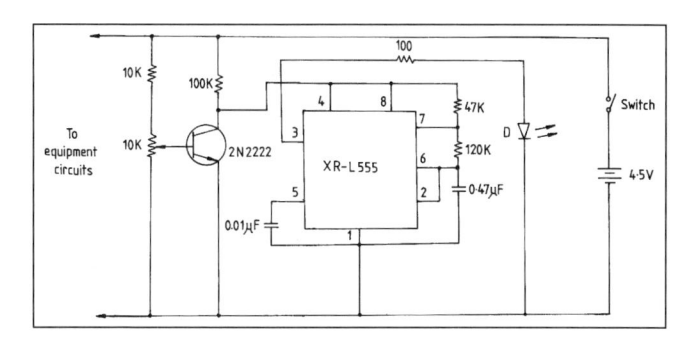

**Figure 10.3.1** Low-cost battery status indicator

arranged as a 10 Hz astable multivibrator. When, therefore, the battery voltage falls to a point set by the potentiometer, the led flickers at 10 Hz.

Power consumed by the XR-L555 is only about 1 mW (1/15 of that for the normal 555), so that it does not seriously affect battery life. The circuit is designed for operation at 4.5 V, but is easily modified for other supplies.

*First published in Electronics and Wireless World, December 1990*
*Reprinted with permission*

## 10.4  9 V battery life extender

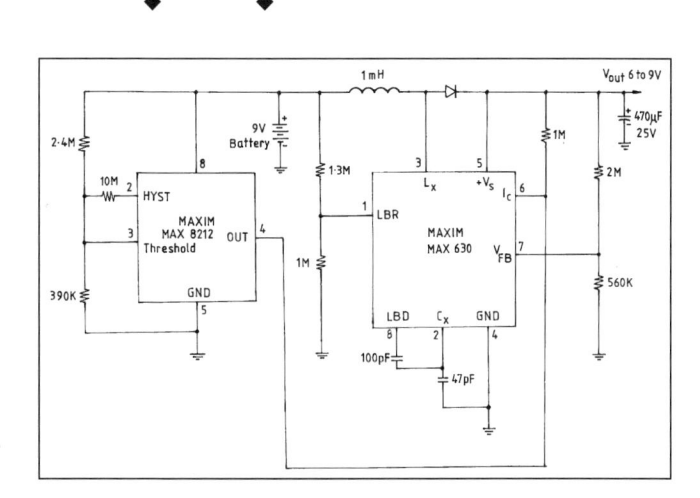

**Figure 10.4.1** Battery life extender

The circuit shown in Figure 10.4.1 provides a minimum of 7 V until the 9 V battery voltage falls to less than 2 V. The circuit uses a MAX 8212 programmable voltage detector and a MAX 630 micropower step-up switching regulator.

The MAX 8212 contains a comparator, a 1.15 V band-gap reference and an open drain N-channel output driver. Two external resistors are used in conjunction with the internal reference to set the trip point voltage to the desired level. A hysteresis output is also included, allowing the user to apply positive feedback for noise-free output switching. The MAX 630 is a low-power step-up switching regulator which can be used in the 5 mW to 5 W range. The chip provides all control and power handling functions in a compact 8 pin package: a 1.31 V bandgap reference, an oscillator, a voltage comparator and a 375 mA N-channel output MOSFET. The operating current is only 70 µA and is nearly independent of output current switch or duty cycle. A logic level input shuts down the regulator to less than 1 µA quiescent current. Low current operation ensures high efficiency even in low power battery operated systems. The MAX 630 can operate from battery voltages ranging from 2 V to 16.5 V.

In the circuit shown in Figure 10.4.1, when the battery voltage is above 7 V the MAX 630's $I_c$ pin is low, putting it into the shutdown mode which draws only 10 nA. When the battery voltage falls to 7 V, the MAX 8212 voltage detector's output goes HIGH, enabling the MAX 630. The MAX 630 then maintains the output voltage at 7 V even as the battery voltage falls below 7 V. The low battery detector (LBD) is used to decrease the oscillator frequency when the battery voltage falls to 3 V, thereby increasing the output current capability of the circuit.

This circuit (with or without the MAX 8212) can be used to provide 5 V from 4 alkaline cells. The initial voltage is approximately 6 V, and the output is maintained at 5 V even when the battery voltage falls to less than 2 V.

*Courtesy of Maxim Integrated Products, Inc., Sunnyvale, CA*

## 10.5 Battery switchover circuit

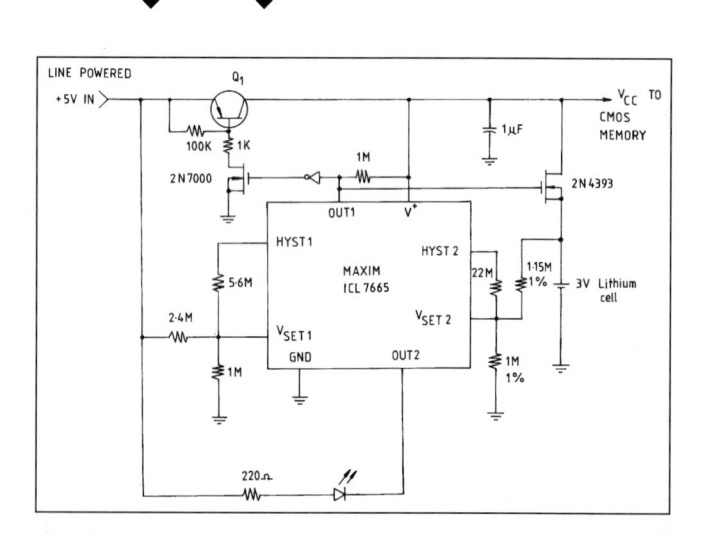

**Figure 10.5.1** Battery switchover circuit

The circuit shown in Figure 10.5.1 using the dual over/under voltage detector 7665 performs two functions: switching the power supply of a CMOS memory backup battery when the line-powered supply is turned off, and lighting a low-battery-warning LED when the backup battery is nearly discharged. The PNP transistor, $Q_1$, connects the line-powered +5 V to the CMOS memory whenever the line-powered +5V supply voltage is greater than 3.5V. The voltage drop across $Q_1$ will only be a couple of hundred millivolts since it will be saturated. Whenever the input voltage falls below 3.5V, OUT1 goes high, turns off $Q_1$ and connects the 3V lithium cell to the CMOS memory.

The second voltage detector of the 7665 monitors the voltage of the lithium cell. If the battery voltage falls below 2.6V, OUT2 goes low and the low-battery-warning LED turns on (assuming that the +5V is present, of course).

*Courtesy of Maxim Integrated Products Inc., Sunnyvale, CA*

## 10.6 Automatic battery back-up switch

**Figure 10.6.1** ICL 7673 battery back-up circuit

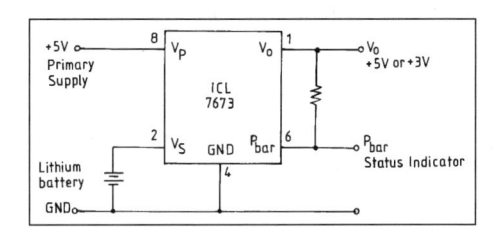

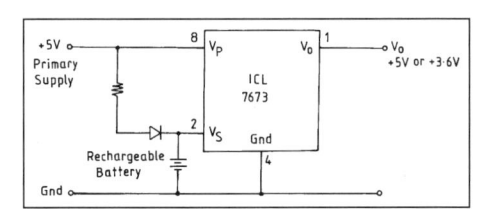

**Figure 10.6.2** Application requiring rechargeable battery back-up

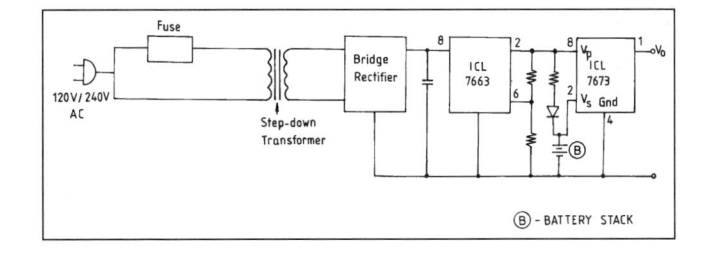

**Figure 10.6.3** Power supply for low power portable AC to DC systems

The ICL 7673 is a monolithic CMOS battery back-up circuit that offers unique performance advantages over conventional means of switching to a backing supply. The ICL 7673 is intended as a low-cost solution for the switching of systems between two power supplies; main and battery backup. The main application is keep-alive-battery power switching for use in volatile CMOS RAM memory systems and real-time clocks. In many applications this circuit will represent a low insertion voltage loss between the supplies and load. This circuit features low current consumption, wide operating voltage range, and exceptionally low leakage

176  Battery circuits

between inputs. Logic outputs are provided that can be used to indicate which supply is connected and can also be used to increase the power switching capability of the circuit by driving external PNP transistors.

A typical application of the ICL 7673 in a battery backup circuit is shown in Figure 10.6.1. A trickle-charge system can be implemented with an additional resistor and a diode as shown in Figure 10.6.2. A complete low power AC to regulated DC system can be implemented using the ICL 7673 and ICL 7663S micropower voltage regulator as shown in Figure 10.6.3.

The input operating voltage range for $V_P$ or $V_S$ is 2.5 V to 15 V. The input supply voltage slew rate should be limited to 2 V/µs to avoid potential damage to the circuit. For battery-operated applications it may be necessary to use a capacitor between the input and ground pins to limit the rate of rise of the supply voltage. A low-impedance capacitor such as a 0.047 µF disc ceramic can be used to reduce the rate of rise.

*Courtesy of Harris Semiconductor, Melbourne, FL*

◆ ◆

## 10.7  Polarity insensitive battery powered supply

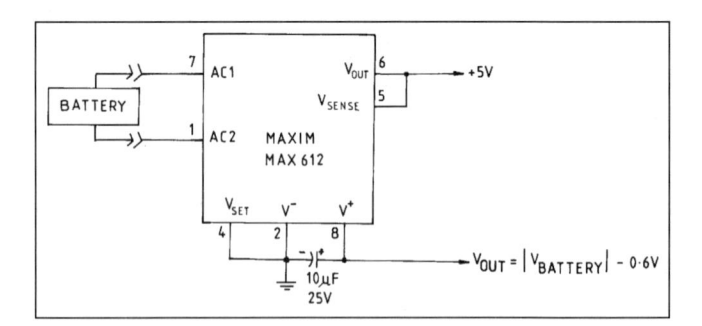

**Figure 10.7.1** Polarity insensitive battery-powered supply

Figure 10.7.1 shows a +5 V power supply which will work even if the battery is installed backwards: the full-wave bridge rectifier of the MAX 612 AC–DC regulator will correct the battery polarity. The MAX 612 is well suited for battery powered circuits since its quiescent current is only 70 µA. The MAX 610 AC–DC regulator can also be used if the battery voltage is less than 10 V.

*Courtesy of Maxim Integrated Products, Inc., Sunnyvale, CA*

◆ ◆

# 10.8 Combination low-battery warning and low-battery disconnect

**Figure 10.8.1** Low-battery warning and low-battery disconnect

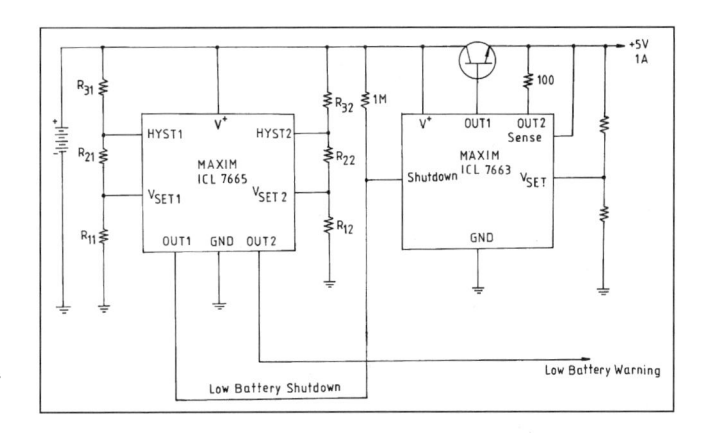

Nickel cadmium (NiCad) batteries are excellent rechargeable power sources for portable equipment, but care must be taken to ensure that NiCad batteries are not damaged by overdischarge. Specifically, a NiCad battery should not be discharged to the point where the polarity of the lowest capacity cell is reversed and that cell is reverse charged by the higher capacity cells. This reverse charging will dramatically reduce the life of a NiCad battery.

The circuit in Figure 10.8.1 both prevents reverse charging and also gives a low battery warning. The circuit consists of an ICL 7665, which is a low power dual over/under voltage detector drawing a typical operating current of only 3 µA, and an ICL 7663, which is a high efficiency programmable positive voltage regulator with a quiescent current of less than 10 µA. The output voltage of ICL 7663 can be set by two external resistors to any voltage in the range 1.3–16 V range, with an input voltage range of 1.5–16 V. The ICL 7663 is well suited for battery powered supplies, featuring low $V_{in}$ to $V_{out}$ differential, 40 mA output current capability, low quiescent current and a logic input level shutdown control.

In the case of a NiCad battery, a typical low battery voltage is 1 V per cell. Since a NiCad 9 V battery is ordinarily made up of six cells with a nominal voltage of 7.2 V, a low battery warning of 6 V is appropriate, with a small hysteresis of 100 mV. To prevent over discharge of a battery the load should be disconnected when the battery voltage is $1 V \times (N - 1)$, where $N$ = number of cells. In this case the low battery load disconnect should occur at 5 V. Since the battery voltage will rise when the load is disconnected, 800 mV of hysteresis is used to prevent repeated on–off cycling.

*Courtesy of Maxim Integrated Products, Inc., Sunnyvale, CA*

# 11 Motor control circuits

## 11.1 DC servomotor phase locked loop

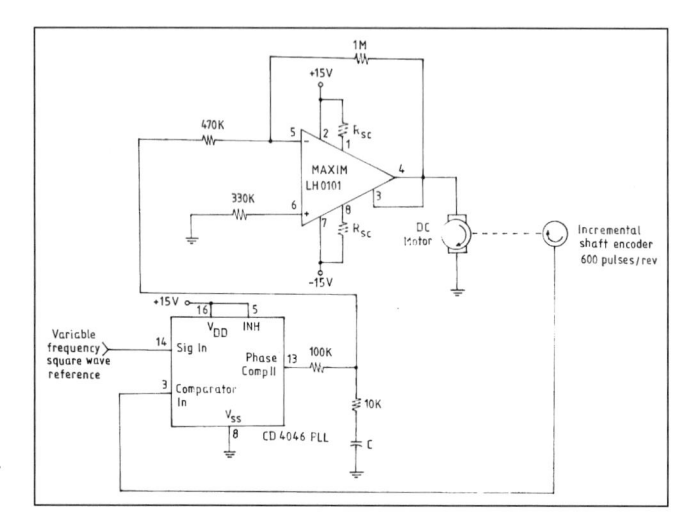

**Figure 11.1.1** Servomotor phase locked loop

In the circuit of Figure 11.1.1, the shaft encoder produces 600 pulses per revolution. These pulses are compared to a reference frequency by the digital phase comparator of the CD 4046. The output of the phase comparator passes through a low-pass filter and drives the input of the LH 0101. The LH 0101 power operational amplifier can deliver up to 5 A peak output current. Packaged in a rugged TO-3 case, the LH 0101 combines the ease of use and performance of a FET input op-amp with the power handling capabilities of a 5 A output stage. The output short-circuit protection makes this device ideal for driving AC and DC motors, large capacitive loads, and electromagnetic actuators. In Figure 11.1.1, the LH 0101 amplifies the signal at its input derived from the low-pass filter, and drives the DC servomotor. The phase-frequency comparator of the CD 4046 increases or decreases the input voltage to the LH 0101 until the shaft encoder output is the same frequency as the reference input.

Motor speed (in RPM) = $F_{in} \times 60/N$

where $F_{in}$ is the frequency of the reference input and $N$ is the number of shaft encoder pulses per revolution.

A single-pulse-per-revolution speed pickup can be used in place of the shaft encoder, but the PLL low-pass filter time constant must be greatly increased.

This circuit is similar to a standard phase locked loop except that the LH 0101, the motor, and the shaft encoder replace the internal VCO of the CD 4046. Unlike the VCO of the CD 4046, the motor adds another pole to the system response and loop stability must be carefully analysed, particularly if the motor and its load has significant inertia. As with most feedback systems, the loop will be stable when there is only one dominant pole. The loop filter time constant should preferably be at least 1 decade higher or lower than the constant of the motor and its load.

*Courtesy of Maxim Integrated Products, Inc., Sunnyvale, CA*

## 11.2 Simple bidirectional DC motor speed controller

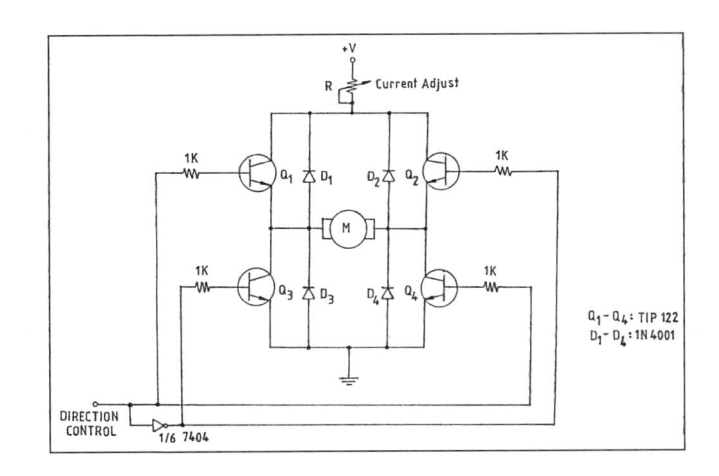

**Figure 11.2.1** Bidirectional DC motor control

The Figure 11.2.1 circuit is a simple scheme for bidirectional control of a DC motor. This circuit is also known as a bridge driver for DC motor control. The controller bridge uses four NPN power Darlington transistors, TIP 122, which have a high DC current gain $h_{FE}$ of 2500 (typical) at $I_C = 4\,A$ DC. The DC motor rotates in one direction if transistors $Q_1$ and $Q_4$ are ON and in the reverse direction if transistors $Q_2$ and $Q_3$ are ON. The direction control line can be used to apply the appropriate logic signal to turn ON the transistors depending on the direction of rotation desired. The potentiometer R helps in controlling the motor speed by allowing variation of motor current. Diodes $D_1$–$D_4$ connected across the collector–emitter terminals of the transistors protect the transistors against collector–emitter breakdown by providing a free-wheeling path for decaying currents when the winding current is switched OFF.

## 11.3 Constant current motor drive

The simple DC motor drive circuit shown in Figure 11.3.1 is particularly useful where there is some likelihood of stalling or lock up; if the motor locks, the current drive remains constant and the system does not destroy itself. Using this approach, two 6 V batteries are sufficient for good performance. A 10 V input will

## 180 Motor control circuits

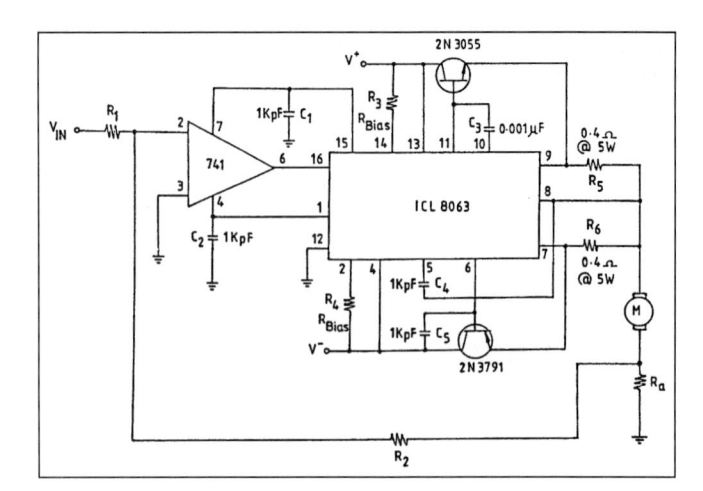

**Figure 11.3.1** Constant current motor drive

produce 1 A of output current to drive the motor, and if the motor is stalled, $I_{OUT}$ remains at 1 A. The circuit uses an ICL 8063 power transistor driver/amplifier.

For example, suppose it is necessary to drive a 24 V DC motor with 1 A of drive current. First make $V_{SUPPLY}$ at least 6 V more than the motor being driven (in this case 30 V). Next select $R_{BIAS}$ according to $V_{SUPPLY}$ from the data sheet of ICL 8063, which indicates $R_{BIAS}$ = 1 Mohm. Then choose $R_1$, $R_2$ and $R_a$ for optimum sensitivity. That means making $R_a$ = 1 ohm to minimize the voltage drop across $R_a$ (the drop will be 1 A × 1 ohm or 1 V). If 1 A/V sensitivity is desirable let $R_2$ = $R_1$ = 10k to minimize feedback current error. Then a ±1 V input voltage will produce a ±1 A current through the motor.

Capacitors should be at least 50 V working voltage and all resistors ½ W, except for those valued at 0.4 ohm. Power across $R_a$ = $I × V$ = 1 A × 1 V = 1 W, so at least a 2 W value should be used. Use large heat sinks for the 2N 3055 and 2N 3791 power transistors. Use a thermal compound when mounting the transistor to the heat sink.

*Courtesy of Harris Semiconductor, Melbourne, FL*

◆ ◆ ◆

## 11.4 Control a bidirectional 4-phase stepper motor

Figure 11.4.1 circuit allows simple control of 4-phase bidirectional stepper motors. The 555 timer configured in the circuit generates a variable-frequency pulse train as high as 10 kHz. The output of the astable oscillator is used as clock input for a modulo-4 binary up/down counter made up of dual J-K flip-flops (7473) and

# Motor control circuits

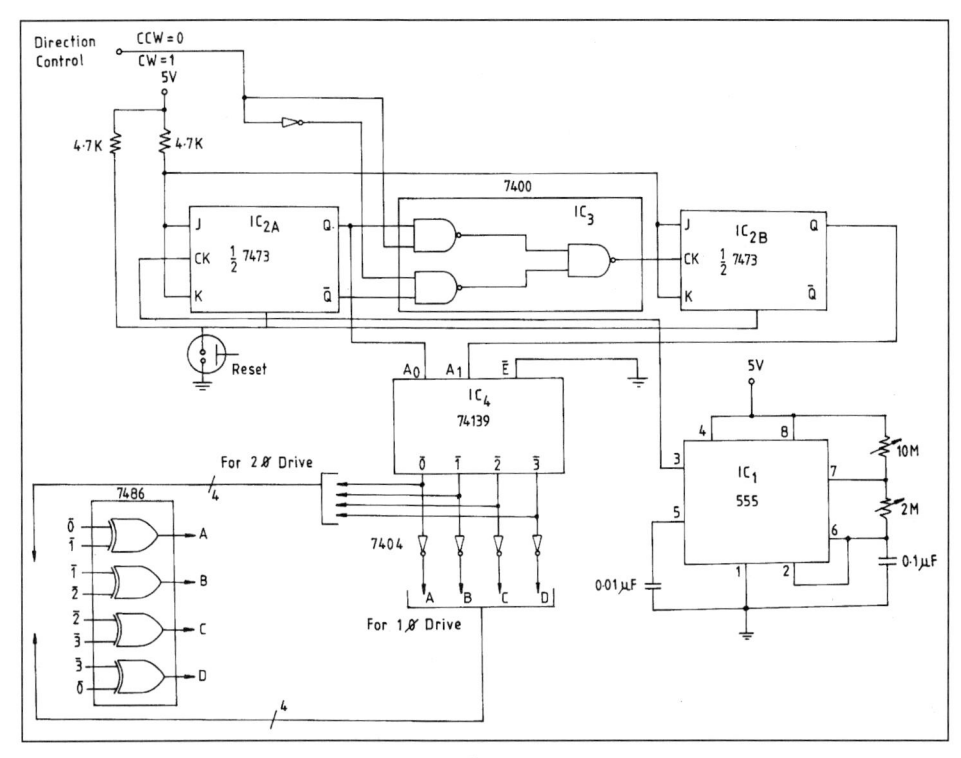

**Figure 11.4.1** 4-phase stepper motor controller

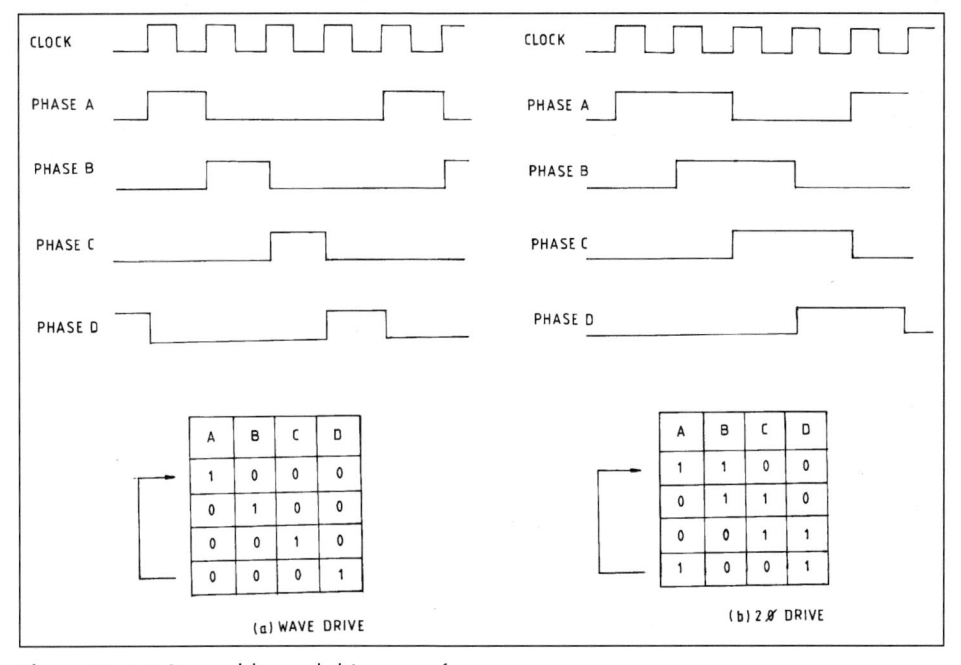

**Figure 11.4.2** State tables and drive waveforms

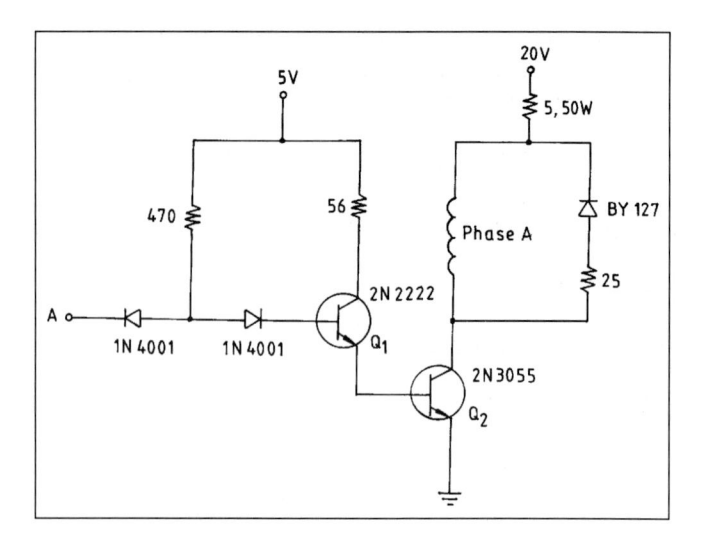

**Figure 11.4.3** Power amplifier for each phase

NAND gates (7400). Connect the reset terminals of the flip-flops as shown to ensure that the count is 00 at power-up. Decode the outputs of the counter with a 2-to-4 line decoder (74139) to select one of the output lines $\overline{0}$ through $\overline{3}$. The outputs of the decoder provide wave-drive or 2-phase drive.

Wave drive has one phase ON at any given time, and the phases are energized sequentially. Figure 11.4.2a shows the state table for wave drive. The outputs of the decoder are inverted to turn on the power amplifiers for each phase.

In 2-phase drive, two phases are on at any given instant. The state table for 2-phase drive is shown in Figure 11.2.2b. To derive the driving sequence for this mode, the outputs are combined two at a time as inputs to the exclusive-OR gates (7486). The gate outputs are connected to the power amplifier inputs. The power amplifier for one phase is shown in Figure 11.4.3.

*Reprinted from EDN (September 5, 1985)*
*© 1992 CAHNERS PUBLISHING COMPANY*
*A Division of Reed Publishing USA*

◆  ◆

## 11.5 PLD adds flexibility to motor controller

The stepper-motor controller shown in Figure 11.5.1 uses no discrete devices and is flexible enough that you can program it for any type of drive. The circuit uses a PLD to generate the necessary logic sequence depending on the type of drive you're using. This PLD approach improves upon the shift register and up/down counter logic-sequencing generation methods because spurious noise pulses can't alter the drive sequence. In the counter and shift-register methods, a single

Motor control circuits 183

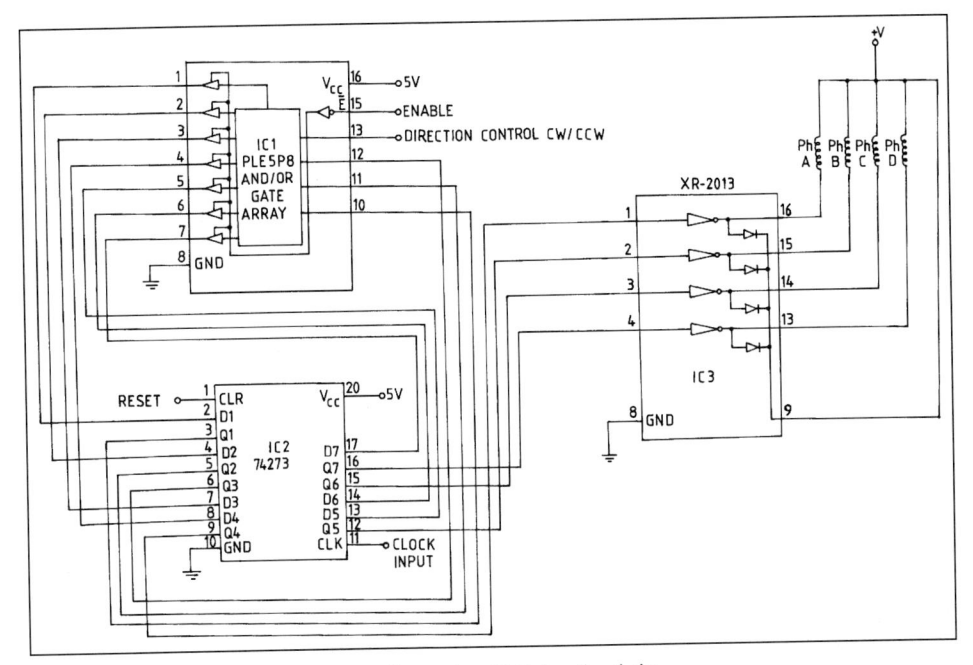

**Figure 11.5.1** Stepper motor controller using PLD for flexibility

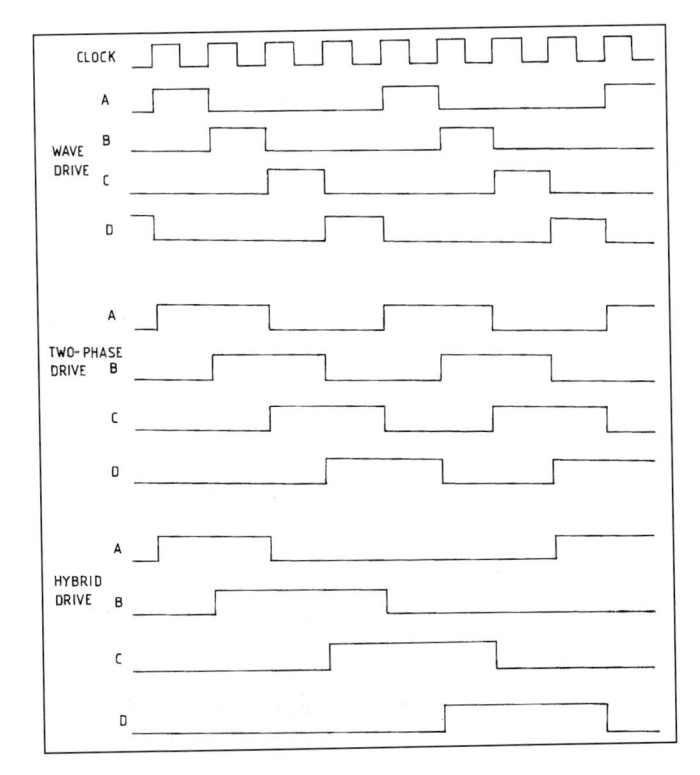

**Figure 11.5.2** Phase switching waveforms for various types of drive

**Table 11.5.1** Timing generator for stepper motor controller: truth table

A2 A1 A0 B2 B1 B0 P A B C D

| State 210 | Next state 210 | P | Timing signals; AAA |   |   |   | BBB comments |   |   |
|---|---|---|---|---|---|---|---|---|---|
|   |   |   | A | B | C | D |   |   |   |
| LLL | LLH | L | H | L | L | L | Phase A | ON | Wave- |
| LLH | LHL | L | L | H | L | L | Phase B | ON | drive |
| LHL | LHH | L | L | L | H | L | Phase C | ON | CW DIR. |
| LHH | HLL | L | L | L | L | H | Phase D | ON |   |
| LLL | LLH | H | H | L | L | L | Phase A | ON | Wave- |
| LLH | LHL | H | L | H | L | L | Phase B | ON | drive |
| LHL | LHH | H | L | L | H | L | Phase C | ON | CCW |
| LHH | HLL | H | L | L | L | H | Phase D | ON | DIR. |
| LLL | LLH | L | H | H | L | L | Phases A,B | ON | Two- |
| LLH | LHL | L | L | H | H | L | Phases B,C | ON | phase |
| LHL | LHH | L | L | L | H | H | Phases C,D | ON | drive |
| LHH | HLL | L | H | L | L | H | Phases D,A | ON | CW DIR. |
| LLL | LLH | H | H | H | L | L | Phases A,B | ON | Two- |
| LLH | LHL | H | L | H | H | L | Phases B,C | ON | phase |
| LHL | LHH | H | L | L | H | H | Phases C,D | ON | drive |
| LHH | HLL | H | H | L | L | H | Phases D,A | ON | CCW DIR. |
| LLL | LLH | L | H | L | L | L | Phase A | ON | Hybrid- |
| LLH | LHL | L | H | H | L | L | Phases A,B | ON | drive |
| LHL | LHH | L | L | H | L | L | Phase B | ON | CW DIR. |
| LHH | HLL | L | L | H | H | L | Phases B,C | ON |   |
| HLL | HLH | L | L | L | H | L | Phase C | ON |   |
| HLH | HHL | L | L | L | H | H | Phases C,D | ON |   |
| HHL | HHH | L | L | L | L | H | Phase D | ON |   |
| HHH | LLL | L | H | L | L | H | Phases A,D | ON |   |
| LLL | LLH | H | H | L | L | L | Phase A | ON | Hybrid- |
| LLH | LHL | H | H | H | L | L | Phases A,B | ON | drive |
| LHL | LHH | H | L | H | L | L | Phase B | ON | CCW DIR. |
| LHH | HLL | H | L | H | H | L | Phases B,C | ON |   |
| HLL | HLH | H | L | L | H | L | Phase C | ON |   |
| HLH | HHL | H | L | L | H | H | Phases C,D | ON |   |
| HHL | HHH | H | L | L | L | H | Phase D | ON |   |
| HHH | LLL | H | H | L | L | H | Phases A,D | ON |   |

change caused by noise in any bit can continually circulate in the logic and alter the drive sequence completely. Using a PLD also adds flexibility to the sequencing logic design (see Figure 11.5.2).

The 74273 octal D flip-flop latches the PLD outputs at the required frequency. The XR-2013, a high-voltage, high-current Darlington transistor array, translates the output logic level of the octal latch to a higher voltage. You can apply voltages as high as 50 V to each phase of the stepper motor in order to force the required current through the motor winding. Each Darlington pair in the array can supply 600 mA on a continuous basis, which is sufficient phase current to drive a small stepper motor. To drive large stepper motors, you can increase the drive-current-per-phase by paralleling two or more of the Darlington pairs in series with each phase's winding.

**Table 11.5.2** PLD equations

Wave drive
Next state generator
B2 = /A2*A1*A0
B1 = /A2*A1*A0 + /A2*/A1*A0
B0 = /A2*/A0
Stepping sequence generator
A = /A2*A1*A0*P + /A2*/A1*/A0*/P
B = /A2*A1*/A0*P + /A2*/A1*A0*/P
C = /A2*A1*/A0*/P + /A2*/A1*A0*P
D = /A2*/A1*/A0*P + /A2*A1*A0*/P

Two-phase drive
Next state generator
B2 = /A2*A1*A0
B1 = /A2*A1*/A0 + /A2*/A1*A0
B0 = /A2*/A0
Stepping sequence generator
A = /A2*/A1*/A0 + /A2*A1*A0
B = /A2*A1*P + /A2*/A1*/P
C = /A2*A1*/A0 + /A2*/A1*A0
D = /A2*A1*/P + /A2*/A1*P

Hybrid drive
Next state generator
B2 = A2*/A0 + A2*/A1 + /A2*A1*A0
B1 = A0 :+: A1
B0 = /A0
Stepping sequence generator
A = /A2*/A1*/P + /A2*/A1*/A0 + A2*A1*P + A2*A1*A0
B = /A2*A0*/P + /A2*A1*/P + A2*/A0*P + A2*A1*P
C = A2*/A1*/P + A2*/A1*/A0 + /A2*A1*P + /A2*/A1*A0
D = /A2*/A1*P + /A2*/A0*P + A2*/A1*/P + A2*A0*/P

The circuit uses the PLE 5P8 PLD, which has five inputs and eight outputs. The circuit uses three of the five inputs for the state-incrementing control function. Another input controls the motor's direction input. The circuit uses three of the eight outputs to generate the next address, and four of the outputs to drive the four motor phases. The circuit doesn't require that you connect any external free-wheeling diodes across the phase windings because they are included in the XR-2013.

Table 11.5.1 lists the timing generator programming details for different types of stepper-motor drivers. The variable $P$ represents the direction control bit (clockwise versus counterclockwise), and the variables $A$ to $D$ represent the four motor phases. You can derive the PLD programming equations shown in Table 11.5.2, using Karnaugh maps. Table 11.5.2 presents the programming details for the PLD for some of the commonly used drive sequences.

*Reprinted from EDN (March 1, 1990)*
*© 1992 CAHNERS PUBLISHING COMPANY*
*A Division of Reed Publishing USA*

# 12 Encoders/decoders

## 12.1 Revolution sensor

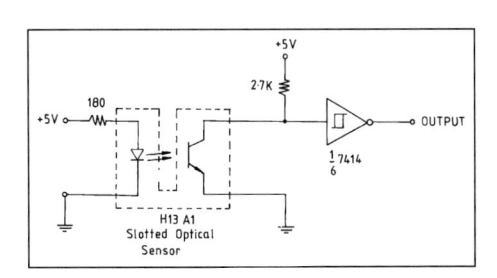

**Figure 12.1.1** Revolution sensor

The circuit shown in Figure 12.1.1 can be used to sense revolutions of a motor shaft or a similar object. The circuit uses a photo-coupled interrupter module, H13 A1. A slotted disc or a disc with fine radial opaque lines can be inserted between the emitter and the detector of H13 A1 through the gap provided. When the IR beam is interrupted in H13 A1 by the revolution of the disc, a logic transition in the output of the interrupter is observed. The Schmitt trigger inverter (1/6 7414) wave-shapes the output into well-defined pulses for the subsequent logic. This monolithic arrangement does away with discrete LEDs/lamps and photo-transistors, thereby improving reliability and simplifying design.

♦ ♦

## 12.2 Photo-diode detector

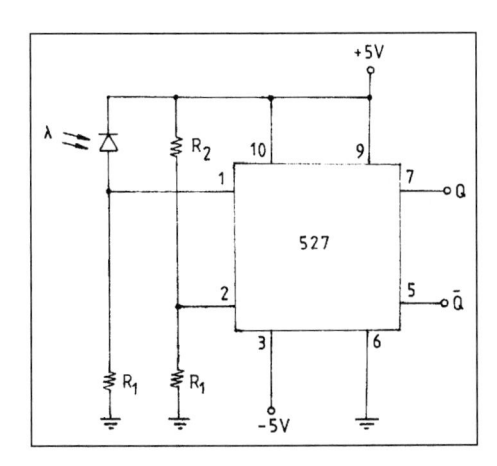

**Figure 12.2.1** Photo-diode detector

Responding to the presence or absence of light, a photo-diode increases or decreases the current through it. Detecting the changes becomes a matter of converting light and dark currents into voltage across a resistor as shown in

Figure 12.2.1. The circuit uses the high-speed comparator NE 527. $R_1$ is selected to be large enough to generate detectable differences between light and dark conditions. Once the signal levels are defined by $R_1$ and the diode characteristics, the average between light and dark signals is used for V-reference and is produced by the resistive divider consisting of $R_1$ and $R_2$. The comparator then produces an output dependent upon the presence or absence of light upon the diode. Logic-compatible digital outputs are obtained at output pins 7 (Q) and 5 ($\overline{Q}$) as shown.

*Courtesy of Philips, The Netherlands*

## 12.3 Detector for magnetic transducer

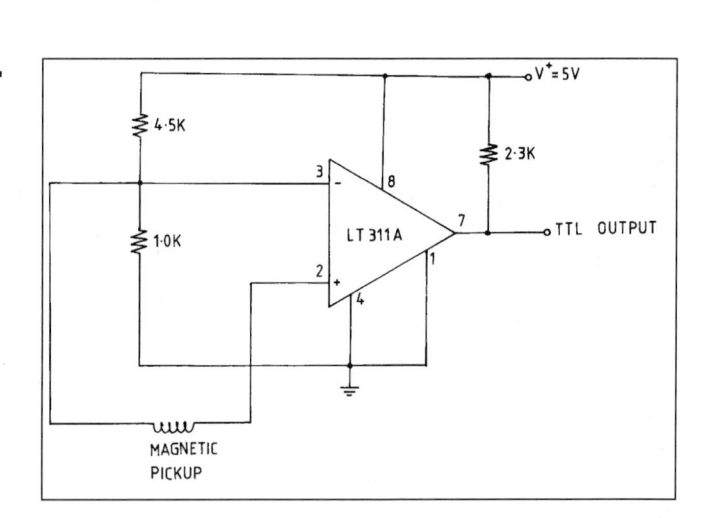

**Figure 12.3.1** Detector for magnetic transducer

The simple circuit shown in Figure 12.3.1 can be used as a detector for a magnetic transducer. It uses a LT 311A voltage comparator operating at 5 V to give a TTL-compatible output. This configuration can be used for detecting the output of any other transducer as well.

*Courtesy of Linear Technology Corporation, Milpitas, CA*

## 12.4 FSK demodulator using PLL 565

Frequency shift keying (FSK) refers to data transmission by means of a carrier which is shifted between two preset frequencies. This frequency shift is usually accomplished by driving a VCO with the binary data signal so that the two

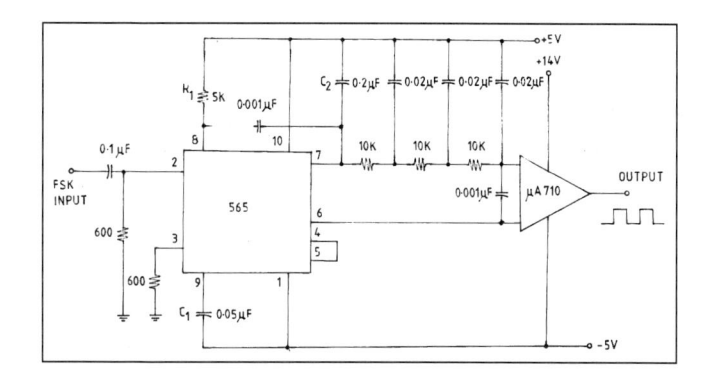

**Figure 12.4.1** FSK decoder using PLL 565

resulting frequencies correspond to the '0' to '1' states (commonly called space and mark) of the binary data signal.

A simple scheme using the 565 to receive FSK signals of 1070 Hz and 1270 Hz is shown in Figure 12.4.1. As the signal appears at the input, the loop locks to the input frequency and tracks it between the two frequencies with a corresponding DC shift at the output (pin 7).

The loop filter capacitor of the PLL ($C_2$ in Figure 12.4.1) is chosen to set the proper overshoot on the output and a three-stage RC ladder filter is used to remove the frequency components. The band edge of the ladder filter is chosen to be approximately half-way between the maximum keying rate (300 baud or 150 Hz). The free-running frequency should be adjusted (with $R_1$) so that the DC voltage level at the output is the same as that at pin 6 of the loop. The output signal can now be made logic-compatible by connecting a voltage comparator between the output and pin 6. The input connection is typical for cases where a DC voltage is present at the source and therefore, a direct connection is not desirable. Both input terminals are returned to ground with identical resistors (in this case, the values are chosen to achieve 600 ohms input impedance).

*Courtesy of Philips, The Netherlands*

◆ ◆

## 12.5 SCA decoder

Some FM stations are authorized by the FCC to broadcast uninterrupted background music for commercial use. To do this a frequency modulated sub-carrier of 67 kHz is used. The frequency is chosen so as not to interfere with the normal stereo or monaural programme; in addition, the level of the sub-carrier is only 10% of the amplitude of the combined signal.

The SCA (subsidiary carrier authorization) signal can be filtered out and demodulated with the 565 phase-locked loop without the use of any resonant circuits. A connection diagram is shown in Figure 12.5.1. The circuit in Figure

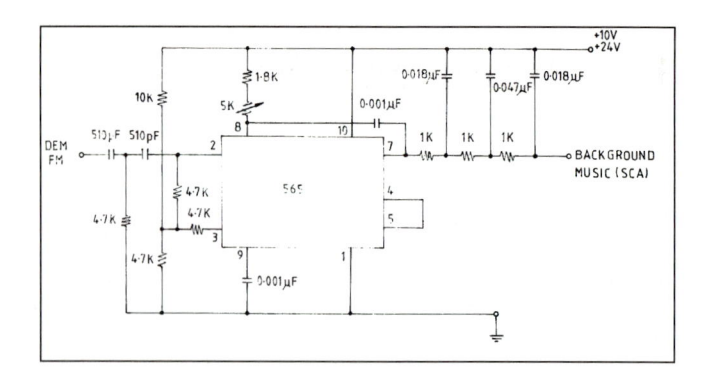

**Figure 12.5.1** SCA decoder

12.5.1 is connected to a point between the FM discriminator and the de-emphasis filter of a commercial band (home) FM receiver. The working of the circuit is as follows:

A resistive voltage divider is used to establish a bias voltage for the input (pins 2 and 3). The demodulated (multiplex) FM signal is fed to the input through a two-stage high-pass filter, both to effect capacitive coupling and to attenuate the strong signal of the regular channel. A total signal amplitude, between 80 mV and 300 mV, is required at the input. Its source should have an impedance of less than 10k. The phase locked loop is tuned to 67 kHz with a 5k potentiometer; only approximate tuning is required, since the loop will seek the signal.

The demodulated output (pin 7) passes through a three-stage low-pass filter to provide de-emphasis and attenuate the high-frequency noise which often accompanies SCA transmission. Note that no capacitor is provided directly at pin 7; thus, the circuit is operating as a first-order loop. The demodulated output signal is in the order of 50 mV and the frequency response extends to 7 kHz. By tuning the receiver to a station which broadcasts an SCA signal, one can obtain hours of commercial-free background music.

*Courtesy of Philips, The Netherlands*

# 13 Tester circuits

## 13.1 On-board transistor tester

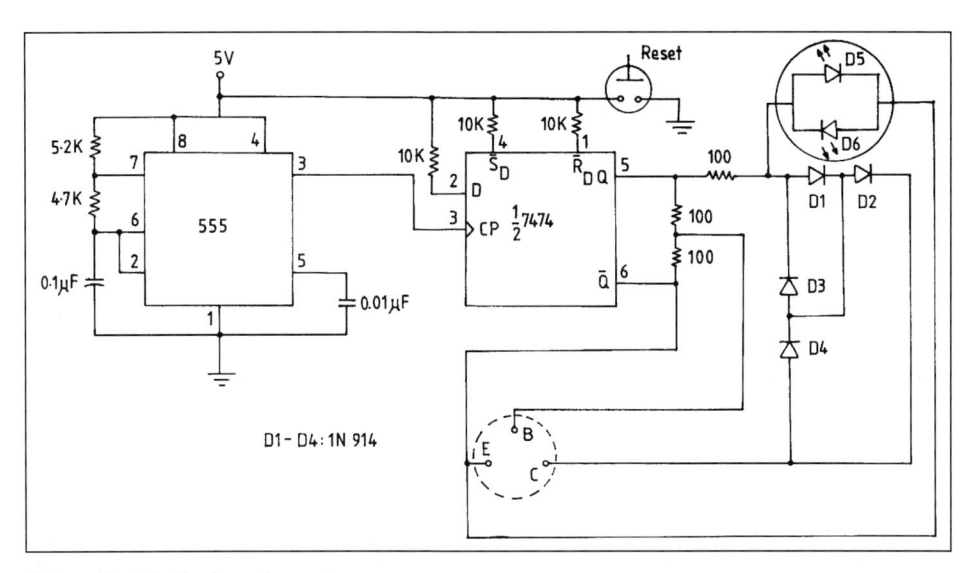

**Figure 13.1.1** On-board transistor tester

This instrument will test transistors without removing them from the printed board. A 555 multivibrator oscillates at a frequency of 1 kHz, its output being taken to a 7474 D-type flip-flop connected as a toggle, which produces complementary square waves (Q and $\overline{Q}$) at 500 Hz. Diodes $D_5$ and $D_6$ are red and green LEDs in the same package. Base drive for the transistor-under-test comes from the mid-point of the Q and $\overline{Q}$ potential divider (Figure 13.1.1).

**Table 13.1.1**

| State | $D_5$ | $D_6$ |
| --- | --- | --- |
| Open C/E | Flicker | Flicker |
| Short C/E | Off | Off |
| Good PNP | On | Off |
| Good NPN | Off | On |

With no transistor connected, the bicolour LED appears amber, since both LEDs switch ON and OFF at 500 Hz. If a good PNP transistor is connected, it is ON when Q is LOW and $\overline{Q}$ is HIGH, since its base–emitter junction is forward-biased; in this condition, neither LED lights since a low Q reverse-biases $D_5$ and the voltage across $D_6$ is equal to $V_{CE(ON)}$ which, for a good transistor is 0.1 V. During the next pulse, Q becomes HIGH and $\overline{Q}$ LOW and a good transistor at the terminals will be OFF; in this condition, $D_6$ is OFF because it is reverse-biased and D₅ is ON. The opposite effect applies if a good NPN transistor is connected.

A good transistor has a collector/emitter voltage of around 0.1 V and a silicon diode drops about 0.6 V. Each of the two loops formed by $D_{1,2}$ and $D_{3,4}$ between pins 5 and 6 of the flip-flop has one collector/emitter drop and two diode drops to impress on the LEDs which, at 0.1 V + 1.2 V = 1.3 V, is insufficient to turn ON one of the LEDs which stays OFF if the transistor is good and ON if the device has one shorted junction to make it behave like a diode; in this case the voltage drop is 1.2 V + 0.6 V = 1.8 V.

Therefore, one LED lights if the transistor is good (both PNP and NPN); both LEDs are OFF for a transistor with shorted C/E junction; and both are ON if the transistor is bad. Diodes $D_{1,2,3,4}$ prevent false indications of normality when a transistor with a B/C short or a B/E short is connected. Table 13.1.1 summarizes the tester operation.

*First published in Electronics and Wirless World, May 1991. Reprinted with permission*

◆ ◆

## 13.2 Zener diode tester and unmarked Zener identifier

**Figure 13.2.1** Zener diode tester and unmarked Zener identifier

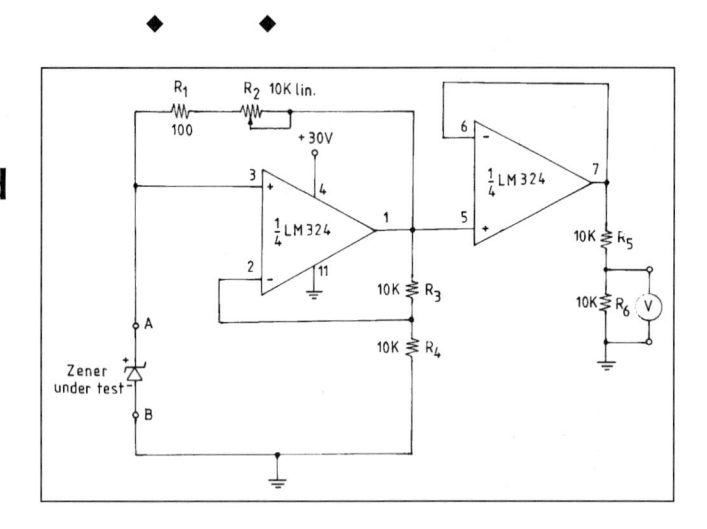

The circuit in Figure 13.2.1 can be used to test Zener diodes and also to identify the breakdown voltage of an unmarked Zener diode. The simple circuit provides a constant current for the Zener under measurement, adjustable by varying $R_2$, independent of changes in power supply voltage. The output voltage of the first op-amp is given by

$$V_o = V_z(R_3 + R_4)/R_4$$

and the Zener current by

$$I_z = (V_o - V_z)/(R_1 + R_2)$$

The output of the first op-amp depends on the breakdown voltage of the Zener diode connected across the terminals A and B and the values of the resistors. The circuit can go into a latch-up state on the application of power if dual supply voltages are used for the op-amp since the circuit is based on positive feedback. Under this condition, the Zener will become forward-biased and function like an ordinary diode. To overcome this problem, the negative supply terminal of the op-amp is grounded and the op-amp is connected to a single positive supply as shown. In the circuit, if $R_3 = R_4$, then $V_o = 2V_z$. The second op-amp in Figure 13.2.1 provides a buffered output which is then halved for reading the Zener breakdown voltage $V_z$ (given by $V_z = V_o/2$) across $R_6$. A voltmeter connected across $R_6$ will directly indicate the breakdown voltage of the Zener under test, if the Zener current is adjusted to the optimum value as specified by the manufacturer.

To test a given Zener diode, the current through the Zener is adjusted to the specified value and a voltmeter connected across $R_6$. If the Zener diode is a good one, the voltmeter reading should equal the breakdown voltage of the Zener; any other reading indicates a faulty Zener.

Conversely, to identify an unmarked Zener diode, connect the Zener across terminals A and B. Vary the Zener current by adjusting $R_2$. At a particular value of Zener current, the voltmeter connected across resistor $R_6$ indicates a fairly constant voltage. The voltmeter reading indicates the Zener breakdown voltage of the Zener under test. A reference to the standard Zener-voltage table will give the exact Zener breakdown voltage, since Zeners are available in certain standard values.

This single chip circuit should prove useful in the lab for testing and identifying Zener diodes.

## 13.3 FET tester

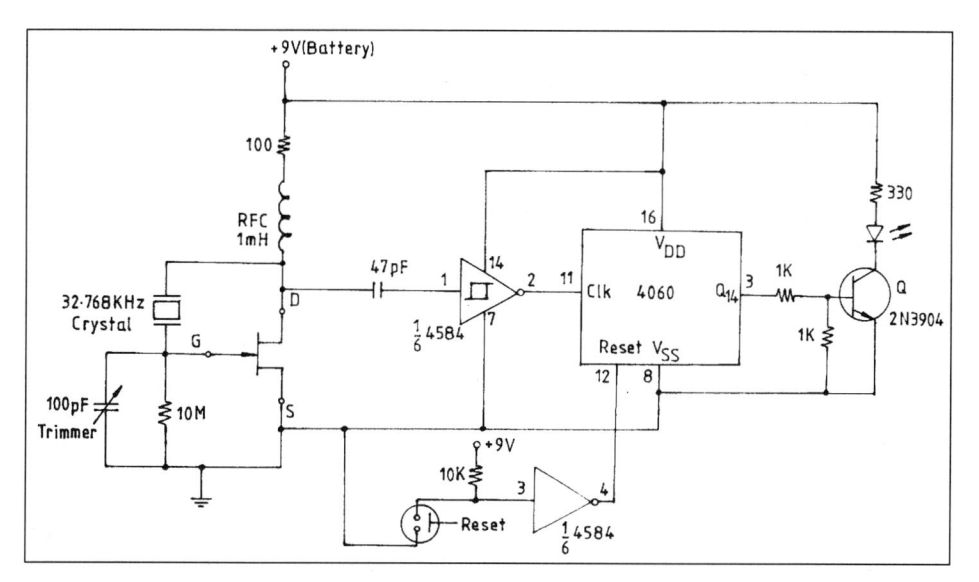

**Figure 13.3.1** FET tester

194  Tester circuits

Often it is required to test Field Effect Transistors in the laboratory. The circuit shown in Figure 13.3.1 can be used for this purpose. The circuit consists of a Pierce crystal oscillator in which the FET under test is used as the active device. The output of the crystal oscillator is wave-shaped by a Schmitt trigger inverter (1/6 4584) and used to drive a 14-stage binary counter and oscillator, 4060. The frequency of the crystal used in this circuit is 32.768 kHz (watch crystal). The output of the frequency divider at $Q_{14}$ is, therefore, $32768/2^{14} = 2\,Hz$. The $Q_{14}$ output drives the transistor Q (2N 3904) ON and OFF at a frequency of 2 Hz. The LED connected at the collector of transistor Q, therefore, becomes ON and OFF at 2 Hz rate if the FET under test is a good one and makes the oscillator work. The 100 pF trimmer capacitor is used for tuning the circuit. The counter can be cleared using the Reset pushbutton provided. This circuit can perform a simple GO/NO GO test on a FET – if the FET is good the LED flickers, otherwise it doesn't.

## 13.4 Crystal tester

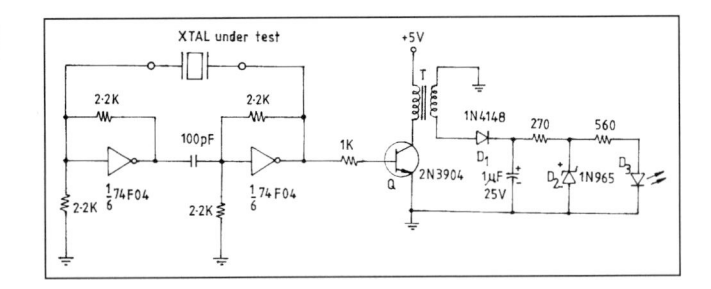

**Figure 13.4.1** Crystal tester

The circuit shown in Figure 13.4.1 can be used as a functional tester for series resonant crystals. The tester uses the crystal under test in an oscillator circuit and gives an LED indication regarding its functionality. The output of the crystal oscillator drives the transistor Q ON and OFF at the oscillator frequency. The primary of a 1 mH pulse transformer is connected as the collector load for the transistor. The secondary of the pulse transformer is connected to a diode rectifier, a capacitor filter and a Zener diode regulator. The final DC output of the voltage regulator is connected to a LED through a current limiting resistor.

If the crystal under test is a good one, the oscillator circuit switches the transistor Q ON and OFF and this switching action produces voltage spikes at the transistor collector due to the inductive load (the primary coil of the pulse transformer). This gets reflected at the secondary of the pulse transformer. The voltage spikes at the secondary are rectified by diode $D_1$, filtered by the 1 µF capacitor and regulated by the Zener diode $D_2$ to about 5 V. A current flows through the LED $D_3$ due to the availability of DC voltage from the regulator output and the LED glows. If, on the other hand, the crystal under test is defective, the oscillator does not give any output and no DC voltage is available for the LED and it is OFF. Thus, the circuit functions as a simple GO/NO GO tester for series resonant crystals. The frequency limit of the crystal that can be tested with this type of circuit depends on the type of digital logic used. This circuit is useful for simple testing of crystals for functionality. A frequency counter can be connected to the output of the oscillator circuit to know the exact frequency of the crystal under test.

## 13.5 Coaxial-cable tester

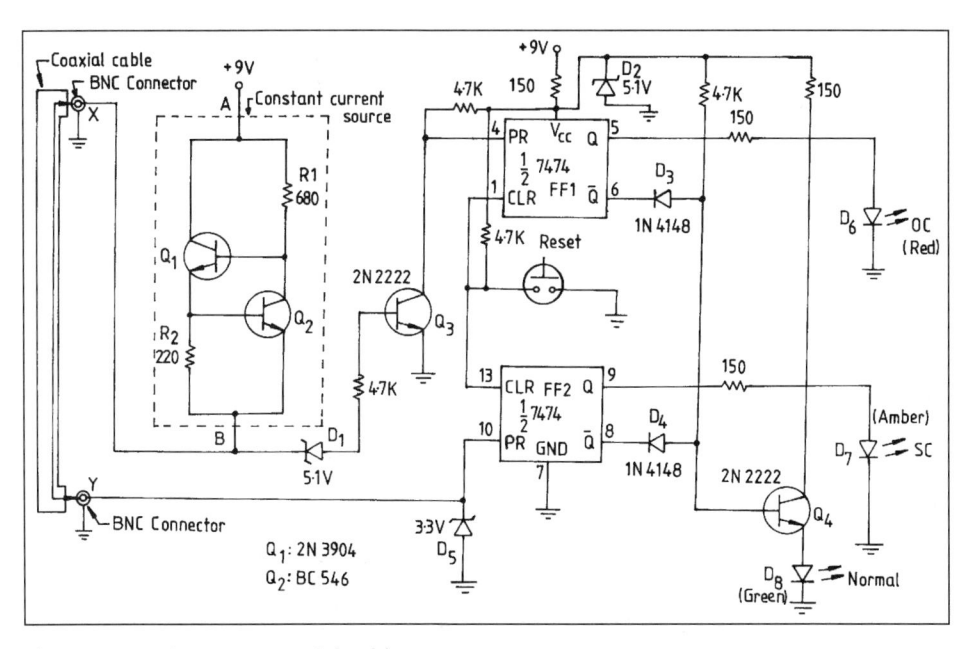

**Figure 13.5.1** Low-cost coaxial-cable tester

Three LEDs indicate the condition of a coaxial cable: whether it is short-circuit, open-circuit or good. A constant-current ring-of-two circuit, Zener diodes and a couple of flip-flops in a 7474 comprise the circuit (Figure 13.5.1).

The current source works as follows. Initially, current flows through the 680 ohm resistor to the base of $Q_1$; as the supply voltage increases beyond $2V_{be}$, voltage across $R_2$ increases and supplies base current to $Q_2$. With a further increase, $Q_2$ conducts more heavily, diverting current from $Q_1$ and keeping a constant current between A and B. In this application, the source supplies 5 mA ($I_1 = V_{be}/R_2$ and $I_2 = (V_{R_1} - 2V_{be})/R_1$).

To test a cable, connect it by BNC connectors or other suitable types across X and Y and reset the flip-flops by the switch. If a good cable is under test, 5 mA flows through the cable to the 3.3 V Zener $D_5$ which, being a lower voltage type than $D_1$, prevents it from conducting. Since the $\overline{Q}$ outputs of the flip-flops are high when reset, $Q_4$ is gated ON and lights the green LED $D_8$.

An open-circuit cable results in current being diverted through $D_1$ and $Q_3$, setting flip-flop 1 and lighting the red LED $D_6$.

Short-circuit cables pull the preset input of flip-flop 2 to ground, setting it and lighting the amber LED $D_7$.

*First published in Electronics World + Wireless World, July 1991. Reprinted with permission*

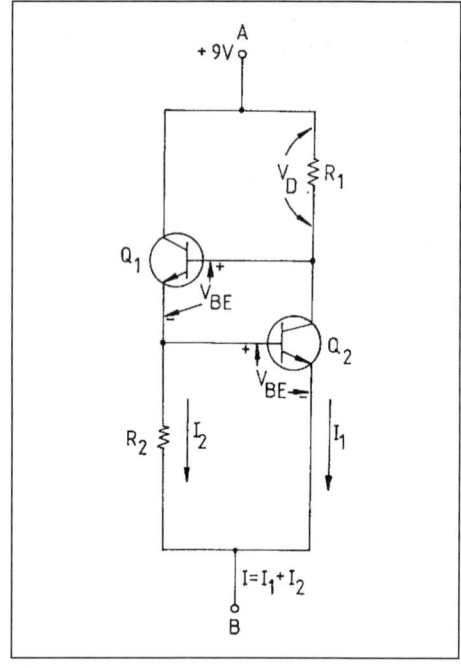

**Figure 13.5.2** Constant current source

# 14 Miscellaneous circuits

## 14.1 High-side switch

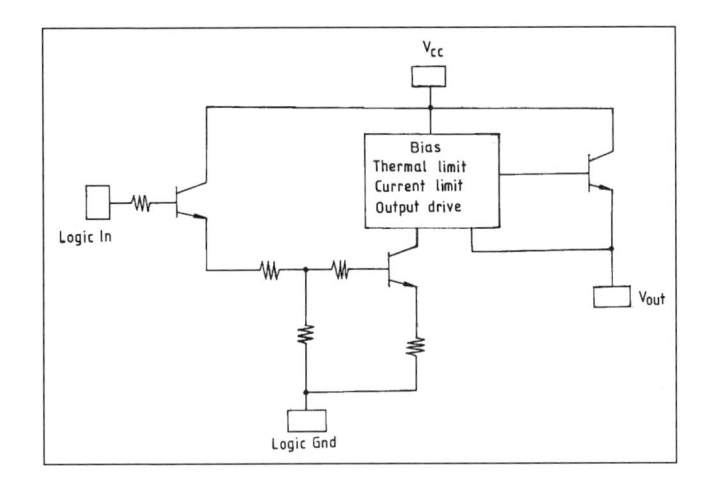

**Figure 14.1.1** Block diagram of high-side switch LT 1089

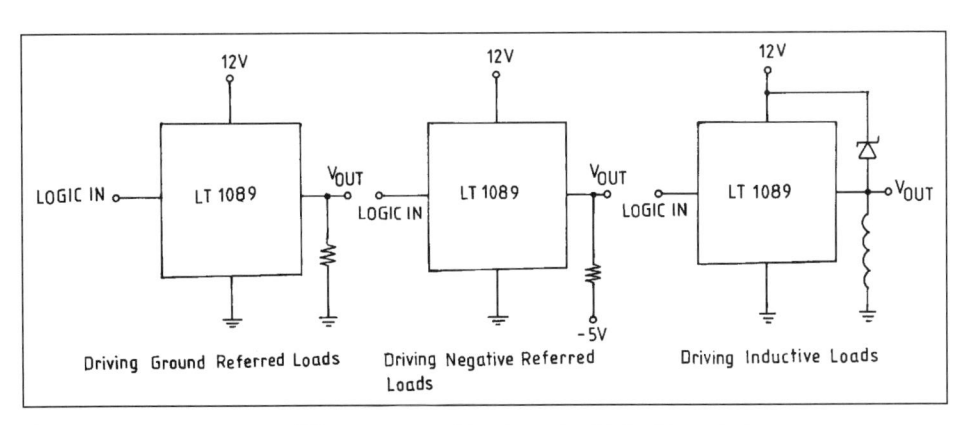

**Figure 14.1.2** Connecting different types of loads to the high-side switch

The LT 1089 from Linear Technology Corporation is a logic-driven, high-current, high-side switch. The device is capable of driving loads up to 7.5 A with a low series drop of only 1.5 V, and the series drop is specified over the full range of switch currents. The device has internal current limiting and thermal overload protection. The input logic is designed so that the output can drive loads referenced either above or below the device ground pin. Either positive or negative logic can be used to drive the input.

The logic and ground pins function as a differential logic input with a common-mode range of $V_{CC}$ to $V_{CC} - 20$ V and a differential threshold voltage ($V_{LOGIC} - GND$) of 1.5 V. If either Logic In or GND is left open the switch remains inactive.

198  Miscellaneous circuits

The following precautions must be taken to protect the device. The LT 1089 must be protected against over-voltage at turn-off when driving inductive loads. The inductive flyback voltage can easily exceed the maximum operating switch voltage ($V_{CC} - V_{OUT}$) of 20 V, potentially damaging the switch. The solution is to clamp the switch voltage to 20 V or less with a Zener diode. The switch can handle 7.5 A and the Zener may be required to handle the same amount of current.

Care must be exercised when operating near the maximum switch voltage. A high-current or capacitive load may trip the current limit circuit at turn-on, thereby adversely affecting the rise-time of $V_{OUT}$. The rise-time is then governed by the current limit divided by the load capacitance, while the fall-time is a function of the complex load. In addition, at switch voltages greater than 18 V the switch current must be less than 0.5 A or the device output will not pull up.

Figure 14.1.1 gives a block diagram of LT 1089 and Figure 14.1.2 shows the method of connecting different types of loads to the switch.

*Courtesy of Linear Technology Corporation, Milpitas, CA*

◆  ◆

## 14.2 Power MOSFET driver

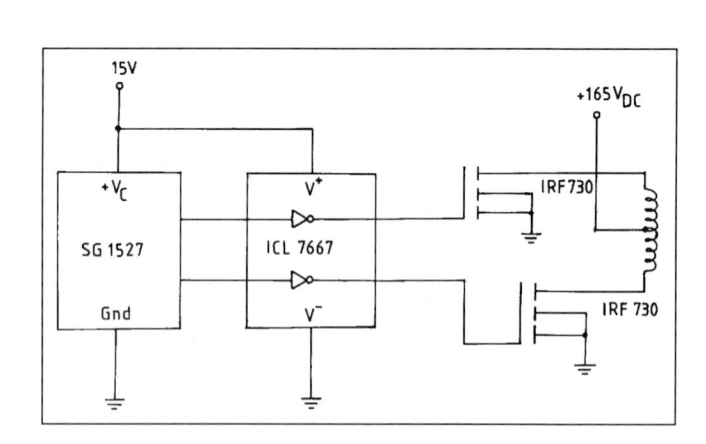

**Figure 14.2.1** Direct drive of MOSFET gates

The ICL 7667 is a dual monolithic high-speed driver designed to convert TTL level signals into high current output at voltages up to 15 V. Its high speed and current output enable it to drive large capacitive loads with high slew rates and low propagation delays. With an output voltage swing only millivolts less than the supply voltage and a maximum supply voltage of 15 V, the ICL 7667 is well suited for driving power MOSFETs in high-frequency switched-mode power converters. The ICL 7667's high current outputs minimize power losses in the power MOSFETs by rapidly charging and discharging the gate capacitances. The ICL 7667's inputs are TTL compatible and can be directly driven by common pulse-width modulation control ICs.

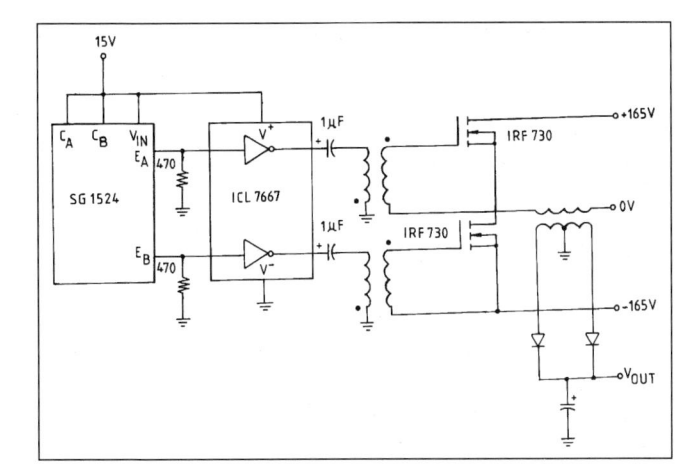

**Figure 14.2.2** Transformer-coupled drive circuit

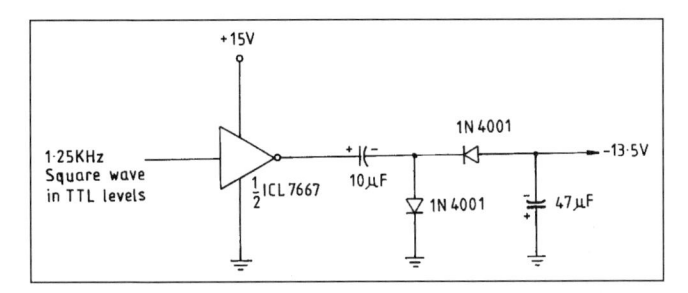

**Figure 14.2.3** Voltage inverter

In addition to power MOS drivers, the ICL 7667, is well suited to other applications such as, control signal, and clock drivers on large memory or microprocessor boards, where the load capacitance is large and low propagation delays are required. Other potential applications include peripheral power drivers and charge-pump voltage inverters.

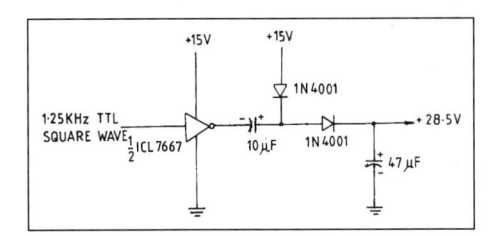

**Figure 14.2.4** Voltage doubler

## Input stage

The input stage is a large N-channel FET with a P-channel constant-current source. This circuit has a threshold of about 1.5 V, relatively independent of the $V_{CC}$ voltage. This means that the inputs will be directly compatible with TTL over the entire 4.5–15 V $V_{CC}$ range. Being CMOS, the input draws less than 1 µA of current over the entire input voltage range of ground to $V_{CC}$. The quiescent current or no-load supply current of the ICL 7667 is affected by the input voltage, going to nearly zero when the inputs are at the 0 logic level and rising to 7 mA maximum when both inputs are at the 1 logic level. A small amount of hysteresis, about 50–100 mV at the input is generated by positive feedback around the second stage.

## Output stage

The ICL 7667 output is a high-power CMOS inverter swinging between ground and $V_{CC}$. At $V_{CC}$ = 15 V, the output impedance of the inverter is typically 7 ohms. The high peak current capability of the ICL 7667 enables it to drive a 1000 pF load with a rise time of only 40 ns. Because the output stage impedance is very low, up to 300 mA will flow through the series N- and P-channel output devices (from $V_{CC}$ to ground) during output transitions. This crossover current is responsible for a significant portion of the internal power dissipation of the ICL 7667 at high frequencies. It can be minimized by keeping the rise and fall times of the input to the ICL 7667 below 1 µs.

## Application hints

Although the ICL 7667 is simply a dual level-shifting inverter, there are several areas to which careful attention must be paid.

*Grounding*  Since the input and the high current output current paths both include the ground pin, it is very important to minimize any common impedance in the ground return. Since the ICL 7667 is an inverter, any common impedance will generate negative feedback, and will degrade the delay, rise and fall times. Use a ground plane if possible, or use separate ground returns for the input and output circuits. To minimize any common inductance in the ground return, separate the input and output circuit ground returns as close to the ICL 7667 as is possible.

*Bypassing*  The rapid charging and discharging of the load capacitance requires very high current spikes from the power supplies. A parallel combination of capacitors that has a low impedance over a wide frequency range should be used. A 4.7 µF tantalum capacitor in parallel with a low inductance 0.1 µF capacitor is usually sufficient bypassing.

*Output damping*  Ringing is a common problem in any circuit with very fast rise or fall times. Such ringing will be aggravated by long inductive lines with capacitive loads. Techniques to reduce ringing include:

(1) Reduce inductance by making printed circuit board traces as short as possible.
(2) Reduce inductance by using a ground plane or by closely coupling the output lines to their return paths.
(3) Use a 10 to 30 ohm resistor in series with the output of the ICL 7667. Although this reduces ringing, it will also slightly increase the rise and fall times.
(4) Use good bypassing techniques to prevent supply voltage ringing.

*Power dissipation* The power dissipation of the ICL 7667 has three main components:

(1) input inverter current loss;
(2) output stage crossover current loss;
(3) output stage $I^2R$ power loss.

The sum of the above must stay within the specified limits for reliable operation.

In cases where the load is a power MOSFET and the gate drive requirements are described in terms of gate charge, the ICL 7667 power dissipation will be

$$P_{AC} = Q_G V_{CC} f$$

where $Q_G$ = Charge required to switch the gate in coulombs and f = frequency.

## Direct drive of MOSFETs

Figure 14.2.1 shows interfaces between the ICL 7667 and typical switching regulator ICs.

## Transformer-coupled drive of MOSFETs

Transformers are often used for isolation between the logic and control section and the power section of a switching regulator. The high output drive capability of the ICL 7667 enables it to directly drive such transformers. Figure 14.2.2 shows a typical transformer-coupled drive circuit. PWM ICs with either active HIGH or active LOW outputs can be used in this circuit, since any inversion required can be obtained by reversing the windings on the secondaries.

## Relay and lamp drivers

The ICL 7667 is suitable for converting low power TTL or CMOS signals into high current, high voltage outputs for relays, lamps and other loads. Unlike many other level translator/driver ICs, the ICL 7667 will both source and sink current. The continuous output current is limited to 200 mA by the $I^2R$ power dissipation in the output FETs.

## Charge pump or voltage inverters and doublers

The low output impedance and wide $V_{CC}$ range of the ICL 7667 make it well suited for charge pump circuits. Figure 14.2.3 shows a typical charge pump voltage inverter circuit. A common use of this circuit is to provide a low current negative supply for analog circuitry or RS 232 drivers. With an input voltage of +15 V, this circuit will deliver 20 mA at −12.6 V. By increasing the size of the capacitors, the current capability can be increased and the voltage loss decreased. The practical range of the input frequency is 500 Hz to 250 kHz. As the frequency goes up, the charge pump capacitors can be made smaller, but the internal losses in the ICL 7667 will rise, reducing the circuit efficiency.

Figure 14.2.4, a voltage doubler, is very similar in both circuitry and performance. A potential use of this circuit would be to supply the higher voltage needed for EEPROM or EPROM programming.

*Courtesy of Harris Semiconductor, Melbourne, FL*

## 14.3 DTMF filter detects tones in exchanges

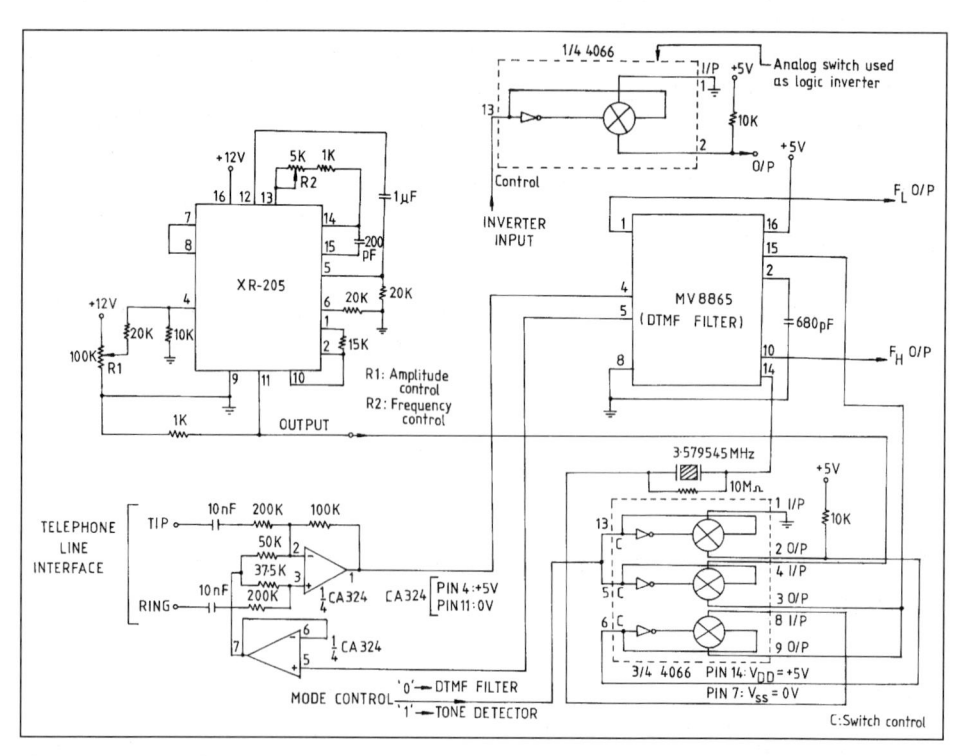

**Figure 14.3.1** A simple interface enables a DTMF filter to detect various exchange tones

A dual-tone multifrequency (DTMF) filter with a decoder detects tone-dialling information and rejects other supervisory exchange tones such as off-hook (dial-tone) ringing and busy signals. DTMF filters built with switched-capacitor techniques achieve good stability, sharp filtering and predictable characteristics, since a single crystal – a low-cost 3.579545 MHz TV crystal – sets the filter's critical passing frequency.

The ability to control the filter's passing frequency can also be used for detecting – instead of rejecting – the telephone-exchange supervisory tones. Reducing the clock frequency to nearly half the usual value shifts the filter's detection frequency to the tones the DTMF filter normally rejects: the 350 and 440 Hz off-hook tone, the 440 and 480 Hz ringing tone, and the 480 and 620 Hz busy tone.

Such a circuit has wide applications for precise detection of the supervisory tones in exchanges. The more traditional techniques of frequency detection, such as phase locked loop tone decoders, tend to give spurious response. But switched-capacitor DTMF integrated circuits eliminate false signals and have good filter characteristics.

Instead of using a special nearly half-frequency crystal or a standard crystal plus a frequency divider, the circuit (see Figure 14.3.1) uses a high-stability monolithic XR-205 waveform generator for the clock frequency. To detect the tones, trimpots $R_1$ and $R_2$ fine-tune the frequency and amplitude required by the switched capacitor filter. A crystal-plus-frequency-divider method cannot achieve the fine-tuning required by such exact frequencies.

The circuit, which includes a 3.579545 MHz crystal, has a quad 4066 analog switch that allows it to operate as a normal DTMF filter. One of the switches in the quad 4066 package acts as an inverter. Switching the mode-control line HIGH or LOW selects, respectively, the XR-205 or the crystal for control-signal frequency detection/DTMF operation.

*Reprinted with permission from Electronic Design (Vol.37, No.4) February 23, 1989. Copyright 1989, Penton Publishing Inc.*

◆ ◆

## 14.4 Four channels on a single-channel oscilloscope

On a single-channel oscilloscope, this circuit, using only four ICs, multiplexes four signal channels for, effectively, simultaneous display (Figure 14.4.1).

The differential 4052 multiplexer works with two sets of four inputs: pins 11, 12, 14 and 15 carry the $y$ signal, while pins 1, 2, 4 and 5 take DC potentials from the four potentiometers to determine the $y$ position on the screen (a DC-coupled oscilloscope is assumed).

Clock pulses variable up to 2 MHz are generated by the 4047 astable and drive the 2-bit Johnson counter, which produces A and B select waveforms for the

204  Miscellaneous circuits

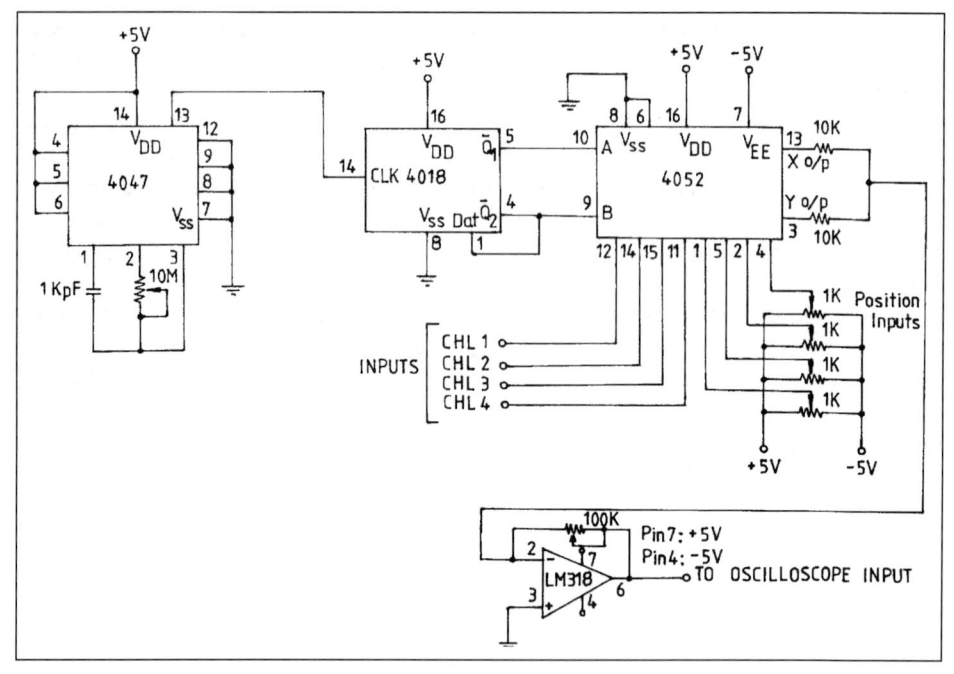

**Figure 14.4.1** Display multiple signals on a single-channel oscilloscope

multiplexer. High switching rates multiplex the *y* inputs at a higher rate than the oscilloscope sweep to give a virtually continuous display.

Output to the single input of the oscilloscope comes from an LM 318 variable-gain op-amp.

*First published in Electronics and Wireless World, January 1992.*
*Reprinted with permission*

## 14.5 Light-activated alarm

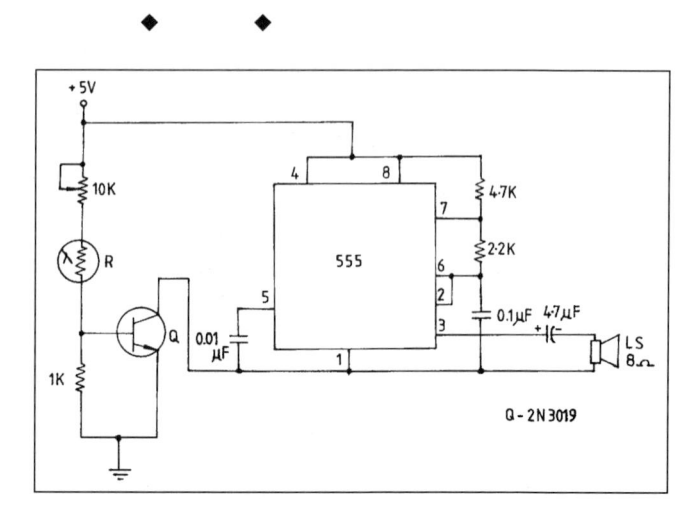

**Figure 14.5.1** Light-activated alarm

The simple circuit shown in Figure 14.5.1 can be used as a light-activated alarm. It consists of an astable multivibrator configured using a 555 timer, the ground path of which is closed and opened by a simple transistor switch. The output of the astable drives a loudspeaker. The switching transistor Q derives its base drive through a 10k potentiometer and a light-dependent resistor connected in series. When no light falls on the LDR its dark resistance is high and sufficient base current does not flow to Q and it remains OFF. The 555 astable does not oscillate since its ground path is open. When light falls on the LDR, its bright resistance falls and there is sufficient base drive to Q to turn it ON. The ON transistor provides a low-resistance ground path for the 555 and it gives a 1.6 kHz tone which drives the loudspeaker giving an audible alarm. The 10k potentiometer is used to set the threshold level of the alarm so that ambient light does not cause spurious operation. The LDR should be enclosed in an opaque casing with a small window for the entry of light.

## 14.6 Remote light monitor

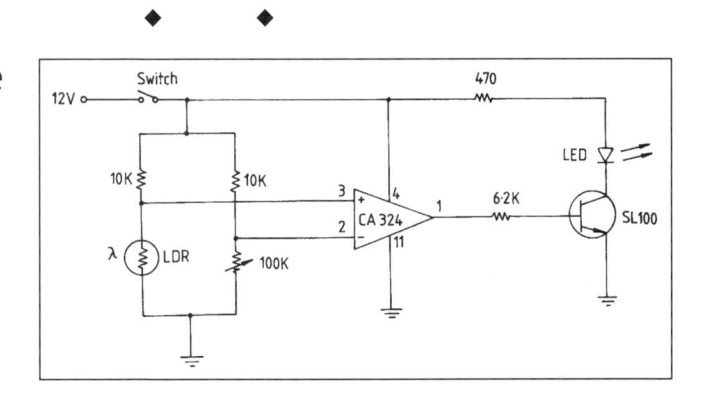

**Figure 14.6.1** Remote light monitor

You can build a very low-cost remote light monitor, for example to monitor the headlight of a vehicle, using the circuit shown in Figure 14.6.1. In this circuit a light-dependent resistor (LDR) is used as one of the four arms of a Wheatstone bridge to sense changes in light intensity. Two 10k resistors and a 100k potentiometer form the other arms of the bridge. One of the four op-amps in the quad op-amp, CA324, is used as a voltage comparator to derive switching action between the headlight ON and OFF conditions. The LDR is mounted such that it faces the headlight reflector. If the headlight is ON, the LDR is illuminated and its resistance is low. The 100k potentiometer is adjusted so that the comparator output is LOW and consequently the transistor (SL 100) is OFF and the LED does not glow. In the event of a fault in the headlight, the LDR is not illuminated, its resistance increases and the bridge becomes unbalanced. This switches the output of the comparator HIGH, turning ON the transistor and the LED glows warning the driver that the headlight is not operational. The 100k potentiometer controls the sensitivity of the circuit and sets the comparator threshold so that ambient light does not cause spurious operation.

## 14.7 Low-cost electronic lamp dimmer

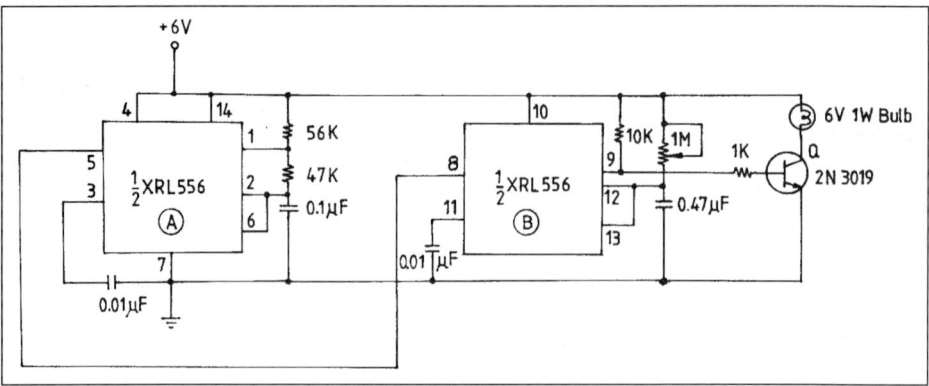

**Figure 14.7.1** Electronic lamp dimmer

You can build a low-cost electronic lamp dimmer using the circuit shown in Figure 14.7.1. The circuit does not use the conventional resistive voltage drop technique which dissipates power in a resistor to control the power delivered to the load such as a lamp. In the circuit shown in Figure 14.7.1, the duty cycle (i.e. the ON/OFF duration) of the voltage pulse train to the lamp is varied by a pulse-width modulator thereby controlling the ON/OFF duration of the lamp. There will be no flicker-effect observed since the ON/OFF frequency is high (100 Hz). The circuit uses a low-power dual timer XR-L556 which consumes less than 1 mW per section, which is less than 1/15 of the power consumed by a normal 556 dual timer, thereby conserving power, especially for battery-operated applications such as emergency-lights, torch-lights, etc.

One half (A) of the dual timer operates as an astable with a frequency of around 100 Hz and feeds the trigger input of the other half (B) which is configured as a monostable with variable pulse width. By varying the 1M potentiometer, the pulse width (ON time) of the monostable output waveform can be adjusted which, in turn, varies the brightness of the bulb connected at the collector of the transistor. In other words, the intensity of the bulb is varied by adjusting the duty cycle of the pulse train turning the transistor ON and OFF. Such a lamp dimmer can be used in emergency-lights, portable torch-lights, etc. to vary the brightness and thereby conserve battery life since full brightness may not be always required. Since the power consumed by the controlling circuits is very small, the drain on the battery is very low.

◆ ◆

## 14.8 Incandescent lamp dimmer and protector

The circuit shown in Figure 14.8.1 can be used to extend the life of an incandescent lamp and also to control its intensity. The circuit is based on a power

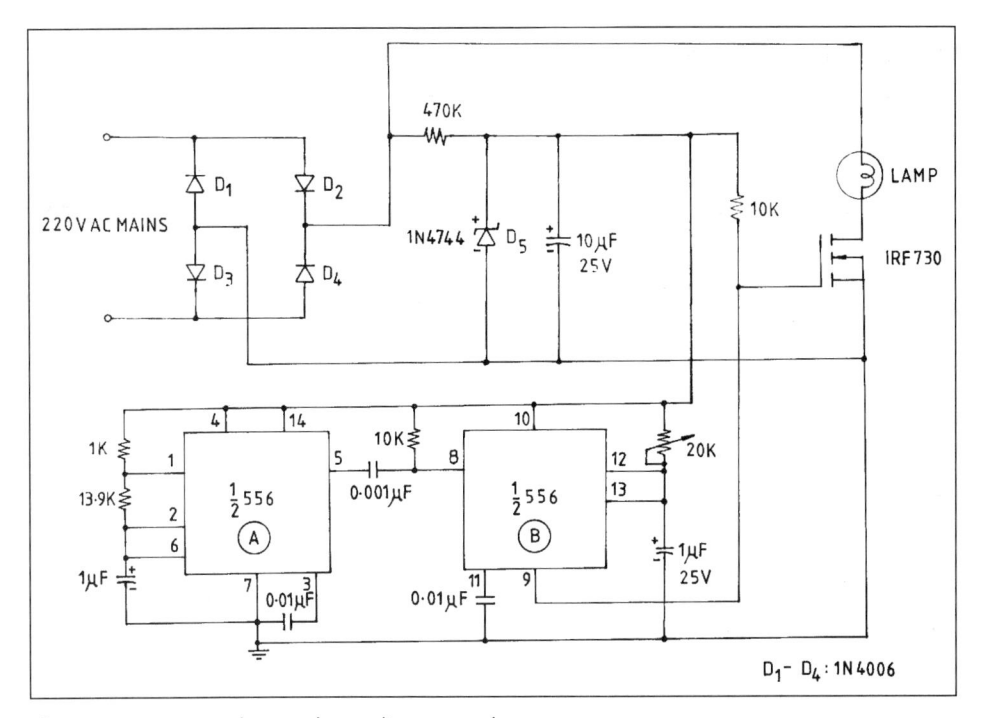

**Figure 14.8.1** Incandescent lamp dimmer and protector

MOSFET and uses a driver circuit for the MOSFET to control the lamp intensity. The life of an incandescent lamp is affected to a large extent by the large inrush current drawn by the lamp at turn-on due to the low resistance of the lamp filament when it is cold compared to its resistance at its normal operating temperature. The driver circuit for the power MOSFET prevents the inrush current at turn-on and extends the life of the lamp. A full-bridge rectifier consisting of diodes $D_1$–$D_4$ provides the DC voltage for the dual timer from the AC mains supply by charging the 10 µF capacitor through the 470k resistor. The Zener diode $D_5$ clamps the capacitor voltage to 15 V. To vary the intensity of the lamp, a duty-cycle control technique is used in the circuit. This is achieved by configuring a fixed frequency variable duty-cycle oscillator using a 556 timer. The 'A' half of the 556 is configured as an astable multivibrator with a free-running frequency of about 50 Hz in this circuit (about equal to the AC mains supply frequency) and the 'B' half as a monostable whose output pulse width can be varied by adjusting the 20k potentiometer. The output of the astable 'A' is used to trigger the monostable 'B'. If the *RC* components of 'A' and 'B' are chosen such that the time-period of oscillation of astable 'A' is greater than the output pulse width of one-shot 'B', then the output of 'B' has the same frequency as the output of 'A' but its duty cycle can be varied from 1% to 99% by varying its *RC* value. The output of the monostable 'B' is used as the gate drive for the MOSFET. The 10k resistor is used to pull-up the monostable output. Adjusting the 20k potentiometer varies the pulse width of the one-shot output and this, in turn, varies the gate drive duty cycle for the MOSFET. This variation in the conduction level of the MOSFET varies the intensity of the incandescent lamp by controlling the current flow through it.

This simple circuit thus performs the dual function of extending the life of an incandescent lamp and also controlling its intensity.

## 14.9 Analog switch needs no supply

An analog switch IC, such as 4016 or 4066, can draw its power requirements from the signal applied at its input without significantly loading the source of the input signal. The diagram shows the scheme for deriving +5 V for $V_{DD}$ and -5 V for $V_{SS}$ supplies needed for the analog switch to handle bipolar input signals without signal clipping.

These quad packages of analog switches need only 1.5 µA under conditions of $V_{in} = V_{SS}$ or $V_{DD}$. The scheme works well for switching frequencies down to 100 Hz. Since the ON resistance of the switch is typically 200 ohms, loading of the source does not occur.

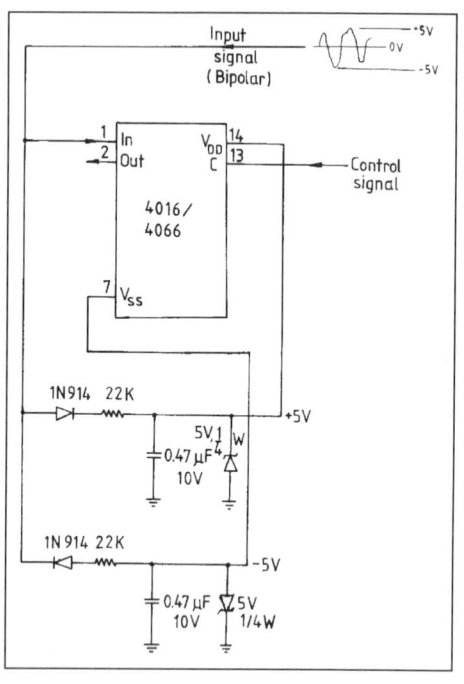

**Figure 14.9.1** Analog switch requires no supply

*First published in Electronics and Wireless World, April 1990.*
*Reprinted with permission*

## 14.10 Simple high/low temperature alarm

The ICL 7665 is a low-power dual over/under-voltage detector drawing a typical operating current of only 3 µA. The trip points and hysteresis of the two voltage detectors are individually programmed via external resistors to any voltage

Miscellaneous circuits 209

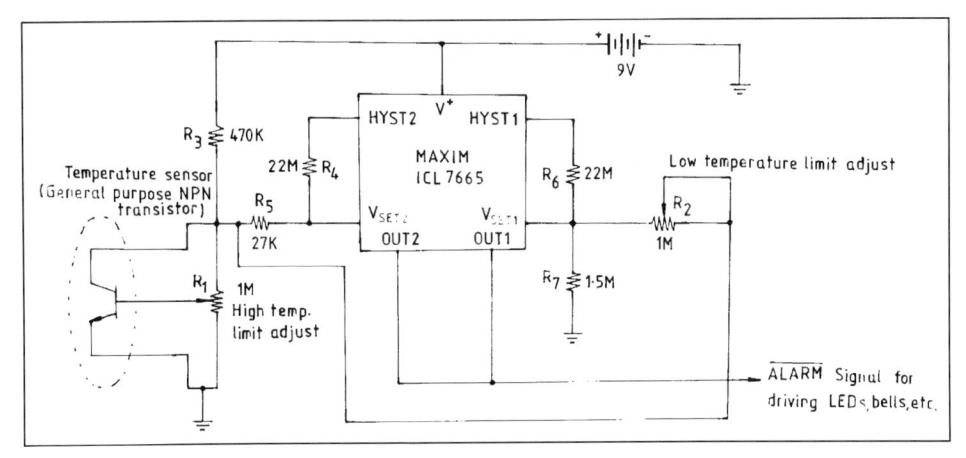

**Figure 14.10.1** Simple high/low temperature alarm

greater than 1.3 V. The ICL 7665 will operate from any supply voltage in the 1.6 V to 16 V range, while monitoring voltages from 1.3 V to several hundred volts.

The circuit shown in Figure 14.10.1 is a simple high/low temperature alarm which uses a low-cost NPN transistor as the sensor and an ICL 7665 as the high/low detector. The NPN transistor and potentiometer $R_1$ form a $V_{be}$ multiplier whose output voltage is determined by the $V_{be}$ of the transistor and the position of $R_1$'s wiper arm. The voltage at the top of $R_1$ will have a temperature coefficient of approximately $-5\,\text{mV}/°C$. $R_1$ is set so that the voltage at $V_{SET2}$ is equal to the $V_{SET2}$ trip voltage when the temperature of the NPN transistor reaches the temperature selected for the high temperature alarm desired. $R_2$ can be adjusted so that the voltage at $V_{SET1}$ is 1.3 V when the NPN transistor's temperature reaches the low temperature limit.

*Courtesy of Maxim Integrated Products, Inc., Sunnyvale, CA*

# Appendix 1

## Reference index of integrated circuits and their sources

| Device | Description | Source code* |
|---|---|---|
| 7400 | Quad 2-input NAND gate | |
| 7402 | Quad 2-input NOR gate | 3,4,7,8,9 |
| 7404 | Hex inverter | |
| 74LS04 | Hex inverter | 3,4,6,7,8,9 |
| 7407 | Hex buffer/driver, open collector | |
| 7408 | Quad 2-input AND gate | |
| 7413 | Dual 4-input NAND Schmitt trigger | |
| 7414 | Hex inverter Schmitt trigger | |
| 7474 | Dual D-type edge triggered flip-flop | |
| 7486 | Quad 2-input EXCLUSIVE-OR gate | |
| 74121 | Monostable multivibrator | |
| 74122 | Retriggerable monostable multivibrator | 3,4,7,8,9 |
| 74123 | Dual retriggerable monostable multivibrator | |
| 74139 | Dual 1-of-4 decoder/demultiplexer | |
| 74161 | Synchronous 4-fit binary counter | |
| 74221 | Dual monostable multivibrator | |
| 74273 | Octal D-type flip-flop with reset | |
| 74367A | Hex buffer/driver (3-state) | |
| 74393 | Dual 4-bit binary ripple counter | |
| 74HC4017 | Decade counter/divider with 10 decoded outputs | 3,4,5,6,7,8,9 |
| 4001 | Quad 2-input NOR gate | |
| 4009 | Hex inverting buffer | |
| 4010 | Hex non-inverting buffer | |
| 4016 | Quad bilateral switch | |
| 4024 | 7 stage binary counter | |
| 4040 | 12-stage binary counter | |
| 4046 | CMOS phase-locked loop | |
| 4047 | Monostable/astable multivibrator | 3,4,5,7,8 |
| 4049 | Hex inverting buffer | |
| 4050 | Hex non-inverting buffer | |
| 4060 | 14 stage ripple-carry binarycounter/divider/oscillator | |
| 4066 | Quad bilateral switch | |
| 4069 | Inverter circuit | |
| 4082 | Dual 4-input AND gate | |
| 4528 | Dual monostable multivibrator | 4,5,7,14 |
| 4538 | Dual retriggerable monostable multivibrator | |

* Source code refers to the serial number in the address list

Appendix 1 211

| Device | Description | Source code* |
|---|---|---|
| 4584 | Hex Schmitt trigger | 7 |
| 40109 | Quad LOW-to-HIGH voltage level shifter | 5,6 |
| 10124 | TTL-to-ECL translator | } 3,4,7,8 |
| 10125 | ECL-to-TTL translator | |
| 1496 | Balanced modulator/demodulator | 4,7 |
| 4N 25 | Opto-coupler | } 3,9 |
| 4N 33 | Opto-coupler | |
| 527 | Voltage comparator | |
| 529 | Voltage comparator | } 4 |
| 530 | High slew rate operational amplifier | |
| 311 | Voltage comparator | 1,3,4,5,6,7,12 |
| 317 | Positive adjustable voltage regulator | 1,3,7,9 |
| 324 | Low-power quad operational amplifier | 4,5,7 |
| 337 | Negative adjustable voltage regulator | 1,7,9 |
| 338 | 5 A positive adjustable voltage regulator | 1,7 |
| 339 | Quad voltage comparator | 4,5,7 |
| TLC 551 | LinCMOS timer | 9 |
| XR-L555 | Low-power timer | 10 |
| 555 | Timer | } 3,4,6,7,8,9,10 |
| 556 | Dual timer | |
| 558 | Quad timer | 4,10,12,15 |
| 565 | Phase locked loop | 3,4,5,7,12 |
| 566 | Function generator | 4,7,12 |
| 567 | Tone decoder | 4,7,10,12 |
| XR-215 | Monolithic phase locked loop | 10 |
| H13 A1 | Photo-coupled interrupter module | 5 |
| 741 | Operational amplifier | 4,5,6,7,9,12 |
| LT 1010 | Fast ±150 mA power buffer | } 1 |
| LT 1011 | Voltage comparator | |
| LT 1032 | Quad low power line driver | |
| LT 1037 | Low noise, high speed precision op-amp | |
| LTC 1042 | Window comparator | |
| LTC 1045 | Programmable micropower hex translator/receiver/driver | |
| LT 1056 | Precision high speed JFET input op-amp | } 1 |
| LT 1080 | Advanced low power 5 V RS 232 dual driver/receiver | |
| LT 1081 | Advanced low power 5 V RS 232 dual driver/receiver | |
| LT 1089 | High-side switch | |
| LT 1101 | Precision micropower single supply instrumentation amplifier | |
| MAX 610 | AC–DC regulator (110/220 V AC to 5 V DC full wave | |
| MAX 612 | AC–DC regulator 8 V RMS to 5 V DC full wave | |
| MAX 630 | CMOS micropower step-up switching regulator | } 2 |
| MAX 634 | CMOS micropower inverting switching regulator | |
| MAX 635 | CMOS –5 V fixed/adjustable output inverting switching regulator | |
| MAX 680 | +5 V to ±10 V voltage converter | |

# Appendix 1

| Device | Description | Source code* |
|---|---|---|
| ICM 7207 | CMOS time base generator | 5 |
| ICM 7208 | 7 digit LED display counter | 5,15 |
| ICM 7217 | 4 digit LED display programmable up/down counter | 2,5,13,14 |
| ICM 7250 | Programmable timer | 3,5 |
| ICM 7555 | General purpose timer | 5.7.12 |
| ICL 7660 | CMOS voltage converter | 2,5 |
| ICL 7664 | Low power programmable negative voltage regulator | 2 |
| ICL 7665 | Micropower under/over voltage detector | 2,5 |
| ICL 7667 | Dual power MOSFET driver | |
| ICL 7673 | Automatic battery back-up switch | |
| ICL 8013 | Four quadrant analog multiplier | 5 |
| ICL 8038 | Precision waveform generator/voltage controlled oscillator | 5,10 |
| ICL 8048 | Logarithmic amplifier | 5 |
| ICL 8049 | Antilog amplifier | |
| ICL 8063 | Power transistor driver/amplifier | 5 |
| 8211 | Programmable voltage detector | 2,5 |
| 8212 | Programmable voltage detector | |
| 8865 | DTMF filter | 18,19 |
| IM 4702 | Baud rate generator | 5,7 |
| IM 4712 | Baud rate generator | |
| NE 5512 | Dual high performance operational amplifier | 4 |
| NE 5535 | Dual high slew rate op-amp | |
| LH 0101 | Power operational amplifier | 2 |
| 93L34 | 8 bit addressable latch | 4 |
| REF 01 | Precision voltage reference | 1,2,12 |
| OP 07 | Precision op-amp | |
| CA 3193 | BiMOS precision op-amp | 5 |
| LF 353 | High slew rate operational amplifier | 3,6,7,9 |
| ULN 2003 | High voltage/high current Darlington transistor array | 3,4,6,9,10,11 |
| ULN 2013 | High voltage/high current Darlington transistor array | 4,10,11 |
| IM 6402 | Universal asynchronous receiver transmitter (UART) | 5 |
| 7805 | 3-terminal 5 V voltage regulator | 7,9 |
| 7812 | 3-terminal 12 V voltage regulator | |
| 1488 | Quad line driver | 3,4,9,10 |
| 1489 | Quad line receiver | 3,4,7,9,10 |
| 3403 | Quad operational amplifier | 3,4,7,9,10 |
| DAC 08 | 8 bit digital to analog converter | 4,12 |

*Note* A few sources are indicated for ICs based on available information. You can refer to a larger directory such as IC Master for information on additional sources. For discrete and passive components, you can contact your local distributor for electronic components. Many countries have their own component manufacturing facilities and if an equivalent discrete device is available from a local source, the same can be used. If any device has been obsoleted by any manufacturer, an equivalent device from another source can be used. Addresses of manufacturers available at the present time are given; you can contact your local agent for the latest address or catalogue of the manufacturer.

# Appendix 2

## Addresses of manufacturers

1. Linear Technology Corporation
   1630 McCarthy Boulevard
   Milpitas, CA-95035
   USA

2. Maxim Integrated Products
   120 San Gabriel
   Sunnyvale, CA-94086
   USA

3. Motorola Inc.,
   P.O. Box 20912
   Phoenix, Arizona-85036
   USA

4. Philips
   International Marketing & Sales
   Building BAF-1
   P.O. Box 218
   5600 MD
   Eindhoven
   The Netherlands

5. Harris Semiconductor
   1301 Woody Burke Road
   Melbourne, Florida-32902
   USA

6. SGS Thomson Microelectronics
   1000 East Bell Road
   Phoenix AZ 85022
   USA

7. National Semiconductor
   2900 Semiconductor Drive
   P.O. Box 58090
   Santa Clara, California – 95052-8090
   USA

8. Hitachi
   Semiconductors & IC Division
   Karukozaka MN Bldg.
   2-1 Ageba-Cho
   Shinjuku-ku
   Tokyo 162
   Japan

9. Texas Instruments
   P.O. Box 809066
   Dallas, Texas-75380
   USA

10. Exar Corporation
    2222 Qume Drive
    P.O. Box 49007
    San Jose, California-95161
    USA

11. Sprague Electric Company
    Semiconductor Division
    70 Pembroke Road
    Concord, NH-03301
    USA

12. Analog Devices
    One Technology Way
    P.O. Box 9106
    Norwood MA-02062
    USA

13. Integrated Device Technology
    2975 Stender Way
    Santa Clara, California-95054
    USA

14. Toshiba Corporation
    International Operations – Electronic Components
    1-1 Shibaura 1- Chome
    Minato-ku, Tokyo-105-01
    Japan

15. Burr-Brown Corporation
    International Airport Industrial Park
    P.O. Box 11400
    Tucson AZ-85734-1400
    USA

16. Cypress Semiconductor
    3901 North First Street
    San Jose, California-95134
    USA

17. Panasonic
    Matsushita Electronics Corporation
    Semiconductor Sales Division
    Nagaokakyo, Kyoto-617
    Japan

18. Plessey Semiconductors Ltd.,
    Cheney Manor
    Swindon, Wiltshire – SN2 2QW
    UK

19. Mitel Semiconductor
    P.O. Box 13320
    360 Legget Drive
    Kanata, Ontario
    Canada

# Appendix 3

# Datasheets of commonly used integrated circuits

Philips Semiconductors–Signetics Linear Products    Product specification

## General purpose operational amplifier    µA741/µA741C/SA741C

### DESCRIPTION
The µA741 is a high performance operational amplifier with high open-loop gain, internal compensation, high common mode range and exceptional temperature stability. The µA741 is short-circuit-protected and allows for nulling of offset voltage.

### FEATURES
- Internal frequency compensation
- Short circuit protection
- Excellent temperature stability
- High input voltage range

### PIN CONFIGURATION
D, F, N Packages

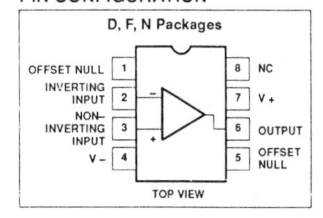

```
OFFSET NULL      1        8  NC
INVERTING INPUT  2        7  V+
NON-INVERTING    3        6  OUTPUT
INPUT
V-               4        5  OFFSET NULL
```
TOP VIEW

### ORDERING INFORMATION

| DESCRIPTION | TEMPERATURE RANGE | ORDER CODE |
|---|---|---|
| 8-Pin Plastic DIP | -55°C to +125°C | µA741N |
| 8-Pin Plastic DIP | 0 to +70°C | µA741CN |
| 8-Pin Plastic DIP | -40°C to +85°C | SA741CN |
| 8-Pin Cerdip | -55°C to +125°C | µA741F |
| 8-Pin Cerdip | 0 to +70°C | µA741CF |
| 8-Pin SO | 0 to +70°C | µA741CD |

### ABSOLUTE MAXIMUM RATINGS

| SYMBOL | PARAMETER | RATING | UNIT |
|---|---|---|---|
| $V_S$ | Supply voltage | | |
|  | µA741C | ±18 | V |
|  | µA741 | ±22 | V |
| $P_D$ | Internal power dissipation | | |
|  | D package | 780 | mW |
|  | N package | 1170 | mW |
|  | F package | 800 | mW |
| $V_{IN}$ | Differential input voltage | ±30 | V |
| $V_{IN}$ | Input voltage[1] | ±15 | V |
| $I_{SC}$ | Output short-circuit duration | Continuous | |
| $T_A$ | Operating temperature range | | |
|  | µA741C | 0 to +70 | °C |
|  | SA741C | -40 to +85 | °C |
|  | µA741 | -55 to +125 | °C |
| $T_{STG}$ | Storage temperature range | -65 to +150 | °C |
| $T_{SOLD}$ | Lead soldering temperature (10sec max) | 300 | °C |

NOTES:
1. For supply voltages less than ±15V, the absolute maximum input voltage is equal to the supply voltage.

Appendix 3    215

Philips Semiconductors–Signetics Linear Products

Product specification

## General purpose operational amplifier    µA741/µA741C/SA741C

### DC ELECTRICAL CHARACTERISTICS
(µA741, µA741C) $T_A = 25°C$, $V_S = \pm15V$, unless otherwise specified.

| SYMBOL | PARAMETER | TEST CONDITIONS | µA741 Min | µA741 Typ | µA741 Max | µA741C Min | µA741C Typ | µA741C Max | UNIT |
|---|---|---|---|---|---|---|---|---|---|
| $V_{OS}$ | Offset voltage | $R_S=10k\Omega$ | | 1.0 | 5.0 | | 2.0 | 6.0 | mV |
| | | $R_S=10k\Omega$, over temp. | | 1.0 | 6.0 | | | 7.5 | mV |
| $\Delta V_{OS}/\Delta T$ | | | | 10 | | | 10 | | µV/°C |
| $I_{OS}$ | Offset current | | | 20 | 200 | | 20 | 200 | nA |
| | | Over temp. | | | | | | 300 | nA |
| | | $T_A=+125°C$ | | 7.0 | 200 | | | | nA |
| | | $T_A=-55°C$ | | 20 | 500 | | | | nA |
| $\Delta I_{OS}/\Delta T$ | | | | 200 | | | 200 | | pA/°C |
| $I_{BIAS}$ | Input bias current | | | 80 | 500 | | 80 | 500 | nA |
| | | Over temp. | | | | | | 800 | nA |
| | | $T_A=+125°C$ | | 30 | 500 | | | | nA |
| | | $T_A=-55°C$ | | 300 | 1500 | | | | nA |
| $\Delta I_B/\Delta T$ | | | | 1 | | | 1 | | nA/°C |
| $V_{OUT}$ | Output voltage swing | $R_L=10k\Omega$ | ±12 | ±14 | | ±12 | ±14 | | V |
| | | $R_L=2k\Omega$, over temp. | ±10 | ±13 | | ±10 | ±13 | | V |
| $A_{VOL}$ | Large-signal voltage gain | $R_L=2k\Omega$, $V_O=\pm10V$ | 50 | 200 | | 20 | 200 | | V/mV |
| | | $R_L=2k\Omega$, $V_O=\pm10V$, over temp. | 25 | | | 15 | | | V/mV |
| | Offset voltage adjustment range | | | ±30 | | | ±30 | | mV |
| PSRR | Supply voltage rejection ratio | $R_S \leq 10k\Omega$ | | | | | 10 | 150 | µV/V |
| | | $R_S \leq 10k\Omega$, over temp. | | 10 | 150 | | | | µV/V |
| CMRR | Common-mode rejection ratio | | | | | 70 | 90 | | dB |
| | | Over temp. | 70 | 90 | | | | | dB |
| $I_{CC}$ | Supply current | | | 1.4 | 2.8 | | 1.4 | 2.8 | mA |
| | | $T_A=+125°C$ | | 1.5 | 2.5 | | | | mA |
| | | $T_A=-55°C$ | | 2.0 | 3.3 | | | | mA |
| $V_{IN}$ | Input voltage range | (µA741, over temp.) | ±12 | ±13 | | ±12 | ±13 | | V |
| $R_{IN}$ | Input resistance | | 0.3 | 2.0 | | 0.3 | 2.0 | | MΩ |
| $P_D$ | Power consumption | | | 50 | 85 | | 50 | 85 | mW |
| | | $T_A=+125°C$ | | 45 | 75 | | | | mW |
| | | $T_A=-55°C$ | | 45 | 100 | | | | mW |
| $R_{OUT}$ | Output resistance | | | 75 | | | 75 | | Ω |
| $I_{SC}$ | Output short-circuit current | | 10 | 25 | 60 | 10 | 25 | 60 | mA |

Philips Semiconductors–Signetics Linear Products

Product specification

## General purpose operational amplifier  µA741/µA741C/SA741C

### DC ELECTRICAL CHARACTERISTICS
(SA741C) $T_A = 25°C$, $V_S = ±15V$, unless otherwise specified.

| SYMBOL | PARAMETER | TEST CONDITIONS | SA741C Min | SA741C Typ | SA741C Max | UNIT |
|---|---|---|---|---|---|---|
| $V_{OS}$ | Offset voltage | $R_S=10k\Omega$ | | 2.0 | 6.0 | mV |
| | | $R_S=10k\Omega$, over temp. | | | 7.5 | mV |
| $\Delta V_{OS}/\Delta T$ | | | | 10 | | µV/°C |
| $I_{OS}$ | Offset current | | | 20 | 200 | nA |
| | | Over temp. | | | 500 | nA |
| $\Delta I_{OS}/\Delta T$ | | | | 200 | | pA/°C |
| $I_{BIAS}$ | Input bias current | | | 80 | 500 | nA |
| | | Over temp. | | | 1500 | nA |
| $\Delta I_B/\Delta T$ | | | | 1 | | nA/°C |
| $V_{OUT}$ | Output voltage swing | $R_L=10k\Omega$ | ±12 | ±14 | | V |
| | | $R_L=2k\Omega$, over temp. | ±10 | ±13 | | V |
| $A_{VOL}$ | Large-signal voltage gain | $R_L=2k\Omega$, $V_O=±10V$ | 20 | 200 | | V/mV |
| | | $R_L=2k\Omega$, $V_O=±10V$, over temp. | 15 | | | V/mV |
| | Offset voltage adjustment range | | | ±30 | | mV |
| PSRR | Supply voltage rejection ratio | $R_S \leq 10k\Omega$ | | 10 | 150 | µV/V |
| CMRR | Common mode rejection ration | | 70 | 90 | | dB |
| $V_{IN}$ | Input voltage range | Over temp. | ±12 | ±13 | | V |
| $R_{IN}$ | Input resistance | | 0.3 | 2.0 | | M$\Omega$ |
| $P_d$ | Power consumption | | | 50 | 85 | mW |
| $R_{OUT}$ | Output resistance | | | 75 | | $\Omega$ |
| $I_{SC}$ | Output short-circuit current | | | 25 | | mA |

### AC ELECTRICAL CHARACTERISTICS
$T_A=25°C$, $V_S = ±15V$, unless otherwise specified.

| SYMBOL | PARAMETER | TEST CONDITIONS | µA741, µA741C Min | µA741, µA741C Typ | µA741, µA741C Max | UNIT |
|---|---|---|---|---|---|---|
| $R_{IN}$ | Parallel input resistance | Open-loop, f=20Hz | 0.3 | | | M$\Omega$ |
| $C_{IN}$ | Parallel input capacitance | Open-loop, f=20Hz | | 1.4 | | pF |
| | Unity gain crossover frequency | Open-loop | | 1.0 | | MHz |
| $t_R$ | Transient response unity gain Rise time | $V_{IN}=20mV$, $R_L=2k\Omega$, $C_L \leq 100pF$ | | 0.3 | | µs |
| | Overshoot | | | 5.0 | | % |
| SR | Slew rate | $C \leq 100pF$, $R_L \geq 2k\Omega$, $V_{IN}=±10V$ | | 0.5 | | V/µs |

Philips Semiconductors–Signetics Linear Products

Product specification

## General purpose operational amplifier

µA741/µA741C/SA741C

**EQUIVALENT SCHEMATIC**

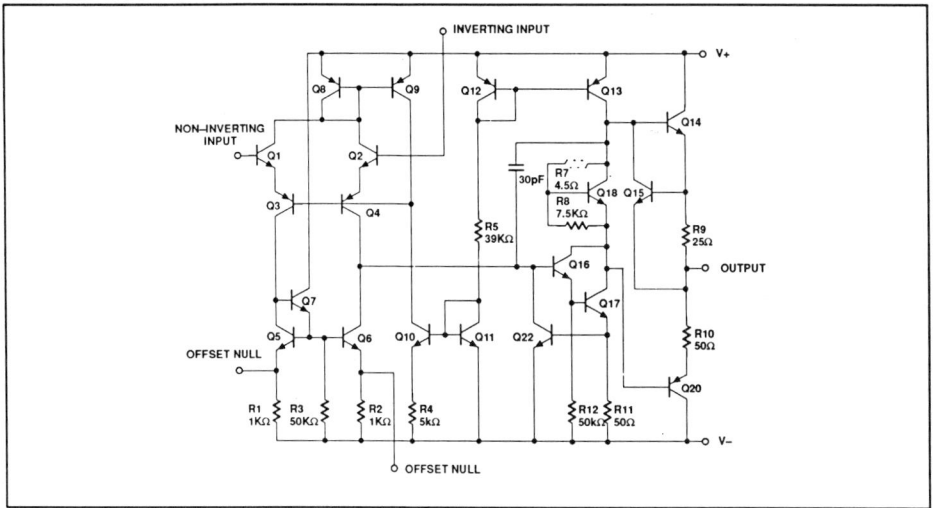

218  Appendix 3

Philips Semiconductors–Signetics Linear Products

Product specification

## Timer

## NE/SA/SE555/SE555C

### DESCRIPTION
The 555 monolithic timing circuit is a highly stable controller capable of producing accurate time delays, or oscillation. In the time delay mode of operation, the time is precisely controlled by one external resistor and capacitor. For a stable operation as an oscillator, the free running frequency and the duty cycle are both accurately controlled with two external resistors and one capacitor. The circuit may be triggered and reset on falling waveforms, and the output structure can source or sink up to 200mA.

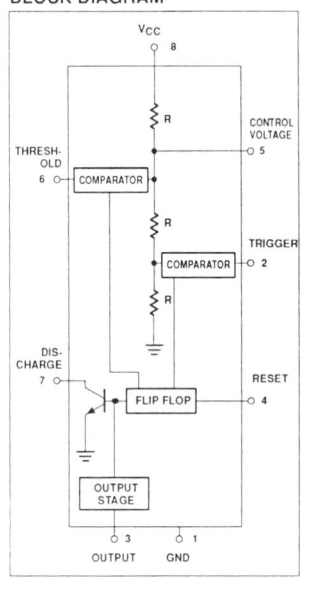

### FEATURES
- Turn-off time less than 2μs
- Max. operating frequency greater than 500kHz
- Timing from microseconds to hours
- Operates in both astable and monostable modes
- High output current
- Adjustable duty cycle
- TTL compatible
- Temperature stability of 0.005% per °C

### APPLICATIONS
- Precision timing
- Pulse generation
- Sequential timing
- Time delay generation
- Pulse width modulation

### PIN CONFIGURATIONS

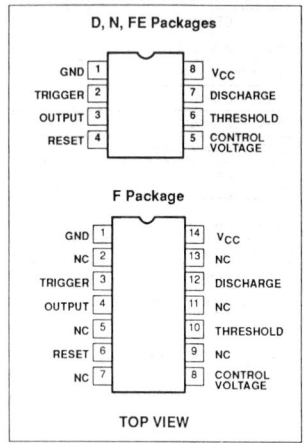

D, N, FE Packages

| | | | |
|---|---|---|---|
| GND | 1 | 8 | V_CC |
| TRIGGER | 2 | 7 | DISCHARGE |
| OUTPUT | 3 | 6 | THRESHOLD |
| RESET | 4 | 5 | CONTROL VOLTAGE |

F Package

| | | | |
|---|---|---|---|
| GND | 1 | 14 | V_CC |
| NC | 2 | 13 | NC |
| TRIGGER | 3 | 12 | DISCHARGE |
| OUTPUT | 4 | 11 | NC |
| NC | 5 | 10 | THRESHOLD |
| RESET | 6 | 9 | NC |
| NC | 7 | 8 | CONTROL VOLTAGE |

TOP VIEW

### BLOCK DIAGRAM

### EQUIVALENT SCHEMATIC

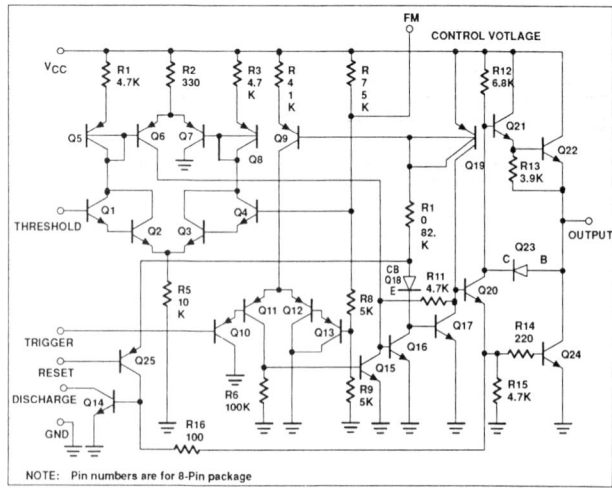

NOTE: Pin numbers are for 8-Pin package

Philips Semiconductors–Signetics Linear Products

Product specification

## Timer

## NE/SA/SE555/SE555C

### ORDERING INFORMATION

| DESCRIPTION | TEMPERATURE RANGE | ORDER CODE |
|---|---|---|
| 8-Pin Plastic SO | 0 to +70°C | NE555D |
| 8-Pin Plastic DIP | 0 to +70°C | NE555N |
| 8-Pin Plastic DIP | -40°C to +85°C | SA555N |
| 8-Pin Plastic SO | -40°C to +85°C | SA555D |
| 8-Pin Hermetic Cerdip | -55°C to +125°C | SE555CFE |
| 8-Pin Plastic DIP | -55°C to +125°C | SE555CN |
| 14-Pin Plastic DIP | -55°C to +125°C | SE555N |
| 8-Pin Hermetic Cerdip | -55°C to +125°C | SE555FE |
| 14-Pin Ceramic DIP | 0 to +70°C | NE555F |
| 14-Pin Ceramic DIP | -55°C to +125°C | SE555F |
| 14-Pin Ceramic DIP | -55°C to +125°C | SE555CF |

### ABSOLUTE MAXIMUM RATINGS

| SYMBOL | PARAMETER | RATING | UNIT |
|---|---|---|---|
| $V_{CC}$ | Supply voltage | | |
| | SE555 | +18 | V |
| | NE555, SE555C, SA555 | +16 | V |
| $P_D$ | Maximum allowable power dissipation[1] | 600 | mW |
| $T_A$ | Operating ambient temperature range | | |
| | NE555 | 0 to +70 | °C |
| | SA555 | -40 to +85 | °C |
| | SE555, SE555C | -55 to +125 | °C |
| $T_{STG}$ | Storage temperature range | -65 to +150 | °C |
| $T_{SOLD}$ | Lead soldering temperature (10sec max) | +300 | °C |

NOTES:
1. The junction temperature must be kept below 125°C for the D package and below 150°C for the FE, N and F packages. At ambient temperatures above 25°C, where this limit would be derated by the following factors:
   D package 160°C/W
   FE package 150°C/W
   N package 100°C/W
   F package 105°C/W

Philips Semiconductors–Signetics Linear Products

Product specification

## Timer

## NE/SA/SE555/SE555C

### DC AND AC ELECTRICAL CHARACTERISTICS

$T_A = 25°C$, $V_{CC} = +5V$ to $+15$ unless otherwise specified.

| SYMBOL | PARAMETER | TEST CONDITIONS | SE555 Min | SE555 Typ | SE555 Max | NE555/SE555C Min | NE555/SE555C Typ | NE555/SE555C Max | UNIT |
|---|---|---|---|---|---|---|---|---|---|
| $V_{CC}$ | Supply voltage | | 4.5 | | 18 | 4.5 | | 16 | V |
| $I_{CC}$ | Supply current (low state)[1] | $V_{CC}=5V$, $R_L=\infty$ | | 3 | 5 | | 3 | 6 | mA |
| | | $V_{CC}=15V$, $R_L=\infty$ | | 10 | 12 | | 10 | 15 | mA |
| $t_M$ $\Delta t_M/\Delta T$ $\Delta t_M/\Delta V_S$ | Timing error (monostable) Initial accuracy[2] Drift with temperature Drift with supply voltage | $R_A=2k\Omega$ to $100k\Omega$ $C=0.1\mu F$ | | 0.5 30 0.05 | 2.0 100 0.2 | | 1.0 50 0.1 | 3.0 150 0.5 | % ppm/°C %/V |
| $t_A$ $\Delta t_A/\Delta T$ $\Delta t_A/\Delta V_S$ | Timing error (astable) Initial accuracy[2] Drift with temperature Drift with supply voltage | $R_A$, $R_B=1k\Omega$ to $100k\Omega$ $C=0.1\mu F$ $V_{CC}=15V$ | | 4 0.15 | 6 500 0.6 | | 5 0.3 | 13 500 1 | % ppm/°C %/V |
| $V_C$ | Control voltage level | $V_{CC}=15V$ | 9.6 | 10.0 | 10.4 | 9.0 | 10.0 | 11.0 | V |
| | | $V_{CC}=5V$ | 2.9 | 3.33 | 3.8 | 2.6 | 3.33 | 4.0 | V |
| $V_{TH}$ | Threshold voltage | $V_{CC}=15V$ | 9.4 | 10.0 | 10.6 | 8.8 | 10.0 | 11.2 | V |
| | | $V_{CC}=5V$ | 2.7 | 3.33 | 4.0 | 2.4 | 3.33 | 4.2 | V |
| $I_{TH}$ | Threshold current[3] | | | 0.1 | 0.25 | | 0.1 | 0.25 | µA |
| $V_{TRIG}$ | Trigger voltage | $V_{CC}=15V$ | 4.8 | 5.0 | 5.2 | 4.5 | 5.0 | 5.6 | V |
| | | $V_{CC}=5V$ | 1.45 | 1.67 | 1.9 | 1.1 | 1.67 | 2.2 | V |
| $I_{TRIG}$ | Trigger current | $V_{TRIG}=0V$ | | 0.5 | 0.9 | | 0.5 | 2.0 | µA |
| $V_{RESET}$ | Reset voltage[4] | $V_{CC}=15V$, $V_{TH}=10.5V$ | 0.3 | | 1.0 | 0.3 | | 1.0 | V |
| $I_{RESET}$ | Reset current | $V_{RESET}=0.4V$ | | 0.1 | 0.4 | | 0.1 | 0.4 | mA |
| | Reset current | $V_{RESET}=0V$ | | 0.4 | 1.0 | | 0.4 | 1.5 | mA |
| $V_{OL}$ | Output voltage (low) | $V_{CC}=15V$ $I_{SINK}=10mA$ | | 0.1 | 0.15 | | 0.1 | 0.25 | V |
| | | $I_{SINK}=50mA$ | | 0.4 | 0.5 | | 0.4 | 0.75 | V |
| | | $I_{SINK}=100mA$ | | 2.0 | 2.2 | | 2.0 | 2.5 | V |
| | | $I_{SINK}=200mA$ | | 2.5 | | | 2.5 | | V |
| | | $V_{CC}=5V$ $I_{SINK}=8mA$ | | 0.1 | 0.25 | | 0.3 | 0.4 | V |
| | | $I_{SINK}=5mA$ | | 0.05 | 0.2 | | 0.25 | 0.35 | V |
| $V_{OH}$ | Output voltage (high) | $V_{CC}=15V$ $I_{SOURCE}=200mA$ | | 12.5 | | | 12.5 | | V |
| | | $I_{SOURCE}=100mA$ | 13.0 | 13.3 | | 12.75 | 13.3 | | V |
| | | $V_{CC}=5V$ $I_{SOURCE}=100mA$ | 3.0 | 3.3 | | 2.75 | 3.3 | | V |
| $t_{OFF}$ | Turn-off time[5] | $V_{RESET}=V_{CC}$ | | 0.5 | 2.0 | | 0.5 | 2.0 | µs |
| $t_R$ | Rise time of output | | | 100 | 200 | | 100 | 300 | ns |
| $t_F$ | Fall time of output | | | 100 | 200 | | 100 | 300 | ns |
| | Discharge leakage current | | | 20 | 100 | | 20 | 100 | nA |

NOTES:
1. Supply current when output high typically 1mA less
2. Tested at $V_{CC}=5V$ and $V_{CC}=15V$.
3. This will determine the max value of $R_A+R_B$, for 15V operation, the max total $R=10M\Omega$, and for 5V operation, the max. total $R=3.4M\Omega$.
4. Specified with trigger input high.
5. Time measured from a positive going input pulse from 0 to $0.8 \times V_{CC}$ into the threshold to the drop from high to low of the output. Trigger is tied to threshold.

Philips Semiconductors–Signetics Linear Products

**Product specification**

## Dual timer

**NE/SA/SE556/NE556-1**

### DESCRIPTION
Both the 556 and 556-1 Dual Monolithic timing circuits are highly stable controllers capable of producing accurate time delays or oscillation. The 556 and 556-1 are a dual 555. Timing is provided by an external resistor and capacitor for each timing function. The two timers operate independently of each other, sharing only $V_{CC}$ and ground. The circuits may be triggered and reset on falling waveforms. The output structures may sink or source 200mA.

### FEATURES
- Turn-off time less than 2µs (556-1, 1C)
- Maximum operating frequency >500kHz (556-1, 1C)
- Timing from microseconds to hours
- Replaces two 555 timers
- Operates in both astable and monostable modes
- High output current
- Adjustable duty cycle
- TTL compatible
- Temperature stability of 0.005%/°C
- SE556-1 compliant to MIL-STD or JAN available from Signetics' Military Division

### APPLICATIONS
- Precision timing
- Sequential timing
- Pulse shaping
- Pulse generator
- Missing pulse detector
- Tone burst generator
- Pulse width modulation
- Time delay generator
- Frequency division
- Touch-Tone® encoder
- Industrial controls
- Pulse position modulation
- Appliance timing
- Traffic light control

### PIN CONFIGURATION

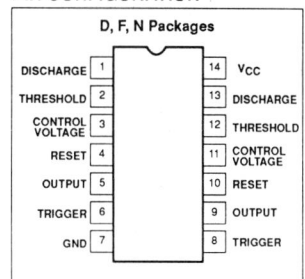

D, F, N Packages

| Pin | Name | | Pin | Name |
|---|---|---|---|---|
| 1 | DISCHARGE | | 14 | $V_{CC}$ |
| 2 | THRESHOLD | | 13 | DISCHARGE |
| 3 | CONTROL VOLTAGE | | 12 | THRESHOLD |
| 4 | RESET | | 11 | CONTROL VOLTAGE |
| 5 | OUTPUT | | 10 | RESET |
| 6 | TRIGGER | | 9 | OUTPUT |
| 7 | GND | | 8 | TRIGGER |

### BLOCK DIAGRAM

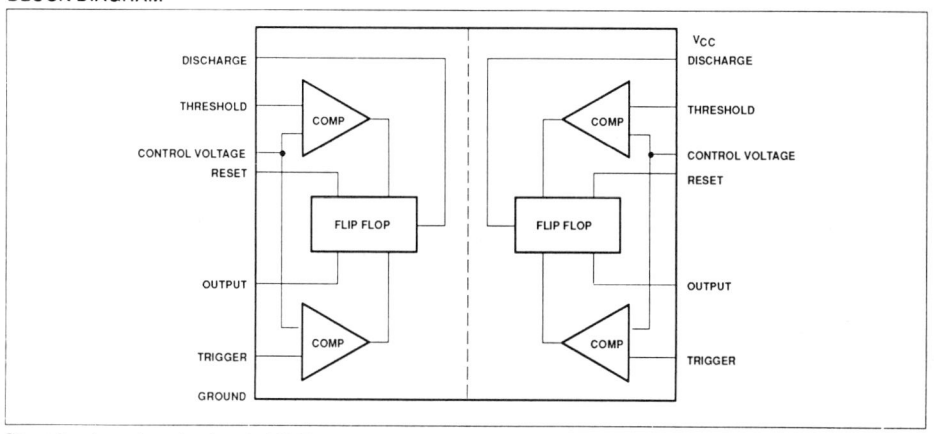

®Touch-Tone is a registered trademark of AT&T

Philips Semiconductors–Signetics Linear Products

Product specification

## Dual timer

NE/SA/SE556/NE556-1

**EQUIVALENT SCHEMATIC** (Shown for one circuit only)

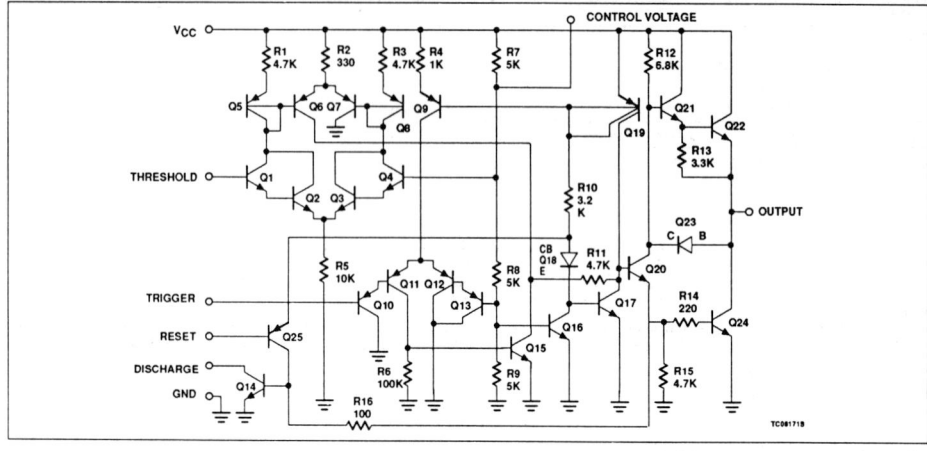

**ORDERING INFORMATION**

| DESCRIPTION | TEMPERATURE RANGE | ORDER CODE |
| --- | --- | --- |
| 14-Pin Plastic SO | 0 to +70°C | NE556D |
| 14-Pin Cerdip | 0 to +70°C | NE556F |
| 14-Pin Plastic DIP | 0 to +70°C | NE556N |
| 14-Pin Cerdip | 0 to +70°C | NE556-1F |
| 14-Pin Plastic DIP | 0 to +70°C | NE556-1N |
| 14-Pin Plastic DIP | -40°C to +85°C | SA556N |
| 14-Pin Cerdip | -55°C to +125°C | SE556F |
| 14-Pin Plastic DIP | -55°C to +125°C | SE556N |

Philips Semiconductors–Signetics Linear Products

Product specification

## Dual timer

**NE/SA/SE556/NE556-1**

### ABSOLUTE MAXIMUM RATINGS

| SYMBOL | PARAMETER | RATING | UNIT |
|---|---|---|---|
| $V_{CC}$ | Supply voltage | | |
| | NE/SA556, 556-1, SE556C, SE556-1C | +16 | V |
| | SE556-1, SE556 | +18 | V |
| $P_D$ | Maximum allowable power dissipation[1] | 800 | mW |
| $T_A$ | Operating temperature range | | |
| | NE556-1, NE556 | 0 to +70 | °C |
| | SA556 | -40 to +85 | °C |
| | SE556 | -55 to +125 | °C |
| $T_{STG}$ | Storage temperature range | -65 to +150 | °C |
| $T_{SOLD}$ | Lead soldering temperature (10sec max) | +300 | °C |

**NOTES:**
1. The junction temperature must be kept below 125°C for the D package and below 150°C for the N and F packages. At ambient temperatures above 25°C, where this limit would be exceeded, the Maximum Allowable Power Dissipation must be derated by the following:
   D package 115°C/W
   N package 80°C/W
   F package 100°C/W

### ELECTRICAL CHARACTERISTICS
$T_A=25°C$, $V_{CC}=+5V$ to $+15V$, unless otherwise specified.

| SYMBOL | PARAMETER | TEST CONDITIONS | SE556/556-1 Min | SE556/556-1 Typ | SE556/556-1 Max | NE/SA556/SE556C NE556-1/SE556-1C Min | NE/SA556/SE556C NE556-1/SE556-1C Typ | NE/SA556/SE556C NE556-1/SE556-1C Max | UNIT |
|---|---|---|---|---|---|---|---|---|---|
| $V_{CC}$ | Supply voltage | | 4.5 | | 18 | 4.5 | | 16 | V |
| $I_{CC}$ | Supply current (low state)[1] | $V_{CC}=5V$, $R_L=\infty$ | | 6 | 10 | | 6 | 12 | mA |
| | | $V_{CC}=15V$, $R_L=\infty$ | | 20 | 24 | | 20 | 30 | mA |
| $t_M$ | Timing error (monostable) Initial accuracy[2] | $R_A=2k\Omega$ to $100k\Omega$ $C=0.1\mu F$ | | 0.5 | 2.0 | | 0.75 | 3.0 | % |
| $\Delta t_M/\Delta T$ | Drift with temperature | $T=1.1$ RC | | 30 | 100 | | 50 | 150 | ppm/°C |
| $\Delta t_M/\Delta V_S$ | Drift with supply voltage | | | 0.05 | 0.2 | | 0.1 | 0.5 | %/V |
| $t_A$ | Timing error (astable) Initial accuracy[2] | $R_A$, $R_B=1k\Omega$ to $100k\Omega$ $C=0\mu F$ | | 4 | 6 | | 5 | 13 | % |
| $\Delta t_A/\Delta T$ | Drift with temperature | $V_{CC}=15V$ | | 400 | 500 | | 400 | 500 | ppm/°C |
| $\Delta t_A/\Delta V_S$ | Drift with supply voltage | | | 0.15 | 0.6 | | 0.3 | 1 | %/V |
| $V_C$ | Control voltage level | $V_{CC}=15V$ | 9.6 | 10.0 | 10.4 | 9.0 | 10.0 | 11.0 | V |
| | | $V_{CC}=5V$ | 2.9 | 3.33 | 3.8 | 2.6 | 3.33 | 4.0 | V |
| $V_{TH}$ | Threshold voltage | $V_{CC}=15V$ | 9.4 | 10.0 | 10.6 | 8.8 | 10.0 | 11.2 | V |
| | | $V_{CC}=5V$ | 2.7 | 3.33 | 4.0 | 2.4 | 3.33 | 4.2 | V |
| $I_{TH}$ | Threshold current[3] | $V_{CC}=15V$, $V_{TH}=10.5V$ | | 30 | 250 | | 30 | 250 | nA |
| $V_{TRIG}$ | Trigger voltage | $V_{CC}=15V$ | 4.8 | 5.0 | 5.2 | 4.5 | 5.0 | 5.6 | V |
| | | $V_{CC}=5V$ | 1.45 | 1.67 | 1.9 | 1.1 | 1.67 | 2.2 | V |
| $I_{TRIG}$ | Trigger current | $V_{TRIG}=0V$ | | 0.5 | 0.9 | | 0.5 | 2.0 | µA |
| $V_{RESET}$ | Reset voltage[5] | | 0.4 | 0.7 | 1.0 | 0.4 | 0.7 | 1.0 | V |
| | Reset current | $V_{RESET}=0.4V$ | 0.4 | 0.1 | 0.4 | 0.4 | 0.1 | 0.6 | mA |

Philips Semiconductors–Signetics Linear Products

Product specification

## Dual timer

## NE/SA/SE556/NE556-1

### ELECTRICAL CHARACTERISTICS (Continued)

| SYMBOL | PARAMETER | TEST CONDITIONS | SE556/556-1 Min | SE556/556-1 Typ | SE556/556-1 Max | NE/SA556/SE556C NE556-1/SE556-1C Min | NE/SA556/SE556C NE556-1/SE556-1C Typ | NE/SA556/SE556C NE556-1/SE556-1C Max | UNIT |
|---|---|---|---|---|---|---|---|---|---|
| $I_{RESET}$ | Reset current | $V_{RESET}=0V$ | | 0.4 | 1.0 | | 0.4 | 1.5 | mA |
| $V_{OL}$ | Output voltage (low) | $V_{CC}=15V$ | | | | | | | |
| | | $I_{SINK}=10mA$ | | 0.1 | 0.15 | | 0.1 | 0.25 | V |
| | | $I_{SINK}=50mA$ | | 0.4 | 0.5 | | 0.4 | 0.75 | V |
| | SE556 | $I_{SINK}=100mA$ | | 2.0 | 2.25 | | | | V |
| | NE/SA556 | | | | | | 2.0 | 3.2 | V |
| | NE556-1 | | | | | | 2.0 | 2.5 | V |
| | | $I_{SINK}=200mA$ | | 2.5 | | | 2.5 | | V |
| | | $V_{CC}=5V$ | | | | | | | |
| | | $I_{SINK}=8mA$ | | 0.1 | 0.2 | | 0.25 | 0.3 | V |
| | | $I_{SINK}=5mA$ | | 0.05 | 0.15 | | 0.15 | 0.25 | V |
| $V_{OH}$ | Output voltage (high) | $V_{CC}=15V$ | | | | | | | |
| | | $I_{SOURCE}=200mA$ | | 12.5 | | | 12.5 | | V |
| | | $I_{SOURCE}=100mA$ | 13.0 | 13.3 | | 12.75 | 13.3 | | V |
| | | $V_{CC}=5V$ | | | | | | | |
| | | $I_{SOURCE}=100mA$ | 3.0 | 3.3 | | 2.75 | 3.3 | | V |
| $t_{OFF}$ | Turn-off time[6] NE556-1 | $V_{RESET}=V_{CC}$ | | 0.5 | 2.0 | | 0.5 | | µs |
| $t_R$ | Rise time of output | | | 100 | 200 | | 100 | 300 | ns |
| $t_F$ | Fall time of output | | | 100 | 200 | | 100 | 300 | ns |
| | Discharge leakage current | | | 20 | 100 | | 20 | 100 | nA |
| | Matching characteristics[4] | | | | | | | | |
| | Initial accuracy[2] | | | 0.5 | 1.0 | | 1.0 | 2.0 | % |
| | Drift with temperature | | | 10 | | | ±10 | | ppm/°C |
| | Drift with supply voltage | | | 0.1 | 0.2 | | 0.2 | 0.5 | %/V |

NOTES:
1. Supply current when output is high is typically 1.0mA less.
2. Tested at $V_{CC}=5V$ and $V_{CC}=15V$.
3. This will determine maximum value of $R_A+R_B$. For 15V operation, the max total R=10MΩ, and for 5V operation, the maximum total R=3.4MΩ.
4. Matching characteristics refer to the difference between performance characteristics for each timer section in the monostable mode.
5. Specified with trigger input high. In order to guarantee reset the voltage at reset pin must be less than or equal to 0.4V. To disable reset function, the voltage at reset pin has to be greater than 1V.
6. Time measured from a positive-going input pulse from 0 to 0.4 $V_{CC}$ into the threshold to the drop from high to low of the output. Trigger is tied to threshold.

Philips Semiconductors–Signetics Linear Products  Product specification

# Quad timer

**NE558**

## DESCRIPTION
The 558 Quad Timers are monolithic timing devices which can be used to produce four independent timing functions. The 558 output sinks current. These highly stable, general purpose controllers can be used in a monostable mode to produce accurate time delays; from microseconds to hours. In the time delay mode of operation, the time is precisely controlled by one external resistor and one capacitor. A stable operation can be achieved by using two of the four timer sections.

The four timer sections in the 558 are edge-triggered; therefore, when connected in tandem for sequential timing applications, no coupling capacitors are required. Output current capability of 100mA is provided in both devices.

## FEATURES
- 100mA output current per section
- Edge-triggered (no coupling capacitor)
- Output independent of trigger conditions
- Wide supply voltage range 4.5V to 18V
- Timer intervals from microseconds to hours
- Time period equals RC
- Military qualifications pending

## APPLICATIONS
- Sequential timing
- Time delay generation
- Precision timing
- Industrial controls
- Quad one-shot

## PIN CONFIGURATION

D[1], N Packages

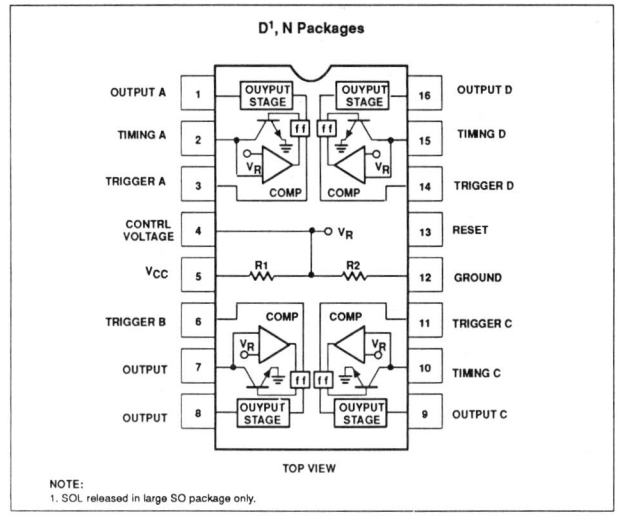

| | Pin | | Pin | |
|---|---|---|---|---|
| OUTPUT A | 1 | | 16 | OUTPUT D |
| TIMING A | 2 | | 15 | TIMING D |
| TRIGGER A | 3 | | 14 | TRIGGER D |
| CONTRL VOLTAGE | 4 | | 13 | RESET |
| V<sub>CC</sub> | 5 | | 12 | GROUND |
| TRIGGER B | 6 | | 11 | TRIGGER C |
| OUTPUT | 7 | | 10 | TIMING C |
| OUTPUT | 8 | | 9 | OUTPUT C |

TOP VIEW

NOTE:
1. SOL released in large SO package only.

Philips Semiconductors–Signetics Linear Products                                                                 Product specification

# Quad timer                                                                                           NE558

## ORDERING INFORMATION

| DESCRIPTION | TEMPERATURE RANGE | ORDER CODE |
|---|---|---|
| 16-Pin Plastic SOL | 0 to +70°C | NE558D |
| 16-Pin Plastic DIP | 0 to +70°C | NE558N |

## ABSOLUTE MAXIMUM RATINGS

| SYMBOL | PARAMETER | RATING | UNIT |
|---|---|---|---|
| $V_{CC}$ | Supply voltage | | |
| | NE/SA558 | +16 | V |
| | SE558 | +18 | V |
| $P_D$ | Maximum power dissipation $T_A$=25°C ambient (still-air)[1] | | |
| | N package | 1450 | mW |
| | D package | 1090 | mW |
| $T_A$ | Operating ambient temperature range | 0 to +70 | °C |
| $T_{STG}$ | Storage temperature range | -65 to +150 | °C |
| $T_{SOLD}$ | Lead soldering temperature (10sec max) | +300 | °C |

NOTES:
1. Derate above 25°C, at the following rates:
   F package at 9.5mW/°C
   N package at 11.6mW/°C
   D package at 8.7mW/°C
$T_A$= 25°C $V_{CC}$=+5V to +15V, unless otherwise specified.

## DC AND AC ELECTRICAL CHARACTERISTICS
$T_A$ = 25°C, $V_{CC}$ = +5V to +15V, unless otherwise specified.

| SYMBOL | PARAMETER | TEST CONDITIONS | NE558 Min | NE558 Typ | NE558 Max | UNIT |
|---|---|---|---|---|---|---|
| $V_{CC}$ | Supply voltage | | 4.5 | | 16 | V |
| $I_{CC}$ | Supply current | $V_{CC}$=Reset=15V | | 16 | 36 | mA |
| | Timing accuracy (t=RC) | R=2kΩ to 100kΩ, C=1µF | | | | |
| $t_A$ | Initial accuracy | | | ±2 | 5 | % |
| $\Delta t_A/\Delta T$ | Drift with temperature | | | 30 | 150 | ppm/°C |
| $\Delta t_A/\Delta V_S$ | Drift with supply voltage | | | 0.1 | 0.9 | %/V |
| $V_{TRIG}$ | Trigger voltage[1] | $V_{CC}$=15V | 0.8 | | 2.4 | V |
| $I_{TRIG}$ | Trigger current | Trigger=0V | | 5 | 100 | µA |
| $V_{RESET}$ | Reset voltage[2] | | 0.8 | | 2.4 | V |
| $I_{RESET}$ | Reset current | Reset | | 50 | 500 | µA |
| $V_{TH}$ | Threshold voltage | | | 0.63 | | $\times V_{CC}$ |
| | Threshold leakage | | | 15 | | nA |
| $V_{OUT}$ | Output voltage[3] | $I_L$=10mA | | 0.1 | 0.4 | V |
| | | $I_L$=100mA | | 1.0 | 2.0 | V |
| | Output leakage | | | 10 | 500 | nA |
| $t_{PD}$ | Propagation delay | | | 1.0 | | µs |
| $t_R$ | Rise time of output | $I_L$=100mA | | 100 | | ns |
| $t_F$ | Fall time of output | $I_L$=100mA | | 100 | | ns |

NOTES:
1. The trigger functions only on the falling edge of the trigger pulse only after previously being high. After reset, the trigger must be brought high and then low to implement triggering.
2. For reset below 0.8V, outputs set low and trigger inhibited. For reset above 2.4V, trigger enabled.
3. The 558 output structure is open-collector which requires a pull-up resistor to $V_{CC}$ to sink current. The output is normally low sinking current.

Philips Semiconductors–Signetics Linear Products

Product specification

# Quad voltage comparator

## LM139A/239A/339A/LM139/239/339/LM2901/MC3302

## DESCRIPTION

The LM139 series consists of four independent precision voltage comparators, with an offset voltage specification as low as 2.0mV max for each comparator, which were designed specifically to operate from a single power supply over a wide range of voltages. Operation from split power supplies is also possible and the low power supply current drain is independent of the magnitude of the power supply voltage. These comparators also have a unique characteristic in that the input common–mode voltage range includes ground, even though they are operated from a single power supply voltage.

The LM139 series was designed to directly interface with TTL and CMOS. When operated from both plus and minus power supplies, the LM139 series will directly interface with MOS logic where their low power drain is a distinct advantage over standard comparators.

## FEATURES

- Wide single supply voltage range 2.0$V_{DC}$ to 36$V_{DC}$ or dual supplies ±1.0$V_{DC}$ to ±18$V_{DC}$
- Very low supply current drain (0.8mA) independent of supply voltage (1.0mW/comparator at 5.0$V_{DC}$)
- Low input biasing current 25nA
- Low input offset current ±5nA and offset voltage
- Input common–mode voltage range includes ground
- Differential input voltage range equal to the power supply voltage
- Low output 250mV at 4mA saturation voltage
- Output voltage compatible with TTL, DTL, ECL, MOS and CMOS logic systems

## APPLICATIONS

- A/D converters
- Wide range VCO
- MOS clock generator
- High voltage logic gate
- Multivibrators

## PIN CONFIGURATION

**D, F, N Packages**

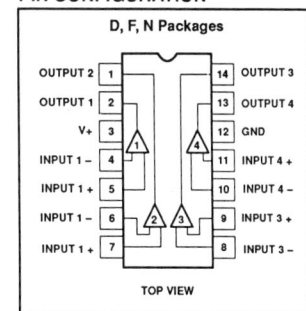

| Pin | | | Pin |
|---|---|---|---|
| OUTPUT 2 | 1 | 14 | OUTPUT 3 |
| OUTPUT 1 | 2 | 13 | OUTPUT 4 |
| V+ | 3 | 12 | GND |
| INPUT 1 – | 4 | 11 | INPUT 4 + |
| INPUT 1 + | 5 | 10 | INPUT 4 – |
| INPUT 1 – | 6 | 9 | INPUT 3 + |
| INPUT 1 + | 7 | 8 | INPUT 3 – |

TOP VIEW

## EQUIVALENT CIRCUIT

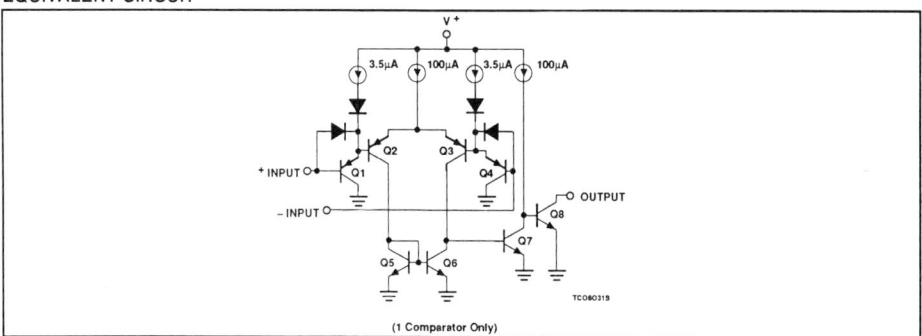

(1 Comparator Only)

228    Appendix 3

Philips Semiconductors–Signetics Linear Products

Product specification

## Quad voltage comparator

LM139A/239A/339A/LM139
/239/339/LM2901/MC3302

### ORDERING INFORMATION

| DESCRIPTION | TEMPERATURE RANGE | ORDER CODE |
| --- | --- | --- |
| 14–Pin Cerdip | –55 to +125°C | LM139F |
| 14–Pin Plastic DIP | –25°C to +85°C | LM239AN |
| 14–Pin Plastic DIP | –25°C to +85°C | LM239N |
| 14–Pin Plastic SO | –25°C to +85°C | LM239D |
| 14–Pin Plastic DIP | –40°C to +85°C | LM2901N |
| 14–Pin Plastic SO | –40°C to +85°C | LM2901D |
| 14–Pin Plastic DIP | 0 to +70°C | LM339AN |
| 14–Pin Plastic SO | 0 to +70°C | LM339D |
| 14–Pin Plastic DIP | 0 to +70°C | LM339N |
| 14–Pin Plastic SO | –40°C to +85°C | MC3302D |
| 14–Pin Cerdip | –40°C to +85°C | MC3302F |
| 14–Pin Plastic DIP | –40°C to +85°C | MC3302N |

### ABSOLUTE MAXIMUM RATINGS

| SYMBOL | PARAMETER | RATING | UNIT |
| --- | --- | --- | --- |
| $V_{CC}$ | $V_{CC}$ supply voltage | 36 or ±18 | $V_{DC}$ |
| $V_{DIFF}$ | Differential input voltage | 36 | $V_{DC}$ |
| $V_{IN}$ | Input voltage | –0.3 to +36 | $V_{DC}$ |
| $P_D$ | Maximum power dissipation, $T_A$=25°C (still–air)[1] | | |
| | F package | 1190 | mW |
| | N package | 1420 | mW |
| | D package | 1040 | mW |
| | Output short–circuit to ground[2] | Continuous | |
| $I_{IN}$ | Input current ($V_{IN}$<–0.3$V_{DC}$)[3] | 50 | mA |
| $T_A$ | Operating temperature range | | |
| | LM139A | –55 to +125 | °C |
| | LM239A | –25 to +85 | °C |
| | LM339A | 0 to +70 | °C |
| | LM2901/MC3302 | –40 to +85 | °C |
| $T_{STG}$ | Storage temperature range | –65 to +150 | °C |
| $T_{SOLD}$ | Lead soldering temperature (10sec max) | 300 | °C |

NOTES:
1. Derate above 25°C, at the following rates:
   F Package at 9.5mW/°C
   N Package at 11.4mW/°C
   D Package at 8.3mW/°C
2. Short circuits from the output to V+ can cause excessive heating and eventual destruction. The maximum output current is aproximately 20mA independent of the magnitude of V+.
3. This input current will only exist when the voltage at any of the input leads is driven negative. It is due to the collector–base junction of the input PNP transistors becoming forward biased and thereby acting as input diode clamps. In addition to this diode action, there is also lateral NPN parasitic transistor action on the IC chip. This transistor action can cause the output voltages of the comparators to go to the V+ voltage level (or to ground for a large overdrive) for the time duration that an input is driven negative. This is not destructive and normal output states will reestablish when the input voltage, which was negative, again returns to a value greater than –0.3$V_{DC}$.

Philips Semiconductors–Signetics Linear Products

Product specification

## Quad voltage comparator

LM139A/239A/339A/LM139 /239/339/LM2901/MC3302

**DC AND AC ELECTRICAL CHARACTERISTICS**
$V+=5V_{DC}$, LM139A/LM139: $-55°C \leq T_A \leq 125°C$; LM239: $-25°C \leq T_A \leq 85°C$; LM339: $0°C \leq T_A \leq 70°C$; LM339A: $0°C \leq T_A \leq 70°C$; LM239A: $-25°C \leq T_A \leq 85°C$; LM2901/LM3302: $-40°C \leq T_A \leq 85°C$, unless otherwise specified.

| SYMBOL | PARAMETER | TEST CONDITIONS | LM139A Min | LM139A Typ | LM139A Max | LM239A/339A Min | LM239A/339A Typ | LM239A/339A Max | UNIT |
|---|---|---|---|---|---|---|---|---|---|
| $V_{OS}$ | Input offset voltage[2] | $T_A=25°C$<br>Over temp. |  | ±1.0 | ±2.0<br>±4.0 |  | ±1.0 | ±2.0<br>±4.0 | mV<br>mV |
| $V_{CM}$ | Input common–mode voltage range[3] | $T_A=25°C$<br>Over temp. | 0<br>0 |  | V+−1.5<br>V+−2.0 | 0<br>0 |  | V+−1.5<br>V+−2.0 | V |
| $V_{IDR}$ | Differential input voltage[1] | Keep all $V_{IN} \geq 0V_{DC}$ (or V− if need) |  |  | V+ |  |  | V+ | V |
| $I_{BIAS}$ | Input bias current[4] | $I_{IN(+)}$ or $I_{IN(-)}$ with output in linear range<br>$T_A=25°C$<br>Over temp. |  | 25 | 100<br>300 |  | 25 | 250<br>400 | nA<br>nA |
| $I_{OS}$ | Input offset current | $I_{IN(+)} - I_{IN(-)}$<br>$T_A=25°C$<br>Over temp. |  | ±3.0 | ±25<br>±100 |  | ±5.0 | ±50<br>±150 | nA<br>nA |
| $I_{OL}$ | Output sink current | $V_{IN(-)} \geq 1V_{DC}$, $V_{IN(+)}=0$, $V_O \leq 1.5V_{DC}$, $T_A=25°C$ $V_O=800mV$, over temp. | 6.0 | 16 |  | 6.0 | 16 |  | mA |
| $I_{OH}$ | Output leakage current | $V_{IN(+)} \geq 1V_{DC}$, $V_{IN(-)}=0$ $V_O=5V_{DC}$, $T_A=25°C$ $V_O=30V_{DC}$, over temp. |  | 0.1 | 1.0 |  | 0.1 | 1.0 | nA<br>µA |
| $I_{CC}$ | Supply current | V+=5V, $R_L=\infty$ on comparators, $T_A=25°C$ V+=30V |  | 0.8 | 2.0 |  | 0.8 | 2.0 | mA |
| $A_V$ | Voltage gain | $R_L \geq 15k\Omega$, V+=15$V_{DC}$ | 50 | 200 |  | 50 | 200 |  | V/mV |
| $V_{OL}$ | Saturation voltage | $V_{IN(-)} \geq 1V_{DC}$, $V_{IN(+)}=0$, $I_{SINK} \leq 4mA$ $T_A=25°C$ Over temp. |  | 250 | 400<br>700 |  | 250 | 400<br>700 | mV<br>mV |
| $t_{LSR}$ | Large–signal response time | $V_{IN}$=TTL logic swing, $V_{REF}=1.4V_{DC}$, $V_{RL}=5V_{DC}$, $R_L=5.1k\Omega$, $T_A=25°C$ |  | 300 |  |  | 300 |  | ns |
| $t_R$ | Response time[5] | $V_{RL}=5V_{DC}$, $R_L=5.1k\Omega$, $T_A=25°C$ |  | 1.3 |  |  | 1.3 |  | µs |

See notes at the end of the Electrical Characteristics.

Appendix 3   229

Philips Semiconductors–Signetics Linear Products

Product specification

## Quad voltage comparator

## LM139A/239A/339A/LM139/239/339/LM2901/MC3302

### DC AND AC ELECTRICAL CHARACTERISTICS

V+=5V$_{DC}$, LM139A/LM139: −55°C ≤ T$_A$ ≤ 125°C; LM239: −25°C ≤ T$_A$ ≤ 85°C; LM339: 0°C ≤ T$_A$ ≤ 70°C; LM339A: 0°C ≤ T$_A$ ≤ 70°C; LM239A: −25°C ≤ T$_A$ ≤ 85°C; LM2901/LM3302: −40°C ≤ T$_A$ ≤ 85°C, unless otherwise specified.

| SYMBOL | PARAMETER | TEST CONDITIONS | LM139 Min | LM139 Typ | LM139 Max | LM239/339 Min | LM239/339 Typ | LM239/339 Max | UNIT |
|---|---|---|---|---|---|---|---|---|---|
| V$_{OS}$ | Input offset voltage[2] | T$_A$=25°C<br>Over temp. | | ±2.0 | ±5.0<br>±9.0 | | ±2.0 | ±5.0<br>±9.0 | mV<br>mV |
| V$_{CM}$ | Input common–mode voltage range[3] | T$_A$=25°C<br>Over temp. | 0<br>0 | | V+−1.5<br>V+−2.0 | 0<br>0 | | V+−1.5<br>V+−2.0 | V |
| V$_{IDR}$ | Differential input voltage[1] | Keep all V$_{IN}$s≥0V$_{DC}$<br>(or V− if need) | | | V+ | | | V+ | V |
| I$_{BIAS}$ | Input bias current[4] | I$_{IN(+)}$ or I$_{IN(−)}$ with output in linear range<br>T$_A$=25°C<br>Over temp. | | 25 | 100<br>300 | | 25 | 250<br>400 | nA<br>nA |
| I$_{OS}$ | Input offset current | I$_{IN(+)}$−I$_{IN(−)}$<br>T$_A$=25°C<br>Over temp. | | ±3.0 | ±25<br>±100 | | ±5.0 | ±50<br>±150 | nA<br>nA |
| I$_{OL}$ | Output sink current | V$_{IN(−)}$≥1V$_{DC}$, V$_{IN(+)}$=0,<br>V$_O$≤1.5V$_{DC}$,<br>T$_A$=25°C<br>V$_O$=800mV,<br>over temp. | 6.0 | 16 | | 6.0 | 16 | | mA |
| I$_{OH}$ | Output leakage current | V$_{IN(+)}$≥1V$_{DC}$, V$_{IN(−)}$=0<br>V$_O$=5V$_{DC}$,<br>T$_A$=25°C<br>V$_O$=30V$_{DC}$,<br>over temp. | | 0.1 | 1.0 | | 0.1 | 1.0 | nA<br>μA |
| I$_{CC}$ | Supply current | V+=5V, R$_L$=∞<br>on comparators,<br>T$_A$=25°C<br>V+=30V | | 0.8 | 2.0 | | 0.8 | 2.0 | mA |
| A$_V$ | Voltage gain | R$_L$≥15kΩ,<br>V+=15V$_{DC}$ | 50 | 200 | | 50 | 200 | | V/mV |
| V$_{OL}$ | Saturation voltage | V$_{IN(−)}$≥1V$_{DC}$, V$_{IN(+)}$=0,<br>I$_{SINK}$≤4mA<br>T$_A$=25°C<br>Over temp. | | 250 | 400<br>700 | | 250 | 400<br>700 | mV<br>mV |
| t$_{LSR}$ | Large–signal response time | V$_{IN}$=TTL logic swing, V$_{REF}$=1.4V$_{DC}$,<br>V$_{RL}$=5V$_{DC}$, R$_L$=5.1kΩ, T$_A$=25°C | | 300 | | | 300 | | ns |
| t$_R$ | Response time[5] | V$_{RL}$=5V$_{DC}$, R$_L$=5.1kΩ,<br>T$_A$=25°C | | 1.3 | | | 1.3 | | μs |

See notes on following page.

Philips Semiconductors–Signetics Linear Products

Product specification

## Quad voltage comparator

### LM139A/239A/339A/LM139 /239/339/LM2901/MC3302

### DC AND AC ELECTRICAL CHARACTERISTICS

V+=5V$_{DC}$, LM139A/LM139: $-55°C \leq T_A \leq 125°C$; LM239: $-25°C \leq T_A \leq 85°C$; LM339: $0°C \leq T_A \leq 70°C$; LM339A: $0°C \leq T_A \leq 70°C$; LM239A: $-25°C \leq T_A \leq 85°C$; LM2901/LM3302: $-40°C \leq T_A \leq 85°C$, unless otherwise specified.

| SYM-BOL | PARAMETER | TEST CONDITIONS | LM2901 Min | LM2901 Typ | LM2901 Max | MC3302 Min | MC3302 Typ | MC3302 Max | UNIT |
|---|---|---|---|---|---|---|---|---|---|
| V$_{OS}$ | Input offset voltage[2] | T$_A$=25°C | | ±2.0 | ±7.0 | | ±3.0 | ±20 | mV |
| | | Over temp. | | ±9 | ±15 | | | ±40 | mV |
| V$_{CM}$ | Input common-mode voltage range[3] | T$_A$=25°C | 0 | | V+−1.5 | 0 | | V+−1.5 | V |
| | | Over temp. | 0 | | V+−2.0 | 0 | | V+−2.0 | |
| V$_{IDR}$ | Differential input voltage[1] | Keep all V$_{IN}$[5]≥0V$_{DC}$ (or V− if need) | | | V+ | | | V+ | V |
| I$_{BIAS}$ | Input bias current[4] | I$_{IN(+)}$ or I$_{IN(−)}$ with output in linear range | | | | | | | |
| | | T$_A$=25°C | | 25 | 250 | | 25 | 500 | nA |
| | | Over temp. | | 200 | 500 | | | 1000 | nA |
| I$_{OS}$ | Input offset current | I$_{IN(+)}$−I$_{IN(−)}$ | | | | | | | |
| | | T$_A$=25°C | | ±5 | ±50 | | ±5 | ±100 | nA |
| | | Over temp. | | ±50 | ±200 | | | ±300 | nA |
| I$_{OL}$ | Output sink current | V$_{IN(+)}$≥1V$_{DC}$, V$_{IN(+)}$=0, V$_O$≤1.5V$_{DC}$, T$_A$=25°C | 6.0 | 16 | | 6 | | | mA |
| | | V$_O$=800mV, over temp. | | | | 2.0 | | | mA |
| I$_{OH}$ | Output leakage current | V$_{IN(+)}$≥1V$_{DC}$, V$_{IN(−)}$=0 V$_O$=5V$_{DC}$, T$_A$=25°C | | 0.1 | | | 0.1 | | nA |
| | | V$_O$=30V$_{DC}$, over temp. | | | 1.0 | | | 1.0 | μA |
| I$_{CC}$ | Supply current | V+=5V, R$_L$=∞ on comparators, T$_A$=25°C | | 0.8 | 2.0 | | .8 | 1.8 | mA |
| | | V+=30V | | 1.0 | 2.5 | | | | |
| A$_V$ | Voltage gain | R$_L$≥15kΩ, V+=15V$_{DC}$ | 25 | 100 | | 2 | 100 | | V/mV |
| V$_{OL}$ | Saturation voltage | V$_{IN(−)}$≥1V$_{DC}$, V$_{IN(+)}$=0, I$_{SINK}$≤4mA T$_A$=25°C | | | 400 | | 150 | 400 | mV |
| | | Over temp. | | 400 | 700 | | | 700 | mV |
| t$_{LSR}$ | Large-signal response time | V$_{IN}$=TTL logic swing, V$_{REF}$=1.4V$_{DC}$, V$_{RL}$=5V$_{DC}$, R$_L$=5.1kΩ, T$_A$=25°C | | 300 | | | 300 | | ns |
| t$_R$ | Response time[5] | V$_{RL}$=5V$_{DC}$, R$_L$=5.1kΩ, T$_A$=25°C | | 1.3 | | | 1.3 | | μs |

NOTES:
1. Positive excursions of input voltage may exceed the power supply level by 17V. As long as the other voltage remains within the common-mode range, the comparator will provide a proper output state. The low input voltage state must not be less than −0.3V$_{DC}$ (or 0.3V$_{DC}$ below the magnitude of the negative power supply, if used).
2. At output switch point, V$_O$ ≈ 1.4V$_{DC}$, R$_S$=0Ω with V+ from 5V$_{DC}$ to 30V$_{DC}$; and over the full input common-mode range (0V$_{DC}$ to V+ − 1.5V$_{DC}$). Inputs of unused comparators should be grounded.
3. The input common-mode voltage or either input signal voltage should not be allowed to go negative by more than 0.3V. The upper end of the common-mode voltage range is V+ − 1.5V, but either or both inputs can go to 30V$_{DC}$ without damage.
4. The direction of the input current is out of the IC due to the PNP input stage. This current is essentially constant, independent of the state of the output so no loading change exists on the reference or input lines.
5. The response time specified is for a 100mV input step with a 5mV overdrive. For larger overdrive signals, 300ns can be obtained (see typical performance characteristics section).

Philips Semiconductors–Signetics Linear Products

Product specification

# Phase-locked loop

# NE/SE564

## DESCRIPTION
The NE/SE564 is a versatile, high guaranteed frequency phase-locked loop designed for operation up to 50MHz. As shown in the Block Diagram, the NE/SE564 consists of a VCO, limiter, phase comparator, and post detection processor.

## FEATURES
- Operation with single 5V supply
- TTL-compatible inputs and outputs
- Guaranteed operation to 50MHz
- External loop gain control
- Reduced carrier feedthrough
- No elaborate filtering needed in FSK applications
- Can be used as a modulator
- Variable loop gain (externally controlled)

## APPLICATIONS
- High speed modems
- FSK receivers and transmitters
- Frequency Synthesizers
- Signal generators
- Various satcom/TV systems
- pin configuration

## PIN CONFIGURATIONS

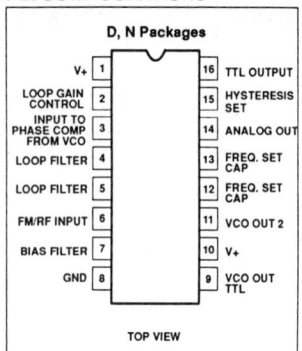

D, N Packages

| Pin | Signal | Pin | Signal |
|---|---|---|---|
| 1 | V+ | 16 | TTL OUTPUT |
| 2 | LOOP GAIN CONTROL | 15 | HYSTERESIS SET |
| 3 | INPUT TO PHASE COMP FROM VCO | 14 | ANALOG OUT |
| 4 | LOOP FILTER | 13 | FREQ. SET CAP |
| 5 | LOOP FILTER | 12 | FREQ. SET CAP |
| 6 | FM/RF INPUT | 11 | VCO OUT 2 |
| 7 | BIAS FILTER | 10 | V+ |
| 8 | GND | 9 | VCO OUT TTL |

TOP VIEW

## ORDERING INFORMATION

| DESCRIPTION | TEMPERATURE RANGE | ORDER CODE |
|---|---|---|
| 16-Pin Plastic SO | 0 to +70°C | NE564D |
| 16-Pin Plastic DIP | 0 to +70°C | NE564N |
| 16-Pin Plastic DIP | -55 to +125°C | SE564N |

## BLOCK DIAGRAM

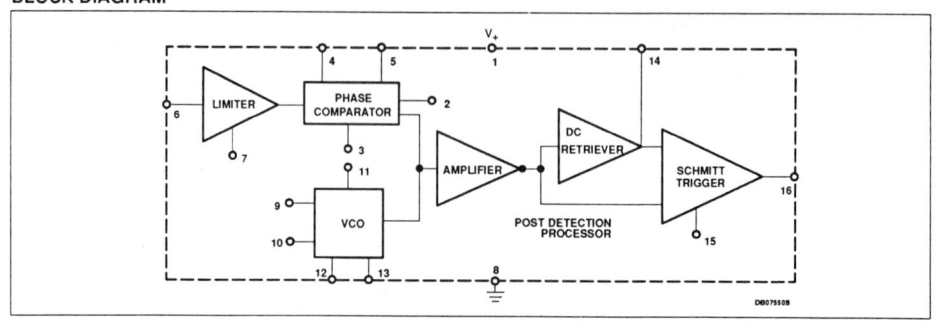

Appendix 3   233

Philips Semiconductors–Signetics Linear Products

Product specification

## Phase-locked loop

## NE/SE564

### ABSOLUTE MAXIMUM RATINGS

| SYMBOL | PARAMETER | RATING | UNITS |
|---|---|---|---|
| V+ | Supply voltage<br>Pin 1<br>Pin 10 | 14<br>6 | V<br>V |
| $I_{OUT}$ | (Sink) Max (Pin 9) | 10 | mA |
| $P_D$ | Power dissipation | 600 | mW |
| $T_A$ | Operating ambient temperature<br>NE<br>SE | 0 to +70<br>-55 to +125 | °C<br>°C |
| $T_{STG}$ | Storage temperature range | -65 to +150 | °C |

**NOTE:**
Operation above 5V will require heatsinking of the case.

### DC AND AC ELECTRICAL CHARACTERISTICS

$V_{CC}$ = 5V; $T_A$ = 0 to 25°C; $f_O$ = 5MHz, $I_2$ = 400µA; unless otherwise specified.

| SYMBOL | PARAMETER | TEST CONDITIONS | LIMITS SE564 MIN | LIMITS SE564 TYP | LIMITS SE564 MAX | LIMITS NE564 MIN | LIMITS NE564 TYP | LIMITS NE564 MAX | UNITS |
|---|---|---|---|---|---|---|---|---|---|
| | Maximum VCO frequency | $C_1$ = 0 (stray) | 45 | 60 | | 45 | 60 | | MHz |
| | Lock range | Input ≥ 200mV$_{RMS}$<br>$T_A$ = 25°C<br>$T_A$ = 125°C<br>$T_A$ = -55°C<br>$T_A$ = 0°C<br>$T_A$ = 70°C | 40<br>20<br>50 | 70<br>30<br>80 | | 40 | 70<br><br><br>70<br>40 | | % of $f_O$ |
| | Capture range | Input ≥ 200mV$_{RMS}$, $R_2$ = 27Ω | 20 | 30 | | 20 | 30 | | % of $f_O$ |
| | VCO frequency drift with temperature | $f_O$ = 5MHz,<br>$T_A$ = -55°C to +125°C<br>$T_A$ = 0 to +70°C<br>= 0 to +70°C<br>$f_O$ = 5MHz,<br>$T_A$ = -55°C to +125°C<br>$T_A$ = 0 to +70°C | | 500<br><br><br>300 | 1500<br><br><br>800 | | 600<br><br><br>500 | | PPM/°C |
| | VCO free-running frequency | $C_1$ = 91pF<br>$R_C$ = 100Ω "Internal" | 4 | 5 | 6 | 3.5 | 5 | 6.5 | MHz |
| | VCO frequency change with supply voltage | $V_{CC}$ = 4.5V to 5.5V | | 3 | 8 | | 3 | 8 | % of $f_O$ |
| | Demodulated output voltage | Modulation frequency: 1kHz<br>$f_O$ = 5MHz, input deviation:<br>2%T = 25°C<br>1%T = 25°C<br>1%T = 0°C<br>1%T = -55°C<br>1%T = 70°C<br>1%T = 125°C | 16<br>8<br><br>6<br><br>12 | 28<br>14<br><br>10<br><br>16 | | 16<br>8 | 28<br>14<br>13<br><br>15 | | mV$_{RMS}$<br>mV$_{RMS}$<br>mV$_{RMS}$<br>mV$_{RMS}$<br>mV$_{RMS}$<br>mV$_{RMS}$ |
| | Distortion | Deviation: 1% to 8% | | 1 | | | 1 | | % |
| S/N | Signal-to-noise ratio | Std. condition, 1% to 10% dev. | | 40 | | | 40 | | dB |
| | AM rejection | Std. condition, 30% AM | | 35 | | | 35 | | dB |
| | Demodulated output at operating voltage | Modulation frequency: 1kHz<br>$f_O$ = 5MHz, input deviation: 1%<br>$V_{CC}$ = 4.5V<br>$V_{CC}$ = 5.5V | 7<br>8 | 12<br>14 | | 7<br>9 | 12<br>14 | | mV$_{RMS}$<br>mV$_{RMS}$ |
| $I_{CC}$ | Supply current | $V_{CC}$ = 5V $I_1$, $I_{10}$ | | 45 | 60 | | 45 | 60 | mA |
| | Output<br>"1" output leakage current<br>"0" output voltage | $V_{OUT}$ = 5V, Pins 16, 9<br>$I_{OUT}$ = 2mA, Pins 16, 9<br>$I_{OUT}$ = 6mA, Pins 16, 9 | <br>0.3<br>0.4 | 1<br>0.6<br>0.8 | 20 | <br>0.3<br>0.4 | 1<br>0.6<br>0.8 | 20 | µA<br>V<br>V |

Philips Semiconductors–Signetics Linear Products  
Product specification

# Function generator

# NE/SE566

## DESCRIPTION
The NE/SE566 Function Generator is a voltage-controlled oscillator of exceptional linearity with buffered square wave and triangle wave outputs. The frequency of oscillation is determined by an external resistor and capacitor and the voltage applied to the control terminal. The oscillator can be programmed over a ten-to-one frequency range by proper selection of an external resistance and modulated over a ten-to-one range by the control voltage, with exceptional linearity.

## FEATURES
- Wide range of operating voltage (up to 24V; single or dual)
- High linearity of modulation
- Highly stable center frequency (200ppm/°C typical)
- Highly linear triangle wave output
- Frequency programming by means of a resistor or capacitor, voltage or current
- Frequency adjustable over 10-to-1 range with same capacitor

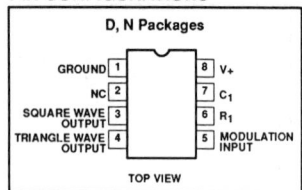

## APPLICATIONS
- Tone generators
- Frequency shift keying
- FM modulators
- Clock generators
- Signal generators
- Function generators

## PIN CONFIGURATIONS

D, N Packages

```
GROUND       1    8  V+
NC           2    7  C1
SQUARE WAVE  3    6  R1
 OUTPUT
TRIANGLE WAVE 4   5  MODULATION
 OUTPUT             INPUT
```
TOP VIEW

## ORDERING INFORMATION

| DESCRIPTION | TEMPERATURE RANGE | ORDER CODE |
|---|---|---|
| 8-Pin Plastic SO | 0 to +70°C | NE566D |
| 14-Pin Cerdip | 0 to +70°C | NE566F |
| 8-Pin Plastic DIP | 0 to +70°C | NE566N |
| 8-Pin Plastic DIP | -55°C to +125°C | SE566N |

## BLOCK DIAGRAM

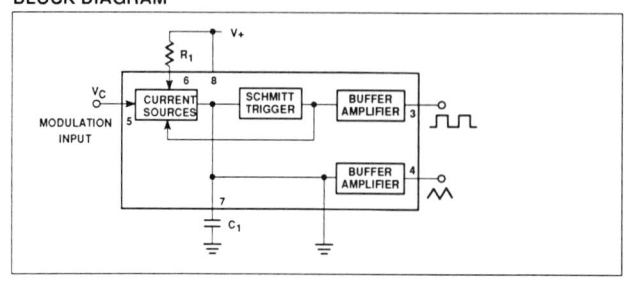

Appendix 3   235

Philips Semiconductors–Signetics Linear Products

Product specification

## Function generator

## NE/SE566

**EQUIVALENT SCHEMATIC**

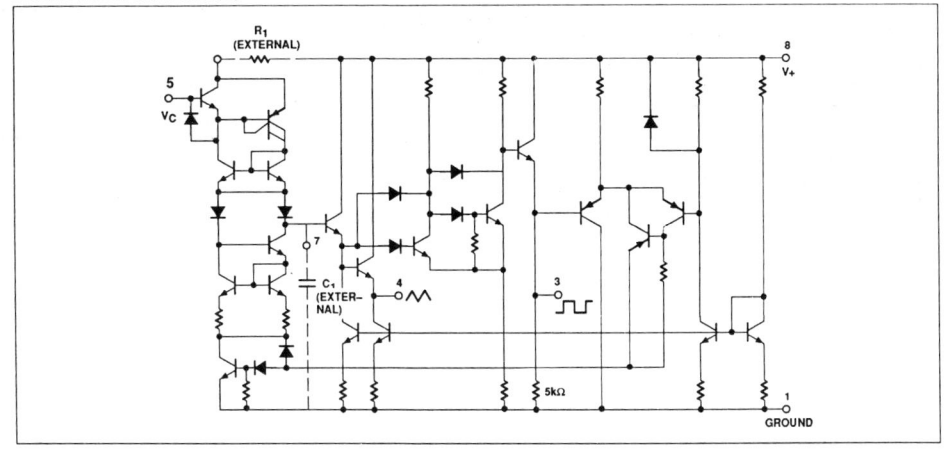

**ABSOLUTE MAXIMUM RATINGS**

| SYMBOL | PARAMETER | RATING | UNIT |
|---|---|---|---|
| $V+$ | Maximum operating voltage | 26 | V |
| $V_{IN}$ | Input voltage | 3 | $V_{P-P}$ |
| $T_{STG}$ | Storage temperature range | -65 to +150 | °C |
| $T_A$ | Operating ambient temperature range | | |
| | NE566 | 0 to +70 | °C |
| | SE566 | -55 to +125 | °C |
| $P_D$ | Power dissipation | 300 | mW |

Philips Semiconductors–Signetics Linear Products                                    Product specification

# Function generator                                                                NE/SE566

## DC ELECTRICAL CHARACTERISTICS
$T_A=25°C$, $V_{CC}=\pm6V$, unless otherwise specified.

| SYMBOL | PARAMETER | SE566 Min | SE566 Typ | SE566 Max | NE566 Min | NE566 Typ | NE566 Max | UNIT |
|---|---|---|---|---|---|---|---|---|
| **General** | | | | | | | | |
| $T_A$ | Operating ambient temperature range | -55 | | 125 | 0 | | 70 | °C |
| $V_{CC}$ | Operating supply voltage | ±6 | | ±12 | ±6 | | ±12 | V |
| $I_{CC}$ | Operating supply current | | 7 | 12.5 | | 7 | 12.5 | mA |
| **VCO[1]** | | | | | | | | |
| $f_{MAX}$ | Maximum operating frequency | | 1 | | | 1 | | MHz |
| | Frequency drift with temperature | | 500 | | | 600 | | ppm/°C |
| | Frequency drift with supply voltage | | 0.1 | 1 | | 0.2 | 2 | %/V |
| | Control terminal input impedance[2] | | 1 | | | 1 | | MΩ |
| | FM distortion (±10% deviation) | | 0.2 | 0.75 | | 0.4 | 1.5 | % |
| | Maximum sweep rate | | 1 | | | 1 | | MHz |
| | Sweep range | | 10:1 | | | 10:1 | | |
| **Output** | | | | | | | | |
| | Triangle wave output | | | | | | | |
| | impedance | | 50 | | | 50 | | Ω |
| | voltage | 1.9 | 2.4 | | 1.9 | 2.4 | | $V_{P-P}$ |
| | linearity | | 0.2 | | | 0.5 | | % |
| | Square wave input | | | | | | | |
| | impedance | | 50 | | | 50 | | Ω |
| | voltage | 5 | 5.4 | | 5 | 5.4 | | $V_{P-P}$ |
| | duty Cycle | 45 | 50 | 55 | 40 | 50 | 60 | % |
| $t_R$ | Rise time | | 20 | | | 20 | | ns |
| $t_F$ | Fall Time | | 50 | | | 50 | | ns |

**NOTES:**
1. The external resistance for frequency adjustment ($R_1$) must have a value between 2kΩ and 20kΩ.
2. The bias voltage ($V_C$) applied to the control terminal (Pin 5) should be in the range $V+ \leq V_C \leq V+$.

Appendix 3   237

Philips Semiconductors–Signetics Linear Products

Product specification

## Tone decoder/phase-locked loop

## NE/SE567

### DESCRIPTION
The NE/SE567 tone and frequency decoder is a highly stable phase-locked loop with synchronous AM lock detection and power output circuitry. Its primary function is to drive a load whenever a sustained frequency within its detection band is present at the self-biased input. The bandwidth center frequency and output delay are independently determined by means of four external components.

### FEATURES
- Wide frequency range (.01Hz to 500kHz)
- High stability of center frequency
- Independently controllable bandwidth (up to 14%)
- High out-band signal and noise rejection
- Logic-compatible output with 100mA current sinking capability
- Inherent immunity to false signals
- Frequency adjustment over a 20-to-1 range with an external resistor
- Military processing available

### APPLICATIONS
- Touch-Tone® decoding
- Carrier current remote controls
- Ultrasonic controls (remote TV, etc.)
- Communications paging
- Frequency monitoring and control
- Wireless intercom
- Precision oscillator

### PIN CONFIGURATIONS

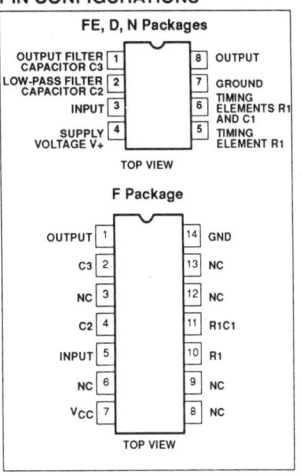

**FE, D, N Packages**

| | | |
|---|---|---|
| OUTPUT FILTER CAPACITOR C3 | 1 | 8  OUTPUT |
| LOW-PASS FILTER CAPACITOR C2 | 2 | 7  GROUND |
| INPUT | 3 | 6  TIMING ELEMENTS R1 AND C1 |
| SUPPLY VOLTAGE V+ | 4 | 5  TIMING ELEMENT R1 |

TOP VIEW

**F Package**

| | | |
|---|---|---|
| OUTPUT | 1 | 14  GND |
| C3 | 2 | 13  NC |
| NC | 3 | 12  NC |
| C2 | 4 | 11  R1C1 |
| INPUT | 5 | 10  R1 |
| NC | 6 | 9   NC |
| V<sub>CC</sub> | 7 | 8   NC |

TOP VIEW

### BLOCK DIAGRAM

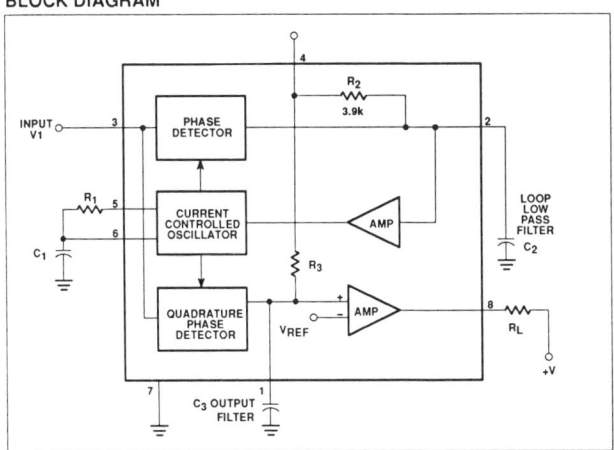

®Touch-Tone is a registered trademark of AT&T.

Philips Semiconductors–Signetics Linear Products

Product specification

## Tone decoder/phase-locked loop

NE/SE567

**EQUIVALENT SCHEMATIC**

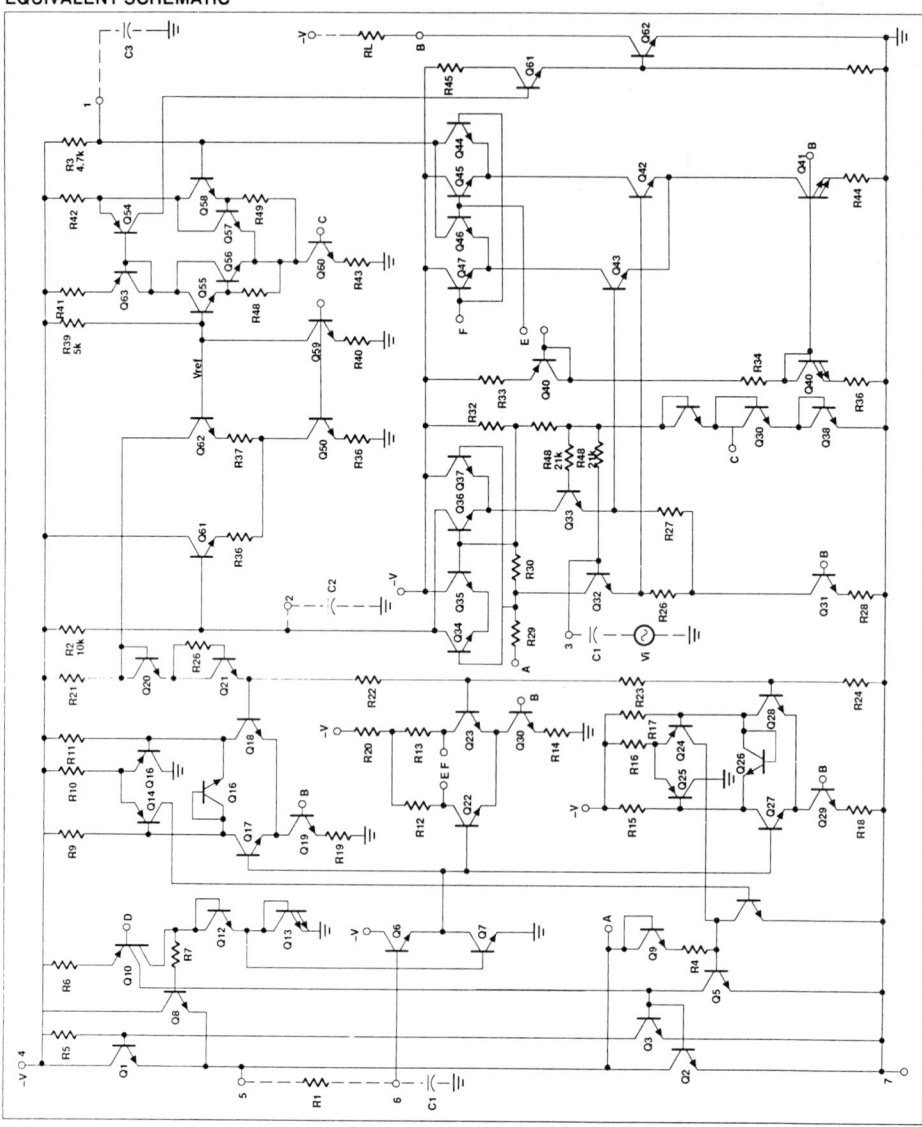

Philips Semiconductors–Signetics Linear Products  Product specification

## Tone decoder/phase-locked loop  NE/SE567

### ORDERING INFORMATION

| DESCRIPTION | TEMPERATURE RANGE | ORDER CODE |
|---|---|---|
| 8-Pin Plastic SO | 0 to +70°C | NE567D |
| 14-Pin Cerdip | 0 to +70°C | NE567F |
| 8-Pin Plastic DIP | 0 to +70°C | NE567N |
| 8-Pin Plastic SO | -55°C to +125°C | SE567D |
| 8-Pin Cerdip | -55°C to +125°C | SE567FE |
| 8-Pin Plastic DIP | -55°C to +125°C | SE567N |

### ABSOLUTE MAXIMUM RATINGS

| SYMBOL | PARAMETER | RATING | UNIT |
|---|---|---|---|
| $T_A$ | Operating temperature | | |
|  | NE567 | 0 to +70 | °C |
|  | SE567 | -55 to +125 | °C |
| $V_{CC}$ | Operating voltage | 10 | V |
| V+ | Positive voltage at input | $0.5 + V_S$ | V |
| V- | Negative voltage at input | -10 | $V_{DC}$ |
| $V_{OUT}$ | 80Output voltage (collector of output transistor) | 15 | $V_{DC}$ |
| $T_{STG}$ | Storage temperature range | -65 to +150 | °C |
| $P_D$ | Power dissipation | 300 | mW |

Philips Semiconductors–Signetics Linear Products

## Tone decoder/phase-locked loop

Product specification

# NE/SE567

### DC ELECTRICAL CHARACTERISTICS
V +=5.0V; $T_A$=25°C, unless otherwise specified.

| SYM-BOL | PARAMETER | TEST CONDITIONS | SE567 Min | SE567 Typ | SE567 Max | NE567 Min | NE567 Typ | NE567 Max | UNIT |
|---|---|---|---|---|---|---|---|---|---|
| **Center frequency[1]** | | | | | | | | | |
| $f_O$ | Highest center frequency | | | 500 | | | 500 | | kHz |
| $f_O$ | Center frequency stability[2] | -55 to +125°C | | 35 ±140 | | | 35 ±140 | | ppm/°C |
| | | 0 to +70°C | | 35 ±60 | | | 35 ±60 | | ppm/°C |
| $f_O$ | Center frequency distribution | $f_O = 100kHz = \frac{1}{1.1 R_1 C_1}$ | -10 | 0 | +10 | -10 | 0 | +10 | % |
| $f_O$ | Center frequency shift with supply voltage | $f_O = 100kHz = \frac{1}{1.1 R_1 C_1}$ | | 0.5 | 1 | | 0.7 | 2 | %/V |
| **Detection bandwidth** | | | | | | | | | |
| BW | Largest detection bandwidth | $f_O = 100kHz = \frac{1}{1.1 R_1 C_1}$ | 12 | 14 | 16 | 10 | 14 | 18 | % of $f_O$ |
| BW | Largest detection bandwidth skew | | | 2 | 4 | | 3 | 6 | % of $f_O$ |
| BW | Largest detection bandwidth—variation with temperature | $V_I$=300mV$_{RMS}$ | | ±0.1 | | | ±0.1 | | %/°C |
| BW | Largest detection bandwidth—variation with supply voltage | $V_I$=300mV$_{RMS}$ | | ±2 | | | ±2 | | %/V |
| **Input** | | | | | | | | | |
| $R_{IN}$ | Input resistance | | 15 | 20 | 25 | 15 | 20 | 25 | kΩ |
| $V_I$ | Smallest detectable input voltage[4] | $I_L$=100mA, $f_i$=$f_O$ | | 20 | 25 | | 20 | 25 | mV$_{RMS}$ |
| | Largest no-output input voltage[4] | $I_L$=100mA, $f_i$=$f_O$ | 10 | 15 | | 10 | 15 | | mV$_{RMS}$ |
| | Greatest simultaneous out-band signal-to-in-band signal ratio | | | +6 | | | +6 | | dB |
| | Minimum input signal to wide-band noise ratio | $B_n$=140kHz | | -6 | | | -6 | | dB |
| **Output** | | | | | | | | | |
| | Fastest on-off cycling rate | | | $f_O$/20 | | | $f_O$/20 | | |
| | "1" output leakage current | $V_8$=15V | | 0.01 | 25 | | 0.01 | 25 | μA |
| | "0" output voltage | $I_L$=30mA | | 0.2 | 0.4 | | 0.2 | 0.4 | V |
| | | $I_L$=100mA | | 0.6 | 1.0 | | 0.6 | 1.0 | V |
| $t_F$ | Output fall time[3] | $R_L$=50Ω | | 30 | | | 30 | | ns |
| $t_R$ | Output rise time[3] | $R_L$=50Ω | | 150 | | | 150 | | ns |
| **General** | | | | | | | | | |
| $V_{CC}$ | Operating voltage range | | 4.75 | | 9.0 | 4.75 | | 9.0 | V |
| | Supply current quiescent | | | 6 | 8 | | 7 | 10 | mA |
| | Supply current—activated | $R_L$=20kΩ | | 11 | 13 | | 12 | 15 | mA |
| $t_{PD}$ | Quiescent power dissipation | | | 30 | | | 35 | | mW |

NOTES:
1. Frequency determining resistor $R_1$ should be between 2 and 20kΩ.
2. Applicable over 4.75V to 5.75V. See graphs for more detailed information.
3. Pin 8 to Pin 1 feedback $R_L$ network selected to eliminate pulsing during turn-on and turn-off.
4. With $R_2$=130kΩ from Pin 1 to V+. See Figure 1.

Philips Semiconductors–Signetics Linear Products

Product specification

## Low power quad op amps

## LM124/224/324/324A/ SA534/LM2902

### DESCRIPTION
The LM124/SA534/LM2902 series consists of four independent, high-gain, internally frequency-compensated operational amplifiers designed specifically to operate from a single power supply over a wide range of voltages.

### UNIQUE FEATURES
In the linear mode, the input common-mode voltage range includes ground and the output voltage can also swing to ground, even though operated from only a single power supply voltage.

The unity gain crossover frequency and the input bias current are temperature-compensated.

### FEATURES
- Internally frequency-compensated for unity gain
- Large DC voltage gain: 100dB
- Wide bandwidth (unity gain): 1MHz (temperature-compensated)
- Wide power supply range Single supply: $3V_{DC}$ to $30V_{DC}$ or dual supplies: $\pm1.5V_{DC}$ to $\pm15V_{DC}$
- Very low supply current drain: essentially independent of supply voltage (1mW/op amp at $+5V_{DC}$)
- Low input biasing current: $45nA_{DC}$ (temperature-compensated)
- Low input offset voltage: $2mV_{DC}$ and offset current: $5nA_{DC}$
- Differential input voltage range equal to the power supply voltage
- Large output voltage: $0V_{DC}$ to $V_{CC}-1.5V_{DC}$ swing

### PIN CONFIGURATION

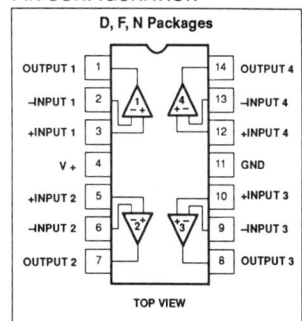

D, F, N Packages

| | | | | |
|---|---|---|---|---|
| OUTPUT 1 | 1 | | 14 | OUTPUT 4 |
| –INPUT 1 | 2 | | 13 | –INPUT 4 |
| +INPUT 1 | 3 | | 12 | +INPUT 4 |
| V+ | 4 | | 11 | GND |
| +INPUT 2 | 5 | | 10 | +INPUT 3 |
| –INPUT 2 | 6 | | 9 | –INPUT 3 |
| OUTPUT 2 | 7 | | 8 | OUTPUT 3 |

TOP VIEW

### ORDERING INFORMATION

| DESCRIPTION | TEMPERATURE RANGE | ORDER CODE |
|---|---|---|
| 14-Pin Plastic DIP | -55°C to +125°C | LM124N |
| 14-Pin Ceramic DIP | -55°C to +125°C | LM124F |
| 14-Pin Plastic DIP | -25°C to +85°C | LM224N |
| 14-Pin Ceramic DIP | -25°C to +85°C | LM224F |
| 14-Pin Plastic DIP | 0°C to +70°C | LM324N |
| 14-Pin Ceramic DIP | 0°C to +70°C | LM324F |
| 14-Pin Plastic SO | 0°C to +70°C | LM324D |
| 14-Pin Plastic DIP | 0°C to +70°C | LM324AN |
| 14-Pin Plastic SO | 0°C to +70°C | LM324AD |
| 14-Pin Plastic DIP | -40°C to +85°C | SA534N |
| 14-Pin Ceramic DIP | -40°C to +85°C | SA534F |
| 14-Pin Plastic SO | -40°C to +85°C | SA534D |
| 14-Pin Plastic SO | -40°C to +85°C | LM2902D |
| 14-Pin Plastic DIP | -40°C to +85°C | LM2902N |

Philips Semiconductors–Signetics Linear Products

Product specification

## Low power quad op amps

LM124/224/324/324A/
SA534/LM2902

**ABSOLUTE MAXIMUM RATINGS**

| SYMBOL | PARAMETER | RATING | UNIT |
|---|---|---|---|
| $V_{CC}$ | Supply voltage | 32 or ±16 | $V_{DC}$ |
| $V_{IN}$ | Differential input voltage | 32 | $V_{DC}$ |
| $V_{IN}$ | Input voltage | -0.3 to +32 | $V_{DC}$ |
| $P_D$ | Maximum power dissipation, $T_A$=25°C (still-air)[1] | | |
| | N package | 1420 | mW |
| | F package | 1190 | mW |
| | D package | 1040 | mW |
| | Output short-circuit to GND one amplifier $V_{CC}$<15$V_{DC}$ and $T_A$=25°C | Continuous | |
| $I_{IN}$ | Input current ($V_{IN}$<-0.3V)[3] | 50 | mA |
| $T_A$ | Operating ambient temperature range | | |
| | LM324/A | 0 to +70 | °C |
| | LM224 | -25 to +85 | °C |
| | SA534/LM2902 | -40 to +85 | °C |
| | LM124 | -55 to +125 | °C |
| $T_{STG}$ | Storage temperature range | -65 to +150 | °C |
| $T_{SOLD}$ | Lead soldering temperature (10sec max) | 300 | °C |

**NOTES:**
1. Derate above 25°C at the following rates:
   F package at 9.5mW/°C
   N package at 11.4mW/°C
   D package at 8.3mW/°C
2. Short-circuits from the output to $V_{CC}$+ can cause excessive heating and eventual destruction. The maximum output current is approximately 40mA, independent of the magnitude of $V_{CC}$. At values of supply voltage in excess of +15$V_{DC}$ continuous short-circuits can exceed the power dissipation ratings and cause eventual destruction.
3. This input current will only exist when the voltage at any of the input leads is driven negative. It is due to the collector-base junction of the input PNP transistors becoming forward biased and thereby acting as input bias clamps. In addition, there is also lateral NPN parasitic transistor action on the IC chip. This action can cause the output voltages of the op amps to go to the V+ rail (or to ground for a large overdrive) during the time that the input is driven negative.

Philips Semiconductors–Signetics Linear Products

Product specification

# Low power quad op amps

**LM124/224/324/324A/ SA534/LM2902**

## DC ELECTRICAL CHARACTERISTICS
$V_{CC}$=5V, $T_A$=25°C unless otherwise specified.

| SYMBOL | PARAMETER | TEST CONDITIONS | LM124/LM224 Min | LM124/LM224 Typ | LM124/LM224 Max | LM324/SA534/LM2902 Min | LM324/SA534/LM2902 Typ | LM324/SA534/LM2902 Max | UNIT |
|---|---|---|---|---|---|---|---|---|---|
| $V_{OS}$ | Offset voltage[1] | $R_S$=0Ω | | ±2 | ±5 | | ±2 | ±7 | mV |
| | | $R_S$=0Ω, over temp. | | | ±7 | | | ±9 | mV |
| $\Delta V_{OS}/\Delta T$ | Temperature drift | $R_S$=0Ω, over temp. | | 7 | | | 7 | | μV/°C |
| $I_{BIAS}$ | Input current[2] | $I_{IN}(+)$ or $I_{IN}(-)$ | | 45 | 150 | | 45 | 250 | nA |
| | | $I_{IN}(+)$ or $I_{IN}(-)$, over temp. | | 40 | 300 | | 40 | 500 | nA |
| $\Delta I_{BIAS}/\Delta T$ | Temperature drift | Over temp. | | 50 | | | 50 | | pA/°C |
| $I_{OS}$ | Offset current | $I_{IN}(+)-I_{IN}(-)$ | | ±3 | ±30 | | ±5 | ±50 | nA |
| | | $I_{IN}(+)-I_{IN}(-)$, over temp. | | | ±100 | | | ±150 | nA |
| $V_{OS}$ | Offset voltage[1] | $R_S$ = 0Ω | | ±2 | ±5 | | ±2 | ±7 | mV |
| | | $R_S$ = 0Ω, over temp. | | | ±7 | | | ±9 | mV |
| $\Delta V_{OS}/\Delta T$ | Temperature drift | $R_S$ = 0Ω, over temp. | | 7 | | | 7 | | μV/°C |
| $I_{BIAS}$ | Input current[2] | $I_{IN}(+)$ or $I_{IN}(-)$ | | 45 | 150 | | 45 | 250 | nA |
| | | $I_{IN}(+)$ or $I_{IN}(-)$, over temp. | | 40 | 300 | | 40 | 500 | nA |
| $\Delta I_{BIAS}/\Delta T$ | Temperature drift | Over temp. | | 50 | | | 50 | | pA/°C |
| $I_{OS}$ | Offset current | $I_{IN}(+) - I_{IN}(-)$ | | ±3 | ±30 | | ±5 | ±50 | nA |
| | | $I_{IN}(+) - I_{IN}(-)$, over temp. | | | ±100 | | | ±150 | nA |
| $\Delta I_{OS}/\Delta T$ | Temperature drift | Over temp. | | 10 | | | 10 | | pA/°C |
| $V_{CM}$ | Common-mode voltage range[3] | $V_{CC}$≤30V | 0 | | $V_{CC}$-1.5 | 0 | | $V_{CC}$-1.5 | V |
| | | $V_{CC}$≤30V, over temp. | 0 | | $V_{CC}$-2 | 0 | | $V_{CC}$-2 | V |
| CMRR | Common-mode rejection ratio | $V_{CC}$=30V | 70 | 85 | | 65 | 70 | | dB |
| $V_{OUT}$ | Output voltage swing | $R_L$=2kΩ, $V_{CC}$=30V, over temp. | 26 | | | 26 | | | V |
| $V_{OH}$ | Output voltage high | $R_L$≤10kΩ, $V_{CC}$=30V, over temp. | 27 | 28 | | 27 | 28 | | V |
| $V_{OL}$ | Output voltage low | $R_L$≤10kΩ, $V_{CC}$=5V, over temp. | | 5 | 20 | | 5 | 20 | mV |
| $I_{CC}$ | Supply current | $R_L$=∞, $V_{CC}$=30V, over temp. | | 1.5 | 3 | | 1.5 | 3 | mA |
| | | $R_L$=∞, $V_{CC}$=5V, over temp. | | 0.7 | 1.2 | | 0.7 | 1.2 | mA |
| $A_{VOL}$ | Large-signal voltage gain | $V_{CC}$=15V (for large $V_O$ swing), $R_L$≥2kΩ | 50 | 100 | | 25 | 100 | | V/mV |
| | | $V_{CC}$=15V (for large $V_O$ swing), $R_L$≥2kΩ, over temp. | 25 | | | 15 | | | V/mV |
| | Amplifier-to-amplifier coupling[5] | f≈1kHz to 20kHz, input referred | | -120 | | | -120 | | dB |
| PSRR | Power supply rejection ratio | $R_S$≤0Ω | 65 | 100 | | 65 | 100 | | dB |
| $I_{OUT}$ | Output current source | $V_{IN}+$=+1V, $V_{IN}-$=0V, $V_{CC}$=15V | 20 | 40 | | 20 | 40 | | mA |
| | | $V_{IN}+$=+1V, $V_{IN}-$=0V, $V_{CC}$=15V, over temp. | 10 | 20 | | 10 | 20 | | mA |
| | sink | $V_{IN}-$=+1V, $V_{IN}+$=0V, V+=15V | 10 | 20 | | 10 | 20 | | mA |
| | | $V_{IN}-$=+1V, $V_{IN}+$=0V, $V_{CC}$=15V, over temp. | 5 | 8 | | 5 | 8 | | mA |
| | | $V_{IN}-$=+1V, $V_{IN}+$=0V, $V_O$=200mV | 12 | 50 | | 12 | 50 | | μA |
| $I_{SC}$ | Short-circuit current[4] | | 10 | 40 | 60 | 10 | 40 | 60 | mA |

Philips Semiconductors–Signetics Linear Products

Product specification

## Low power quad op amps

LM124/224/324/324A/
SA534/LM2902

**DC ELECTRICAL CHARACTERISTICS** (Continued)

| SYMBOL | PARAMETER | TEST CONDITIONS | LM124/LM224 ||| LM324/SA534/LM2902 ||| UNIT |
|---|---|---|---|---|---|---|---|---|---|
| | | | Min | Typ | Max | Min | Typ | Max | |
| GBW | Unity gain bandwidth | | | 1 | | | 1 | | MHz |
| SR | Slew rate | | | 0.3 | | | 0.3 | | V/μs |
| $V_{NOISE}$ | Input noise voltage | f=1kHz | | 40 | | | 40 | | nV/√Hz |
| $V_{DIFF}$ | Differential input voltage[3] | | | | $V_{CC}$ | | | $V_{CC}$ | V |

Philips Semiconductors–Signetics Linear Products

Product specification

## Low power quad op amps

**LM124/224/324/324A/ SA534/LM2902**

### DC ELECTRICAL CHARACTERISTICS (Continued)

| SYMBOL | PARAMETER | TEST CONDITIONS | LM324A Min | LM324A Typ | LM324A Max | UNIT |
|---|---|---|---|---|---|---|
| $V_{OS}$ | Offset voltage[1] | $R_S=0\Omega$ | | ±2 | ±3 | mV |
| | | $R_S=0\Omega$, over temp. | | | ±5 | mV |
| $\Delta V_{OS}/\Delta T$ | Temperature drift | $R_S=0\Omega$, over temp. | | 7 | 30 | μV/°C |
| $I_{BIAS}$ | Input current[2] | $I_{IN}(+)$ or $I_{IN}(-)$ | | 45 | 100 | nA |
| | | $I_{IN}(+)$ or $I_{IN}(-)$, over temp. | | 40 | 200 | nA |
| $\Delta I_{BIAS}/\Delta T$ | Temperature drift | Over temp. | | 50 | | pA/°C |
| $I_{OS}$ | Offset current | $I_{IN}(+)-I_{IN}(-)$ | | ±5 | ±30 | nA |
| | | $I_{IN}(+)-I_{IN}(-)$, over temp. | | | ±75 | nA |
| $\Delta I_{OS}/\Delta T$ | Temperature drift | Over temp. | | 10 | 300 | pA/°C |
| $V_{CM}$ | Common-mode voltage range[3] | $V_{CC} \leq 30V$ | 0 | | $V_{CC}$-1.5 | V |
| | | $V_{CC} \leq 30V$, over temp. | 0 | | $V_{CC}$-2 | V |
| CMRR | Common-mode rejection ratio | $V_{CC}=30V$ | 65 | 85 | | dB |
| $V_{OUT}$ | Output voltage swing | $R_L=2k\Omega$, $V_{CC}=30V$, over temp. | 26 | | | V |
| $V_{OH}$ | Output voltage high | $R_L \leq 10k\Omega$, $V_{CC}=30V$, over temp. | 27 | 28 | | V |
| $V_{OL}$ | Output voltage low | $R_L \leq 10k\Omega$, $V_{CC}=5V$, over temp. | | 5 | 20 | mV |
| $I_{CC}$ | Supply current | $R_L=\infty$, $V_{CC}=30V$, over temp. | | 1.5 | 3 | mA |
| | | $R_L=\infty$, $V_{CC}=5V$, over temp. | | 0.7 | 1.2 | mA |
| $A_{VOL}$ | Large-signal voltage gain | $V_{CC}=15V$ (for large $V_O$ swing), $R_L \geq 2k\Omega$ | 25 | 100 | | V/mV |
| | | $V_{CC}=15V$ (for large $V_O$ swing), $R_L \geq 2k\Omega$, over temp. | 15 | | | V/mV |
| | Amplifier-to-amplifier coupling[5] | f=1kHz to 20kHz, input referred | | -120 | | dB |
| PSRR | Power supply rejection ratio | $R_S \leq 0\Omega$ | 65 | 100 | | dB |
| $I_{OUT}$ | Output current source | $V_{IN}+=+1V$, $V_{IN}-=0V$, $V_{CC}=15V$ | 20 | 40 | | mA |
| | | $V_{IN}+=+1V$, $V_{IN}-=0V$, $V_{CC}=15V$, over temp. | 10 | 20 | | mA |
| | sink | $V_{IN}-=+1V$, $V_{IN}+=0V$, $V+=15V$ | 10 | 20 | | mA |
| | | $V_{IN}-=+1V$, $V_{IN}+=0V$, $V_{CC}=15V$, over temp. | 5 | 8 | | mA |
| | | $V_{IN}-=+1V$, $V_{IN}+=0V$, $V_O=200mV$ | 12 | 50 | | μA |
| $I_{SC}$ | Short-circuit current[4] | | 10 | 40 | 60 | mA |
| $V_{DIFF}$ | Differential input voltage[3] | | | | $V_{CC}$ | V |
| GBW | Unity gain bandwidth | | | 1 | | MHz |
| SR | Slew rate | | | 0.3 | | V/μs |
| $V_{NOISE}$ | Input noise voltage | f=1kHz | | 40 | | nV/√Hz |

**NOTES:**
1. $V_O \approx 1.4 V_{DC}$, $R_S=0\Omega$ with $V_{CC}$ from 5V to 30V and over full input common-mode range ($0V_{DC}+$ to $V_{CC}$ -1.5V).
2. The direction of the input current is out of the IC due to the PNP input stage. This current is essentially constant, independent of the state of the output so no loading change exists on the input lines.
3. The input common-mode voltage or either input signal voltage should not be allowed to go negative by more than 0.3V. The upper end of the common-mode voltage range is $V_{CC}$ -1.5, but either or both inputs can go to +32V without damage.
4. Short-circuits from the output to $V_{CC}$ can cause excessive heating and eventual destruction. The maximum output current is approximately 40mA independent of the magnitude of $V_{CC}$. At values of supply voltage in excess of +15$V_{DC}$, continuous short-circuits can exceed the power dissipation ratings and cause eventual destruction. Destructive dissipation can result from simultaneous shorts on all amplifiers.
5. Due to proximity of external components, insure that coupling is not originating via stray capacitance between these external parts. This typically can be detected as this type of coupling increases at higher frequencies.

Philips Semiconductors–Signetics Linear Products

Product specification

## Low power quad op amps

LM124/224/324/324A/
SA534/LM2902

**EQUIVALENT CIRCUIT**

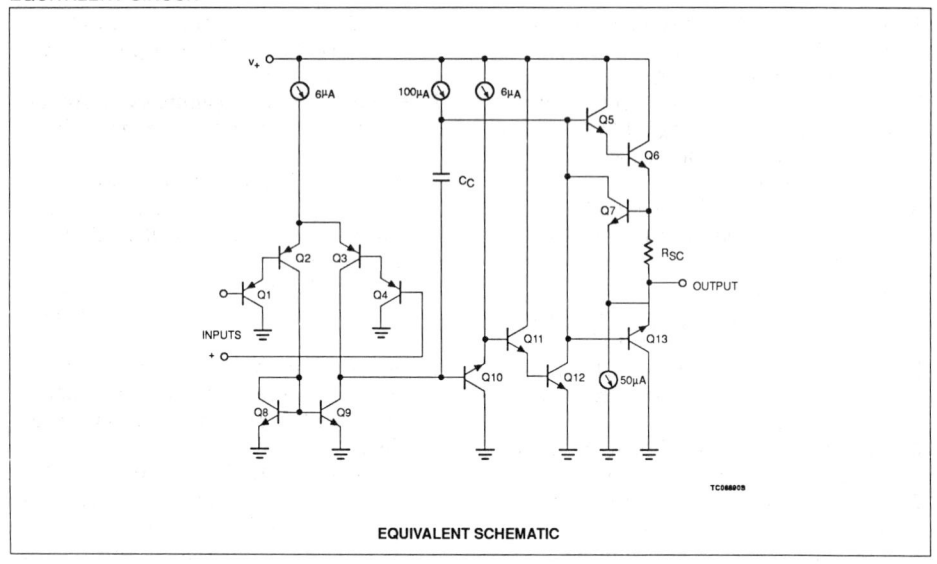

**EQUIVALENT SCHEMATIC**

# Bibliography

1. McMenamin, Michael J., *Linear Integrated Circuits: Operation and Applications*, Prentice Hall, 1985.
2. Prossen, Franklin P. and Winkel, David E., *The Art of Digital Design – An Introduction to Top-down Design*, Prentice Hall, 1987.
3. Wobschall, Darold, *Circuit Design for Electronic Instrumentation – Analog and Digital Devices from Sensor to Display*, McGraw Hill, 1987.
4. Helms, Harry L., *High-speed (HC/HCT) CMOS Guide*, Prentice-Hall, 1989.
5. Bonebreak, Robert L., *Practical Techniques of Electronic Circuit Design*, John Wiley & Sons, 1987.
6. Glasord, Glenn M., *Analog Electronic Circuits*, Prentice-Hall, 1986.
7. Collins, T. H., *Analog Electronics Handbook*, Prentice-Hall, 1989.
8. Dowding, Barry, *Principles of Electronics*, Prentice-Hall, 1988.
9. Johnson, E. L. and Karim, M. A., *Digital Design – A Pragmatic Approach*, PWS Publishers, USA, 1987.
10. Middleton, Robert G., *New Digital Troubleshooting Techniques*, Prentice-Hall, 1984.
11. Shepard, Jeffrey D., *Power Supplies*, Reston Publishing Company, 1984.
12. Dungan, Frank R., *Linear Integrated Circuits for Technicians*, PWS Publishers, USA, 1984.
13. Burger, Peter, *Digital Design: A Practical Course*, John Wiley & Sons, 1988.
14. Ulrich Tietze, Schenk, Christoph and Schmid, Eberhard, *Electronic Circuits*, Springer-Verlag, 1991.
15. Hnatek, Eugene R., *User's Guidebook to Digital CMOS Integrated Circuits*, McGraw Hill, 1981.
16. O'Dell, T. H., *Circuits for Electronic Instrumentation*, Cambridge University Press, 1991.
17. Soclof, Sidney, *Applications of Analog Integrated Circuits*, Prentice Hall, 1985.
18. Coughlin, Robert F. and Driscoll, Frederick F., *Operational Amplifiers and Linear Integrated Circuits*, Prentice Hall, 1982.
19. Clayton, G. B., *Operational Amplifiers*, ELBS Edition, Butterworth-Heinemann, 1986.
20. Dailey, Denton J., *Operational Amplifiers and Linear Integrated Circuits – Theory and Applications*, McGraw Hill, 1989.
21. Thackray, Philip C. and Meiksin Z. H., *Electronic Design with Off-the-shelf Integrated Circuits*, Prentice Hall, 1984.
22. Horowitz, Paul and Hill, Winfield, *The Art of Electronics*, Cambridge University Press, 1986.
23. Middleton, Robert G., *Designing Electronic Circuits*, Prentice Hall, 1986.
24. Millman, Jacob and Grabel, Arvin, *Microelectronics*, McGraw Hill, 1987.
25. Walter G. Jung, *IC Timer Cookbook*, Howard W. Sams & Co., 1981.
26. *Data Handbook – Linear Products*, Vol. IC11, 1989, Philips, The Netherlands.
27. *Intersil Component Data Catalog*, 1987.
28. *Linear Databook 1990*, Linear Technology Corporation, USA.
29. *Integrated Circuits Databook*, 1990, Maxim Integrated Products, USA.
30. *TTL Catalog*, Signetics, 1982.
31. *Digital Integrated Circuits*, CMOS HE4000B family, Part 4, July 1983, Philips, The Netherlands.

# Circuits index

Absolute value
  amplifier, using op-amp, 85
AC
  line synchronizer, 28
  variable input selector, 150
Adjustable
  voltage regulator, negative, 161
  voltage regulator, positive, 155, 158
Alarm
  light-activated, 204
  low-battery, 177
  power fail, 151–4
  temperature, high/low, 208, 209
Amplifier
  absolute value, 85
  anti-log, 90
  audio, 91, 94
  audio, 50 W, 94
  bridge transducer, 86
  differential, 85
  high-speed, 92
  instrumentation, 98
  inverting, 83
  logarithmic, 87
  low-distortion, 91
  low-noise, 92
  non-inverting, 84
  phonograph, pre-, 93
  power, 95
  programmable gain, 92
  quasilinear, 122
  summing, 84
  tape-head, pre-, 93
  variable gain, 81
Analog
  switch, no supply, 208
Antilog
  amplifier, 90
Astable
  multivibrator, 59
Attenuator
  digital, 13
Audio
  amplifier, 91, 94
  oscillator, 102
Automobile
  voltage regulator, 164

Battery
  back-up switch, 175
  charger, simple, 171
  charger, lead-acid, 156
  charger, wind-powered, 171
  disconnect, low, 117
  life extender, 173
  powered DC–DC converter, 146
  powered split supply, 145
  powered supply, polarity insensitive, 176
  status indicator, 172
  switchover circuit, 174
  warning, low, 177
Binary
  frequency divider, 15
Booster
  DC voltage, 147
  high current, 69
  power, op-amp, 68
Bridge
  audio power amplifier, 91
  transducer amplifier, 86
Buffer
  based switch debouncer, 25

Channel
  single-oscilloscope, 203
Clock
  generator, 3-phase, 116
  generator, 4-phase, 117
  generator, phase-shifted, 118
CMOS
  timer, programmable, 63
  TO-TTL interface, 46
Coaxial
  cable driver, 53
  cable receiver, 53
  cable tester, 195
Comparator
  frequency, analog, 27
  frequency, digital, 25
  multi-level, 73
  phase, digital, 12
  voltage, with hysteresis, 71
  window, 70

Constant
    current motor drive, 179
    DC output, variable AC input, 150
Controller
    DC motor speed, 179
    stepper motor, 4-phase, 180, 182
Converter
    DC–DC, +5 V to +10 V, 143
    +3 V to +5 V, 146
    step-up negative DC–DC, 149
    step-up positive DC–DC, 147
    voltage to current, 72
Counter
    frequency, 17, 36
    Johnson, 116, 203
    revolution, 187
    unit, 36
Crystal
    oscillator, 122
    tester, 194

DC
    motor controller, 179
    servomotor PLL, 178
DC–DC converter
    +3 V to +5 V, 146
    +5 V to +10 V, 143
Debouncer
    switch, 23
Decoder
    chatter-free, tone, 133
    dual-tone, 130, 131
    FSK, 188
    high speed narrow band, tone, 133
    SCA, 189
    tone, 125
Delay
    line, digital, 10
Detector
    AC power brownout, 152
    AC power power-fail, 152
    blown-fuse, 154
    digital-edge, 15
    glitch, digital, 37
    low-battery, 177
    magnetic transducer, for, 188
    missing pulse, 18
    phase, 139
    phase, digital, 12
    photo-diode, 187
Differential
    amplifier, 85
Differentiator
    op-amp based, 66

Digital
    AC line synchronizer, 28
    alternate-action switch, 41
    attenuator, 13
    bit rate generators, 30
    decade counter, 34
    delay-line, 10
    edge-detector, 15
    frequency comparator, 25
    frequency counter, 17, 36
    frequency divider, 15
    frequency doubler, 16
    frequency mixer, 12
    glitch detector, 37
    logic interfacing techniques, 44
    missing pulse detector, 18
    monostable, 1–7
    multiplexer, 23
    noise canceller, 9
    phase-detector, 12
    phase lead–lag indicator, 38
    power-on reset circuit, 21
    pulse-stretcher, 8
    quadrature phase-shifter, 19
    switch debouncer, 23
    synchronizer, 39
    tachometer, 17
    time-base generator, 29
Dimmer
    electronic, for battery operated
        applications, 206
    incandescent lamp, 206
Discriminator
    pulse-width, 20
Distortion
    audio amplifier, low-, 91
    reduction in mod-demod circuit, 137
Doubler
    frequency, 140
    frequency, digital, 16
    voltage, 199
Driver
    coaxial cable, 53
    FET, 56
    flat ribbon cable, 54
    lamp, 201
    power MOSFET, 198
    programmable micropower, line, 52
    relay, monolithic, 57
    RS 232, line, 47
    RS 232, 5 V, line, 50
    RS 232, optically isolated, line, 49
    RS 423, line, 55
    stepper motor, PLD based, 182
    stepper motor, simple, 180

DTMF
  filter based exchange tones detector, 202
Dual
  tone decoder, 130

ECL
  to TTL interface, 46
Exchange
  tones detector, 202

Filter
  DTMF, based exchange tones detector, 202
  supply-frequency reject, 73
Frequency
  counter, 17, 26
  doubler, 140
  meter, 132
  multiplier, using PLL, 136
  shift-keying decoder, 188
Function
  generator, single supply, 104

Gated
  oscillator, 123
Generator
  bit rate, 30
  digital phase-shifted clock, 118
  function, single-supply, 104
  linear ramp, 110
  nanoseconds pulse, 115
  programmable pulse, 114
  ramp, 107
  square wave, 106
  square wave, 3-phase, 116
  square wave, 4-phase, 117
  staircase waveform, 111
  time-base, 29
  tone-burst, 109
  triangle wave, 106
  two-phase sine wave, 112
  waveform, 101

Hysteresis
  comparator with, 71

Identifier
  unmarked Zener, 192
Incandescent
  lamp dimmer and protector, 206

Indicator
  blown-fuse, 154
  visible-voltage, 73
Inductor
  simulated, 68
Instrumentation
  amplifier, 98
Integrator
  op-amp based, 66
Interfacing
  techniques for digital logic, 44
Inverter
  using spare EX–OR gate, 33
Inverting
  amplifier, using op-amp, 83
  switching regulator, 165

Lamp
  dimmer, electronic, 206
  dimmer, incandescent, 206
  flasher, 159
  protector, incandescent, 206
Light
  activated alarm, 204
  monitor, remote, 205
Linear
  ramp voltage generator, 110
Logarithmic
  amplifier, 87
Logic
  interfacing techniques, digital, 44
Long
  duration timer, 62
Low
  power monostable using 555 timer, 65

Meter
  frequency, 132
Mixer
  digital frequency, 12
Modulator
  phase, 138
Monostable
  multivibrator, *see under* Multivibrator
  using 555 timer, 2
  using 74121, 3
  using 74122, 3
  using 74123, 4
  using 74221, 4
  using 4047, 4
  using 4528, 5
  using 4538, 5
MOSFET
  based lamp dimmer, 206
  driver for power-, 198

Motor
  constant current drive, 179
  controller, DC, 179
  controller, stepper motor, 180, 182
Multiplier
  analog, 76
Multivibrator
  astable, using 555 timer, 59

Nanoseconds
  pulse generator, 115
Noise
  canceller, digital, 9
Non-inverting
  amplifier, using op-amp, 84

Op-amp
  absolute value amplifier, 85
  differentiator, 66
  high current booster, 69
  integrator, 66
  inverting amplifier, 83
  multilevel comparator, 73
  non-inverting amplifier, 84
  power booster, 68
  precision full-wave rectifier, 75
  precision half-wave rectifier, 74
  simulated inductor, 68
  summing amplifier, 84
  supply frequency reject filter, 73
  supply splitter, 75
  variable gain amplifier, 81
  voltage comparator with hysteresis, 71
  voltage follower, 67
  voltage reference, programmable, 81
  voltage to current converter, 72
  window comparator, 70
Opto-coupler
  based revolution sensor, 187
Oscillator
  crystal, 122
  gated, 123
  precision, sine-wave, 93
  RC phase-shift, 121
  Wien-bridge, 119, 121
Over
  voltage detector, 42

Phase
  detector, 139
  detector, digital, 12
  lead–lag indicator, 38

lock indicator, 134
locked loop, 124
shifted clock generator, 118
shifter, quadrature, 19
Phonograph
  pre-amplifier, 92
Photodiode
  detector, 187
Polarity
  sensitive battery-powered supply, 176
Power
  amplifier, 95
  fail alarm, 151
  fail and brownout detector, 152
  fail warning, 153
  on reset circuit, 21
  supply, uninterruptible, +5 V, 142
Precision
  op-amp, low-noise, high-speed, 92
  rectifier, full-wave, 75
  rectifier, half-wave, 74
  sine-wave generator, 93
  waveform generator, 101
Programmable logic device
  based monostable multivibrator, 6
  based stepper motor controller, 182
Protector
  incandescent lamp, 206
Pulse
  detector, missing-, 18
  extractor, 32
  generator, 108
  generator, nanoseconds-, 115
  generator, programmable-, 114
  stretcher, using counter, 8
Push-button switch
  debouncer, 23

Quad
  timer, 558, 62
Quadrature
  phase-shifter, digital, 19
  waveform decoder, digital, 22
Quasilinear
  amplifier, 122

Ramp
  generator, 107
Receiver
  coaxial cable, 53
  line, RS 232, 47, 49, 50
  programmable, 52

Rectifier
  full-wave, precision, 75
  half-wave, precision, 74
Reference
  voltage, negative, 81
  voltage, negative using positive, 82
  voltage, positive, 81
Regulator
  voltage, automobile, 164
  voltage/current, 165
  voltage, inverting switching, 165
  voltage, negative adjustable, 161
  voltage, positive adjustable, 155, 158
Relay
  driver, 57
  supply, 24 V from 12 V/15 V, 53
Reset
  circuit, power-on, 21

Sawtooth
  generator, 108
SCA
  decoder, 189
Sensor
  revolution, 187
Sine wave
  generator, 93, 101
  generator, two-phase, 112
  generator, ultra-pure, 93
Square
  -root circuit, 78, 80
  squaring circuit, 78, 80
  wave generator, 106
  wave generator, +/−15 V from +5 V, 113
  wave generator, 3-phase, 116
  wave generator, 4-phase, 117
  wave tone-burst, 108
Staircase
  waveform generator, 111
Summing
  amplifier, using op-amp, 84
Supply
  splitter, op-amp based, 75
Switch
  alternate-action, electronic, 41
  analog, 208
  battery back-up, 175
  debouncers, 23
  high-side, 197
Switchover
  circuit, battery-, 174
Synchronizer
  AC line, 28
  digital, 39

Tachometer, 17
Tape-head
  pre-amplifier, 93
Temperature
  alarm, high/low, 208
Tester
  coaxial cable, 195
  crystal, 194
  field-effect transistor, 193
  transistor, on-board, 191
  Zener diode, 192
Time-base
  generator, 29
Timer
  appliance, 60
  astable multivibrator using, 59
  long duration, 62
  low-power monostable, using 555, 65
  programmable, 63
  retriggerable monostable, using 555, 64
  sequential, 64
Tone
  burst generator, single, 109
  decoder, chatter-free, 133
  decoder, dual, 130
  decoder, high speed narrow band, 133
  decoder, monolithic, 125
Transducer
  amplifier, bridge, 86
  magnetic, detector for, 188
Transformer
  isolated power supply, 141
Translator
  level, 52
Triangle
  wave generator, 106
TTL
  to CMOS interface, 46
  to ECL interface, 47

Under
  voltage detector, 42
Uninterruptible
  power supply, +5 V, 142

Visible
  voltage indicator, 73
Voltage
  doubler, 202
  follower, 67
  over-, detector, 42
  under-, detector, 42
Voltage regulators
  *See under* Regulator

Waveform decoder
  quadrature, 22
Waveform generator
  monolithic precision, 101
  pulse, 108
  sawtooth, 108
  sine, two-phase, 112
  square, 106
  staircase, 111
  ramp, 107
  triangle, 106

Wien-bridge
  oscillator, 119
  oscillator, using spare login inverters, 121
Wind
  powered battery charger, 171

Zener diode
  identifier, for unmarked, 192
  tester, 192